SPE REPRINT SERIES NO. 51

CO_2 FLOODING

1999 Edition

Published by the
Society of Petroleum Engineers
Richardson, Texas

CO_2 FLOODING REPRINT COMMITTEE

ISBN 1-55563-083-9

FOREWORD

This is the fifth Reprint Series volume that addresses improved oil recovery by miscible-gas injection. The first four volumes were *Miscible Processes* published in 1963, *Phase Behavior* published in 1981, *Miscible Processes II* published in 1985, and *EOR Field Case Histories* published in 1987. All but the first of these volumes include papers on CO_2 flooding. In addition, the SPE Monograph *Miscible Displacement* reviews and discusses literature on CO_2-flooding technology that appeared before 1983. Because the subject was well covered before 1983, the committee limited its deliberations to literature published after that date.

The most significant development in CO_2 flooding in the 1980's and 1990's has been the implementation of large-scale commercial floods. Although the majority of these floods are ongoing and are not in a mature stage, numerous papers have been published that describe performance results to date, operational issues, project monitoring, and analysis. This Reprint volume gives particular emphasis to field case histories that give insights into how the floods are working, what problems have been encountered, and what actions have been taken. We have also tried to address other subjects where new findings have been published that are important for CO_2-flood design and prediction, such as simulation, phase behavior, displacement mechanisms, relative permeability, and injectivity.

The papers selected are grouped into the following categories: Field Case Histories—Performance; Field Case Histories—Operations; Analysis, Prediction, and Optimization; Phase Behavior; and Injectivity, Displacement Mechanisms, and Relative Permeability. In addition to the 20 papers reproduced in their entirety in this volume, an additional 20 are included in the Annotated Bibliography. The Bibliography cites all the papers the committee reviewed to make its final selections.

Fred I. Stalkup
Chairman

CONTENTS

Field Case Histories—Performance

Field Case Histories—Operations

Analysis, Prediction, and Optimization

Phase Behavior

Injectivity, Displacement Mechanisms, and Relative Permeability

SPE 35190

Dollarhide Devonian CO2 Flood: Project performance review 10 years later.

by J.F.R. Bellavance, UNOCAL Corp.
SPE Member

This paper was prepared for presentation at the SPE Permian Basin Oil & Gas Recovery Conference held in Midland, TX, USA., 27-29 March, 1996.

ABSTRACT

The Dollarhide Devonian CO2 miscible flood has demonstrated substantial production increases in Phases 1, 2, 3 and 4 of the 5 Phase project. To date, incremental recovery is estimated at 15.8 MMBBLS. The field has now produced 63.7 MMBO and is presently under a program of accelerated depletion. In view of the detrimental effects of the WAG cycles on injectivity, water injection has been limited to areas of severe breakthrough. Production currently averages 3080 bbls/d. A total of 66.5 BCF or a 11.2% HCPV of CO2 has been injected into the reservoir and less than 15% has been produced. CO2 breakthrough has not been a problem in the Devonian hence, a WAG process is not warranted at Dollarhide.

INTRODUCTION

The Dollarhide Devonian Unit was formed in 1959. It is located in the Dollarhide Field, Andrews County, Texas, 70 miles west of Midland, Texas. (fig. 1 & 2) Discovered in June 1945 by Magnolia Petroleum and Humble Oil & Refining Co., the Devonian is one of 4 productive intervals within the Dollarhide Field. Subsequent to primary depletion (1947-61) and a successful waterflood (1962-85), a patented tertiary CO2-hybrid WAG miscible process was engineered to maximize recovery. The Dollarhide Devonian Field is one of two Devonian miscible CO2 floods currently operating in West Texas and produces a 40 API light crude oil at a depth of 7800 ft. Production peaked in the vicinity of 8000 BOPD under primary, 9000 BOPD under waterflood and is currently at 3080 BOPD and climbing under miscible CO2 injection. Original oil in place is estimated at 145.8 million bbls of oil. The estimated primary and secondary recoveries from the Unit total 62.6 MMBBLS or 43% of the OOIP. Approximately 19.5 MMBBLS or 13.4% of this volume are attributable to primary recovery. Ultimate tertiary recovery is expected to be 20.4 MMBO with an additional 7.3 MMBO captured through infill drilling. Primary and secondary performances are discussed in greater detail by Poole (1).

This CO2 tertiary project was implemented over the course of 10 years (Fig. 3). Phase 1 was the first of 5 phases to be completed in this field in 1985-86. Phase 2 then followed in 1987-88 completing tertiary development over an area called the southern fault block. Although simulation studies (2,3) predicted recoveries in the order of 14% under tertiary development, mixed results were obtained from CO2 injection in this block. There was no question however that additional oil was being captured through miscible displacement. By May 1988, production in phases 1 and 2 had increased to 1672 bbls/d up from 924 bbls/d in August 1985. This prompted the continuation of the development into the Phase 3 area, located in the northern fault block. By the end of 1990, the entire Phase was under CO2 injection and showing earlier than anticipated response in several wells. Finally, 1993 and 1994 saw the addition of 15 infill producers and 24 well conversions to CO2 injectors to bring the total well count to 101 active producers and 80 injectors. The field was originally drilled on 40 acre spacing and is currently producing on 20 acres after an extensive infill drilling program. Performance to date has been less than anticipated and current efforts are being made to alleviate operational difficulties causing this shortcoming.

This paper presents an update of the flood performance and a cursory evaluation of its behavior over the past 10 years of existence. A comparison of the different phases will be provided to emphasize the influence of the geology and operation strategies on sweep efficiency and response to CO2 flooding. Modifications, to maximize sweep efficiency, flood response time and improve project economics, were made to the later development phases based on observed behavior in Phases 1, 2, and 3. The findings obtained from these are currently being used to optimize the overall production strategy for this tertiary recovery project.

GEOLOGY

The Dollarhide Devonian reservoir is productive from the Thirtyone Formation. Saller et al. (4) divided the Thirtyone Formation at Dollarhide field into 5 main lithologies (fig. 4). These facies may be simplified into two main reservoir facies separated by a tight limestone. They will be referred to in this paper as the upper and lower porosity of the Unitized interval. The Thirtyone formation has a conformable contact with the underlying Silurian Wristen Formation and an unconformable contact with the overlying strata. Regionally, the Upper Devonian Woodford Shale is above this unconformity, however, locally the Woodford Shale has been removed by erosion.(5) This has occurred at Dollarhide on the crest of the anticline.

The Dollarhide Devonian field is a faulted, asymmetric anticline subdivided into four major fault blocks by a series of transverse faults which act as flow barriers. Located within the Dollarhide Devonian Unit are three of the four blocks, the fourth being located in North Dollarhide. The top of the Devonian structure is truncated by the Pennsylvanian unconformity which removed the Mississipian Limestone, Mississipian Woodford shale, and the Upper Devonian reservoir.(6). The middle and southern fault blocks show areas where all Devonian has been eroded altogether.(fig. 5)

The northern fault block, which houses phases 3 and 4, possesses the highest reservoir quality. It is characterized by good reservoir continuity and homogeneity both in the Upper and Lower porosity intervals. Faulting in this block is relatively minor and mostly concentrated on the edges. This fault block shows the best Lower porosity pay and has the best recorded recoveries under waterflood. The southern fault block, which includes the Phase 1 and 2 areas is slightly inferior in quality to the northern fault block. This is mainly due to the elaborate system of small faults throughout the pay sections. Although continuous and of good quality, the Upper and Lower porosity pays are highly variable which accounts for the lower recoveries under waterflood. Lastly, the middle fault block, which consists of Phase 5 and small portions of phases 2 and 4, is highly discontinuous and variable in nature. Lower reservoir quality and heavy faulting have resulted in lower than anticipated recoveries. The interpretation of a 3-D seismic survey conducted in 1988 revealed considerably more faulting than originally thought and provided great insight into problem well behavior and infill drilling location selection.

The Lower porosity interval accounts for 83% of the total oil in place and is therefore the main target for tertiary recovery in this reservoir. The Upper porosity shows permabilities and porosities and net pay averaging 12 mD, 8.5 percent and 28 ft as compared to the Lower porosity's 9 mD, 17 percent, and 48 ft.

ORIGINAL CO2 FLOOD PROPOSAL

The original tertiary flood project was to be implemented in four phases over an 11 year period. This was done to reduce early peak CO2 demand, maximize the utilization of recycled CO2, minimize manpower requirements, and provide flexibility for acceleration or deceleration of project development depending on market conditions. More importantly, however, this would provide enough time to evaluate early Phase 1 behavior and adjust the subsequent phases accordingly to maximize recovery. The southern fault block was selected for the Phase 1 area because of its secondary performance, its geological characteristics, and its remaining oil in place. This Phase was to have a 25% hydrocarbon pore volume slug of pure CO2 injected followed by chase water, all at a reservoir pressure of 3600 Psia, well above the 1600 Psia minimum miscibility pressure. A WAG process would be instituted only if premature breakthrough became a problem. CO2 was trucked in for an 8 month period during which the necessary CO2 transportation pipelines construction and supply contracts were finalized.

Additional computer modeling of the Phase 1 area showed that WAG injection provided the benefit of greater mobility control at the expense of rate acceleration.(2) Optimization of the model showed that a large slug of CO2 followed by WAG injection could capitalize on the benefits of both mechanisms. Using this process, a 14% incremental recovery could be obtained, however, the slug size had to be increased to 30%. This process was subsequently patented by UNOCAL under the name HYBRID-WAG process where a 9% pore volume of CO2 was to be injected followed by the remaining 21% at a 1:1 WAG ratio.

Phases 1 and 2 were developed according to this design, seeing 8.9% and 9.0% HCPV of CO2 injection prior to going to WAG. The Phase 3 area was put on CO2 injection in May 1990 and saw a 6.6% HCPV of CO2 injected prior to going to WAG. It was felt that the higher permeability of the northern fault block would precipitate breakthrough and therefore require mobility control at an earlier stage. Phases 4 and 5, located in the middle fault block of the Dollarhide Field, were developed in 1993-94 through a series of infill wells and conversions. CO2 injection commenced in November 1993 in both phases.

FLOOD BEHAVIOR
PHASE 1

Tertiary response, for Phase 1, was first experienced at the 47-9 well after only 9 months , considerably sooner than the anticipated 18 months predicted in the original computer simulation study. This was mainly due to the irregularity of the pattern as the 47-26 injector was drilled on 10 acre spacing. Other wells in the area, such as 47-6 and 47-7 responded in 22 and 21 months respectively. On average, response time for wells in the Phase 1 area was 35 months and production peaked at 350% of the pre-flood rate. CO2 production was first detected 17 months after first injection and only in minimal quantities. Peak CO2/oil ratios were recorded in November 1993 at 3.7 MCF/BBL and have since decreased to 1.8 MCF/BBL.

When examined individually, a number of wells outperformed the Phase average response. More specifically, wells 46-6, 47-8, and 48-4 showed production increases of 475%, 500%, and 910% respectively. This is explained by the high reservoir quality of the area and above average injectivity of the supporting injectors. Also, the WAG process, once believed to enhance conformance and recovery has been shown to impair well productivity and oil cuts.(7) Several patterns have shown decreases in oil production within 1 month after CO_2 injection was reestablished. Once the pattern was put back on CO_2 injection, a period of 2 to 3 months would be required to regain the rate loss. The fact that cycling, within the Phase 1, 2 and 3 areas had been scheduled in such a manner that at least 2 injectors were on CO_2 injection and two on water injection for each pattern, obscured the adverse injectivity effects caused by water. This conclusion became obvious once continuous CO_2 injection was re-established in mid-1994. (7)

Primary and secondary recoveries for Phase 1 were estimated from production decline at 15.8% and 25.7% respectively, giving a secondary to primary recovery ratio of 1.63. The decline rate under waterflood was estimated to be 11.9 %. Under CO_2 flood operations, the production rate, after peaking at 1240 BOPD, declined at 6.9% until 1994. An internal study showed that, in certain areas of the reservoir, CO_2 injection had been predominantly going into the Upper porosity where only 18.2% of the OOIP is located.(8) As of August 1994, Phase 1 had seen on average in excess of 24% of the CO_2 going into the Upper porosity and some individual patterns posted numbers as high as 84%. Pattern optimization and profile modification have, over the past two years, helped eliminate the decline, redirect injection, and turned the trend around to see an increase in the oil rate (fig. 6). At this time, tertiary recovery for this Phase is projected to be 16.3% using exponential decline. To date a total of 34.8 BCF equating to a 23.1% HCPV of CO_2 has been injected and only 0.3 BCF produced over the course of 10 years. CO_2 breakthrough and vertical migration have not been a problem. As anticipated, watercut in this Phase has decreased to 70.8% from 84.8% prior to initiation of CO_2 injection.

PHASE 2

Tertiary development of the Phase 2 area commenced as the bottom fell out of the oil market. With a less than attractive perspective on the horizon, UNOCAL chose to delay the project until some firming of the WTI price. As of July 1988, CO_2 was being injected at a rate of 5.5 MMCFD into 12 wells and development of the southern fault block was complete. When taking into consideration the poorer reservoir quality of Phase 2, this rate compares favorably with the 15 MMCFD injection in the 21 injectors of Phase 1. First response to CO_2 injection in the Phase 2 area was recorded at the 44-7D well some 7 months later. On average, however, production response in the Phase 2 area averaged 34 months. CO_2 breakthrough occurred 18 months after start of injection and only in minimal quantities. Maximum recorded CO_2/oil ratios were measured in December of 1991 at 8.47 MCF/BBL and have since settled at 2.3 MCF/BBL.

A detailed examination of the individual well behavior reveals some interesting factors. 40-35 and 40-36, which hold the best Upper porosity in the middle fault block, have shown the greatest production increase of the Phase 2 wells coming in at 880% and 710% increases respectively. This is explained by the relatively higher permeability of the Upper porosity in relation to the Lower porosity. This, in combination with the fact that over 55% of the CO_2 injection within the pattern has gone into the Upper porosity explains the high increase of short duration. Unlike phase 1, the wells of Phase 2 were only slightly affected by the WAG cycles. The relatively limited occurrences of good Lower porosity development in this Phase minimized the influence of the loss of injectivity caused by the reduced relative permeabilities encountered following a WAG cycle.

Ultimate primary and secondary recoveries for this Phase were estimated at 13.1% and 22.2% respectively for a secondary to primary ratio of 1.69. A decline rate of 19.8 % under waterflood was slowed considerably once miscible operations were initiated (fig. 7). Peak rate for the Phase 2 area was recorded in August of 1993, at 546 BOPD, after the drilling of the 20 acre infill 40-38D. On a 2.7% tertiary decline from 1989 until the drilling of 40-38, phase 2 production has stabilized 350 BOPD and is currently on an incline. Much like in the Phase 1 area, on an injected pore volume basis, the Upper porosity's 13.7% is substantially more CO_2 than the Lower porosity's 9%. For the Phase 2 area, 24.2% of the OOIP is located in the Upper porosity and it has received 34% of the CO_2 injection to date. As in Phase 1, profile modifications and pattern optimization are being used to maximize and possibly accelerate production. Under current operations, tertiary recovery is expected to be 16.8% using exponential decline. To date, a total of 14.2 BCF of CO_2 (17.4% HCPV) has been injected and only 0.3 BCF of it produced since 1988. CO_2 breakthrough has once again not been a problem and continuous CO_2 injection has been implemented throughout the Phase. Consequently, watercuts in Phase 2 have dropped from 84.4% down to 73.4% currently.

PHASE 3

Phase 3 development followed immediately after Phase 2 completion. With a portion of the infill drilling program initiated in 1986, the phase was infill drilled to 20 acres in 1989 and 1990. A total of 20 producers was drilled and 18 wells converted to CO_2 injection. Oil production prior to CO_2 injection was 590 BOPD and currently averages 1100+ BOPD. The first indication of tertiary response was recorded at the 13-1 and 14-1 wells within 3 months as compared to the phase average value of 27 months. This was highly unusual and after closer examination it became obvious that the fault system associated with the area surrounding the two wells was responsible for the rapid response. Tertiary peak oil rate to pre-CO_2 production rate for this phase calculates to 1.61. Currently, phase 3 is seeing the highest injection rate of all phases at over 18 MMCFD. In order to prevent CO_2 migration to outside acreage, water injection was maintained at the northern and western edges of the Unit. CO_2 production was first recorded in minimal volumes 13 months after first injection. CO_2/oil ratios peaked at 6.0 MCF/BBL in February 1994 and are currently at 3.9 MCF/BBL.

On a well by well basis, 10-106, 15-5, and 15-4 really stand out. They show respective production increases of 470%, 690%, and 390% as compared to a phase individual well average Peak to pre-flood rate of 250%. This anomaly is due mainly to the fact that adjacent injection wells are either open hole completions or have 3 or more high shot density injectors (6+ SPF) within the pattern. Once again, this phase is still in the preliminary stages of CO_2 flooding and several wells are inclining at the current time. Of interest is the occurrence of a breakdown in the CO_2 injection system in December of 1994 for a period of 3 weeks. Production in the phase 3 area, which had been on a steady incline for over 3 years, was almost immediately affected resulting in an estimated 200+ BOPD production loss. CO_2 injection was restored shortly after this failure but production rates lagged in resuming their incline. Production has only recently recovered and now tracks the pre-WAG incline rate.(fig 8)

Decline predicted primary and secondary recoveries for phase 3 calculate to be 12.2% and 39.9% respectively for a secondary to primary ratio of 3.27. The high secondary recovery in this phase is attributable to superior reservoir quality and the absence of minor faulting present in all other phases. A decline rate of 9.4% under waterflood compares favorably with phases 1 and 2. Under miscible flood operations, the oil production is on an increasing trend and currently averages over 1100 BOPD. As for the other two phases, phase 3 has seen a substantially greater pore volume of upper porosity CO_2 flooded. Historical injection profiles show a cumulative 39.4% of the injection going into the upper zone. In certain wells having both upper and lower pay present, up to 65% of the injection has gone into the upper section. This problem is being alleviated by an aggressive profile modification workover program. This is doubly important as over 84% of the OOIP of phase 3 is located in the lower pay interval. Predicted tertiary recovery for this phase is currently estimated at 16.0% using exponential decline. To date, a total of 27.5 BCF (13.9% PV) of CO_2 has been injected into phase 3 and 3.8 BCF produced. CO_2 production has escalated in two of the 29 producers but has stabilized at 1.1 MMCFD and 0.75 MMCFD. Watercuts have decreased from 91% in 1989 to below 80% currently.

PHASES 4 & 5

Phases 4 and 5 were developed simultaneously in 1993 and 1994 with the drilling of 15 infill producers and conversion of 27 injectors. Currently, phase 4 has 14 producers and 13 active injectors as compared to phase 5's 14 producers and 11 injectors. Encompassing an area of poorer reservoir quality, these two phases were anticipated to have lackluster responses to tertiary miscible CO_2 flooding. Oil production prior to tertiary development was 189 BOPD and 267 BOPD for phases 4 and 5 respectively. These figures have increased to what is today 595 BOPD and 276 BOPD. Several operational problems have plagued these two phases, from severe scale formation to injection string leaks and poor downhole conditions. Injection started in the third quarter of 1993 and CO_2 volumes total 3.8 BCF and 3.1 BCF for phases 4 and 5 respectively. At this stage, CO_2 production has been measured only in very small volumes. (fig. 9,10)

Only six wells have demonstrated significant response to date and all are part of the phase 4 area. Strangely enough, the responses have been significant ranging from a 125% increase in 27-1 to a 950% increase in the 26-12 well. This is not surprising when one considers that the well 26-4, which is the common injector to the wells showing high responses, initially had the Unit's highest IP rate. On average, however, phases 4 and 5 have had increases of 144% and 66% respectively. Limited profile information is available at this time to make an accurate evaluation of the distribution of injected CO_2 and zonal displacement. Unlike the first three phases, phases 4 & 5 have remained on continuous injection and will continue in this manner until breakthrough problems are encountered.

Ultimate primary and secondary recoveries calculated to 12.4% & 27.6% and 12.9% & 22.1% for phases 4 and 5 respectively giving secondary to primary ratios of 2.23 and 1.71. Both phases showed strong secondary decline rates coming in at 11.5% and 14.0%. Phase 4 is currently producing at its highest level since flood initiation (595 BOPD) and 4 wells are responsible for 90% of the phase's production. Phase 5 peaked in early 1994 during the infill drilling phase at 397 BOPD and is currently holding steady at 275 BOPD, slightly above the pre-flood rate of 265 BOPD. Watercuts have seen a significant drop in phase 4 going from 89+ % in 1993 to less than 70% currently; Phase 5 has yet to see a decrease.

PRODUCTION CHALLENGES

The major challenge faced by UNOCAL in optimizing this flood was to increase the displacement efficiency, accelerate the reservoir depletion, and increasing project economic viability without sacrificing recovery. Injection profile analysis confirmed that considerable improvement in displacemēnt efficiency could be achieved by redirecting the majority of the injection into the lower porosity zone. This, combined with higher injection rates would offset pattern withdrawals which had historically been below unity. For this purpose, a proposal was put forward to increase perforation density in the injection wells. This was based on the belief that the magnitude of the response to CO_2 flooding was directly correlatable to the cumulative shot density of the pattern injectors. New information is currently being gathered to allow presentation of this hypothesis in a subsequent SPE paper. To date, however, production gains in excess of 600 BOPD have been achieved by using this approach and additional increases are expected as the work is completed.

In order to have an effective CO_2 flood, a CO_2-Hydrocarbon miscible solvent bank has to be formed and maintained to maximize displacement. The introduction of water through the WAG process hinders this mechanism and severely reduces displacement efficiency. Similarly, an adverse effect of the WAG process is the resulting oil trapping from the introduction of water. Laboratory verified that simultaneous injection of solvent and water into waterflooded cores results in trapping of both oil and solvent. Experiments using long Berea cores demonstrated that WAG ratios between 1 and 3 severely reduced oil recovery.(9,10,11,12,13,14). These conditions closely resemble that of the Devonian Formation at Dollarhide and confirm what has been witnessed in the field. This was again reinforced by Huang & Holm in their Dollarhide

Devonian core flooding tests where recoveries under continuous CO_2 injection were 12 to 16% greater than under WAG.(15) Hadlow states that on average, CO_2 projects will show a 20% loss in injectivity in the WAG process, hence slowing recovery.(16)

OBSERVATIONS

In all phases, water production and watercut increased prior to the oil response indicating a more efficient and uniform displacement front due to reduced capillary forces at the pore throat level. This was more prominent in areas where watercuts had reached 90%+ prior to CO_2 injection. Beliveau & Payne(17) have documented similar behavior in Canada.

Long before the implementation of the CO_2 miscible flood, scaling has been a major concern at Dollarhide. During the waterflood operations, produced waters from the Clearfork, Devonian, Fusselman and Ellenburger were mixed and reinjected. This, in combination with the introduction of CO_2, has contributed to considerably worsen the scale occurrences in the Devonian. In order to minimize this problem, several modifications to the water handling facilities were made resulting in the elimination of 98+% of the fines and particulate from the injection stream. The resulting effect on scale formation has yet to be noticed, however, the amount of fill present in the injectors has noticeably decreased.

Of interest is the progression of scale with various stages of the CO_2 flood. In the early stages of the flood, the majority of the scale recovered from producers and injector conversions is calcium sulfate. This stage is felt to be limited to an average of two cleanouts. The second stage shows more iron sulfide concentration with traces of calcium sulfate and calcium carbonate and approximately lasts from 4 to 12 cleanouts. The majority of the scale recovered from the third stage is calcium carbonate and appears to occur for 4 to 6 cleanouts. Asphaltenes come next for a very short period, usually 2 to 3 cleanouts over a period of 12 to 18 months. Finally, once all these stages have been attained, few, if any at all, scale problems are encountered again. It is still too early at this point to tell if scale occurrences have been totally eliminated as few wells have passed the asphasltene stage, but results are encouraging. Proximity to faults and fractures seem to play a key role in the amount of scale encountered in producers.

CO2 UTILIZATION FACTOR

The original design for the Dollarhide Devonian CO_2 Miscible Flood called for a project CO_2 utilization factor of 6. Currently, the five phases show higher utilization rates as they are still in the early stages of the flood. Figure 11 shows the various utilization rates as a function of time for all five phases. Phase 1, which started in 1985 shows a steady decrease in the instantaneous utilization factor until mid-1994. Phase 2 has behaved in a similar fashion since commencing injection in 1988. Phase 3 was on a steady increase until 1993 and has since leveled at the 14 MCF/BBL level. Phases 4 and 5 being at the early stages of the flood, are still expected to rise over the next several years before a downward trend is seen. Currently, the instantaneous factors average 14 MCF/BBL for 4 out of the 5 phases and is decreasing on a yearly basis. At this time, projections for the full-life project utilization factor are between 6 and 8 MMCF/BBL.

CONCLUSIONS

The WAG process is detrimental to productivity and recovery at Dollarhide.

Tertiary response for the Dollarhide Devonian CO_2 flood is expected to average 250% of the pre-flood rates.

Maximizing CO_2 injection by increasing shot density in the Lower porosity interval can significantly increase well productivity.

The high variability of the reservoir quality and the faulted nature of this reservoir make individual pattern monitoring essential in the optimization of this tertiary flood.

Scaling occurrences are highly dependent on fluid throughput, flood maturity and proximity to faults and fracture systems.

ACKNOWLEDGMENT

The author wishes to thank UNOCAL and the other co-owners of the Dollarhide Devonian Unit for permission to publish this paper.

REFERENCES

1. Poole, E.S.: "Evaluation and implementation of CO2 injection at the Dollarhide Devonian Unit". SPE paper 17277 presented at the SPE Permian Basin Oil and Gas Recovery Conference, Midland, Texas, (March 10-11, 1988).

2. Lin, E.C., Poole, E.S.: " A numerical evaluation of various CO2 injection processes for the Dollarhide Devonian Reservoir". UNOCAL Science and Technology Division, Brea, California, Jan 13, 1987.

3. Murray, L.E.: "Field scale simulation study for the estimation of tertiary oil recovery by CO2 injection in the Dollarhide Devonian Field". UNOCAL Science and Technology Division, Brea, California. April 27, 1984.

4. Saller, A.H., D. VanHorn, J. Miller, B.T. Guy, Reservoir Geology of Devonian Carbonates and Chert --- Implications for Tertiary Recovery, Dollarhide Field, Andrews County, Texas". AAPG Bulletin V.75 No.1 P.86-107.

5. Ruppel, S.C., Hovorka, S.D.:"Controls on Reservoir Development in Devonian Chert: Permian Basin,

Texas. AAPG Bulletin, V.79, No. 12, December 1995, P.1757-1785.

6. McPherson, B.D.: "Dollarhide Devonian CO2 Flood --- Summary and Observations". UNOCAL Corp. Midland Office, April, 1995.

7. Huang, E.T.S., Holm, L.W.: "Core Labs Report on Displacement Experiments Using Dollarhide Devonian Cores". UNOCAL Science and Technology Division, Brea California. April, 1986.

8. Jackson, V. :"CO2 Flood Pattern Analysis for Upper and Lower Devonian --- Dollarhide Devonian, Andrews, Texas. October 2, 1994.

9. Raimondi, P. and Torcaso, M.A.: "Distribution of the Oil Phase Obtained Upon Imbibition of Water", SPE Journal (March 1964) P. 49-55.

10. Thomas, G.H., Countryman, G.R., and Fatt, I.: "Miscible Displacement in a Multiphase System", SPE Journal (Sept. 1963) p. 189-196.

11. Stalkup, F.I.: "Displacement of Oil by Solvent at High Water Saturations", SPE Journal (Dec. 1970) P.337-348.

12. Jones, S.C.: "Some Surprises in the Transport of Miscible Fluids in the Presence of a Second Immiscible Phase". , SPE paper 12125 presented at the 58th Annual Technical Conference in San Francisco, Oct. 5-8, 1983.

13. Campbell, B.T., and Orr, F.M. Jr.: "Flow Visualization for CO2/Crude Oil Displacements". SPE paper 11958, presented at the 58th Technical Conference in San Francisco, Oct. 5-8, 1983.

14. Eiffin, D.L. and Yellig, W.F.: "Effects of Mobile Water on Multiple-Contact Miscible Gas Displacement" SPE Journal, June 1983.

15. Holm, L.W.: "Mobilization of Waterflood Residual Oil by Miscible Fluids". SPE paper 13216 presented at the 59th Annual Technical Conference and Exhibition, Houston, Texas. Sept. 16-19, 1984.

16. Hadlow, R.E.: "Update of Industry Experience with CO2 Injection." SPE paper 24928 presented at the 67th Annual Technical Conference in Washington, D.C., October 4-7, 1992.

17. Beliveau, D and Payne, D.A: "Analysis of a Tertiary CO2 Flood in a Naturally Fractured Reservoir". SPE paper 22947 presented at the 66th Annual Technical Conference and Exhibition in Dallas, Texas. Oct. 6-9, 1991.

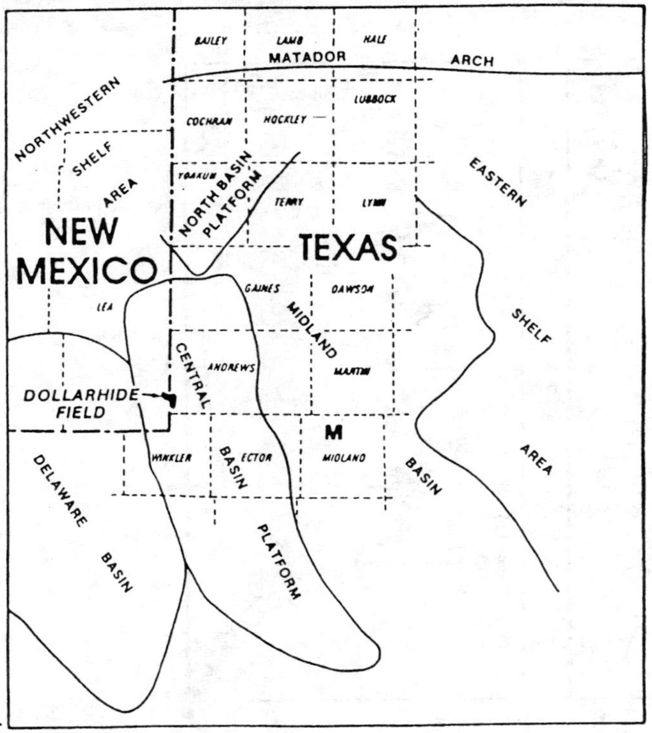

FIGURE 1
LOCATION OF DOLLARHIDE FIELD

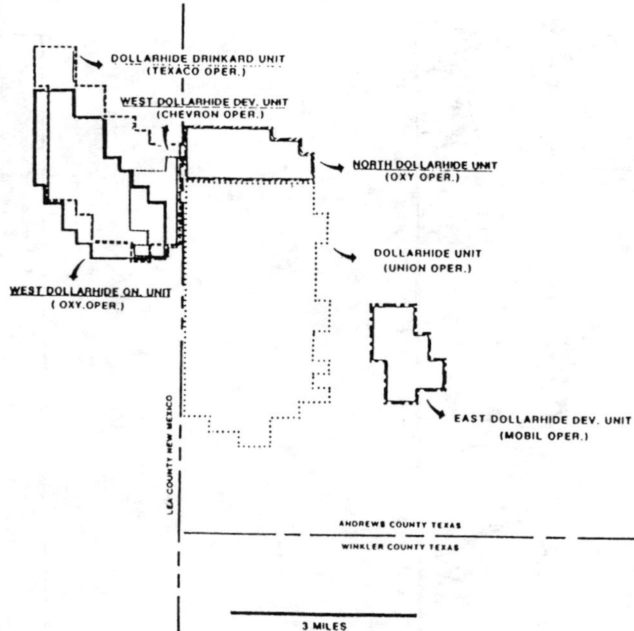

FIGURE 2

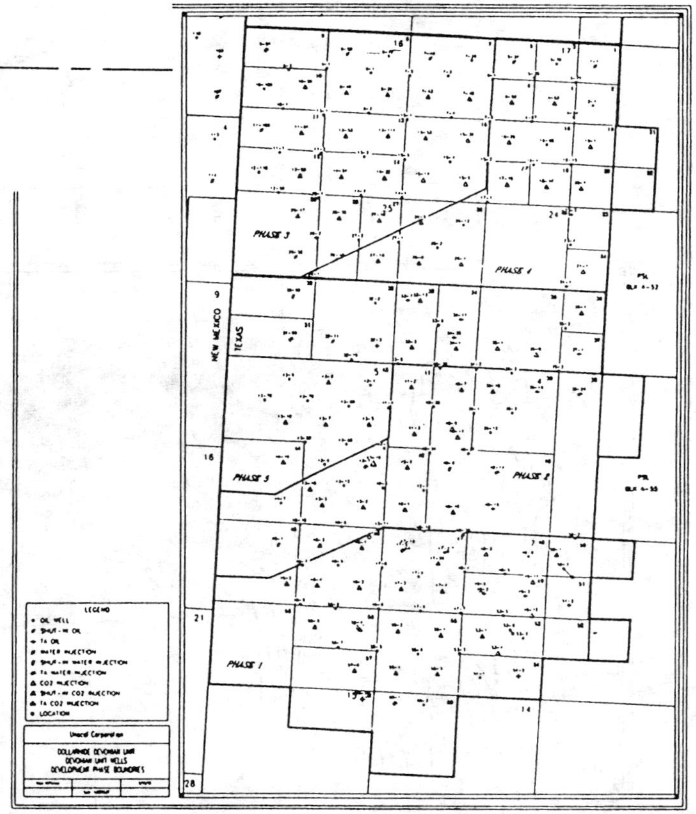

FIGURE 3

DOLLARHIDE 46 - 5D

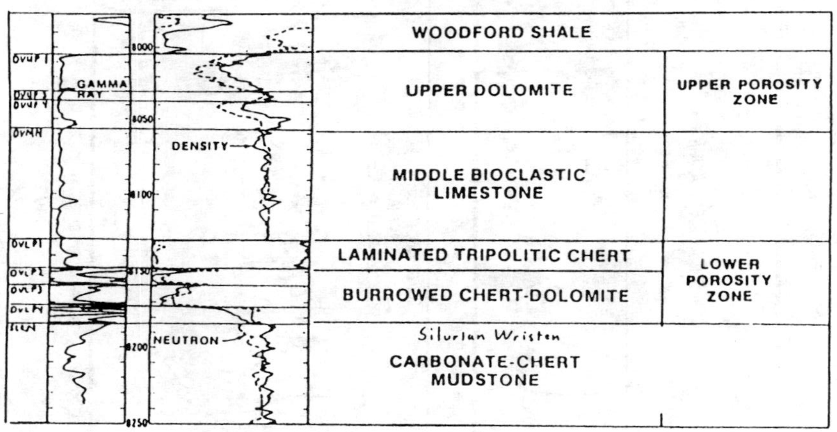

FIGURE 4

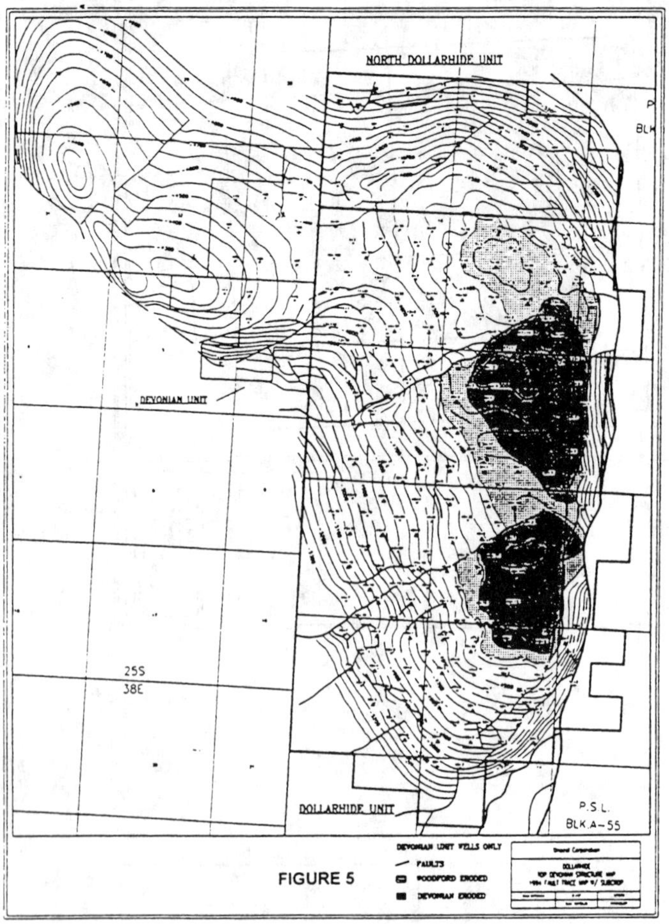

FIGURE 5

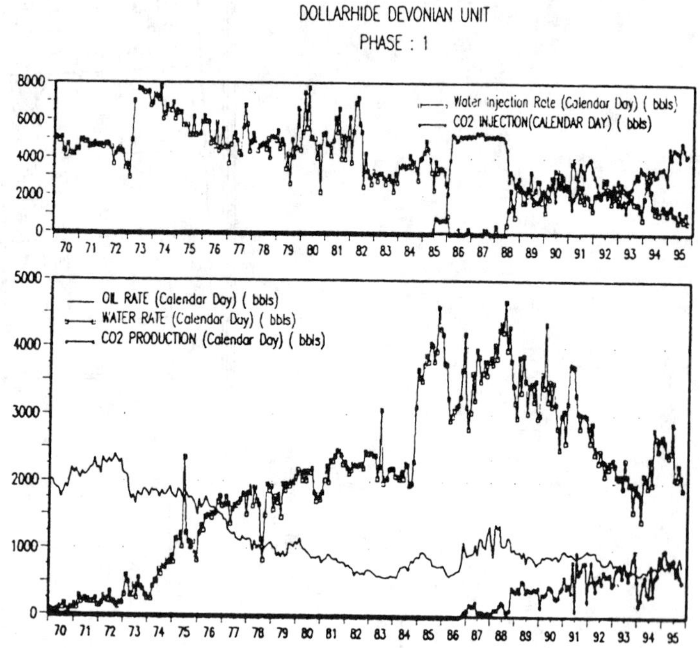

FIGURE 6

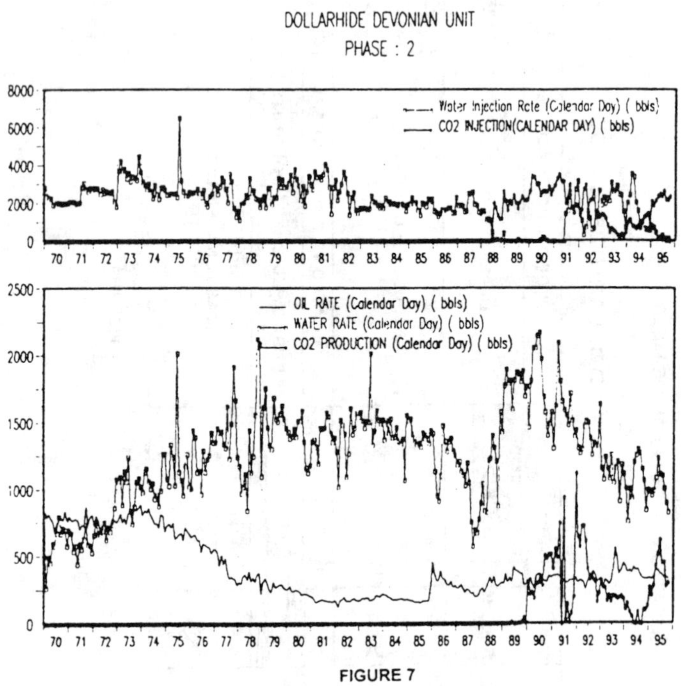

FIGURE 7

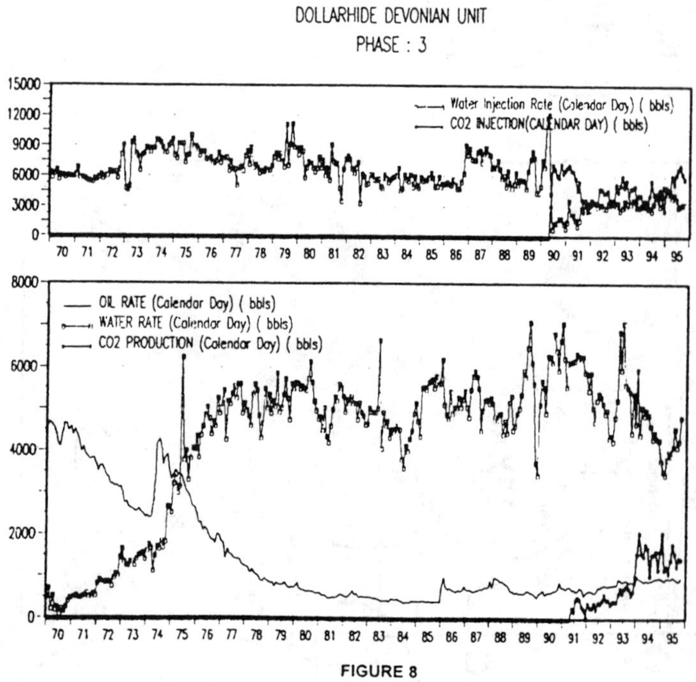

FIGURE 8

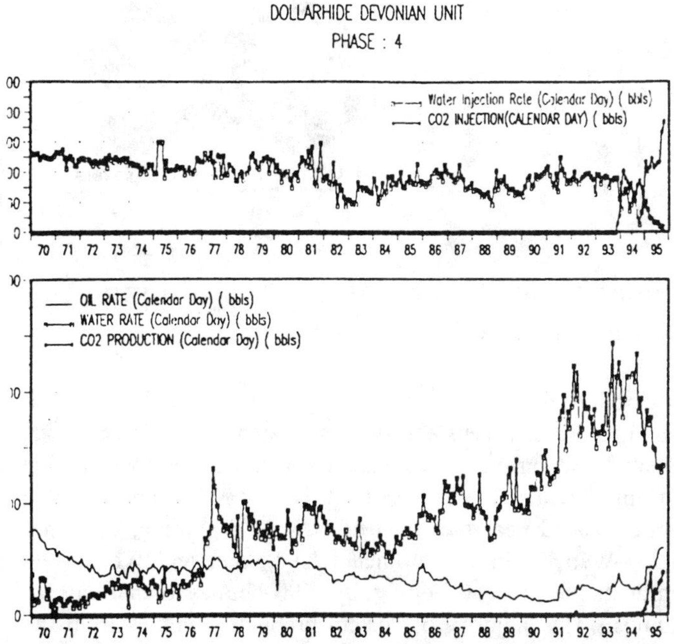

FIGURE 9

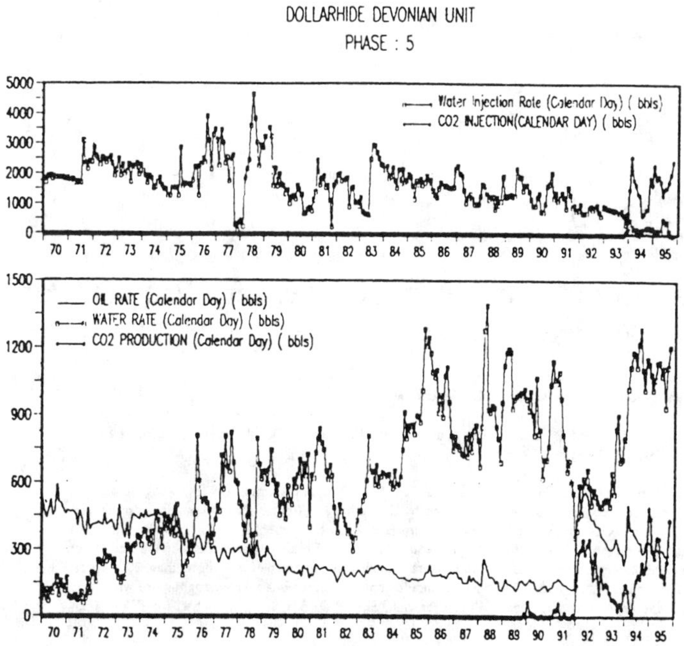

FIGURE 10

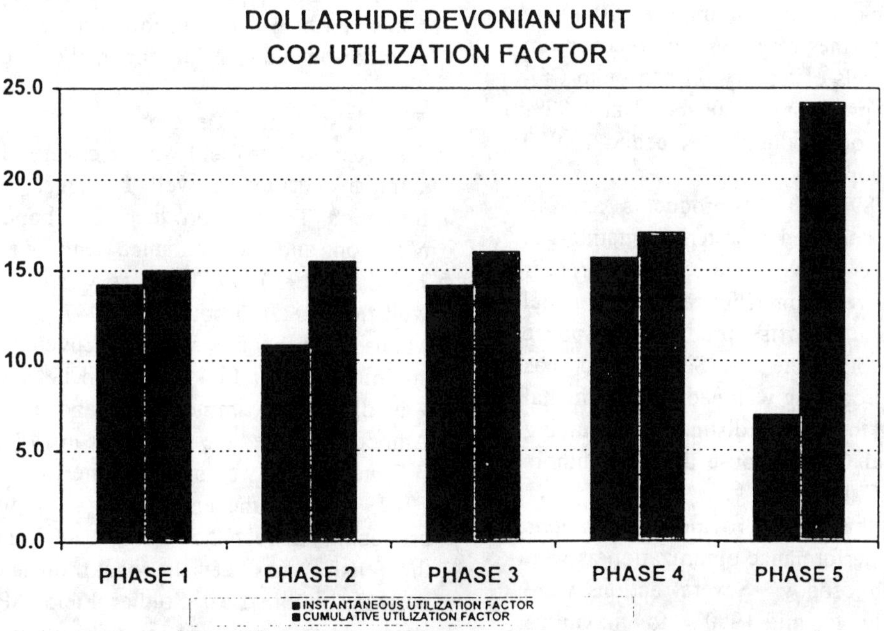

FIGURE 11

SPE 35191

Lost Soldier Tensleep CO_2 Tertiary Project, Performance Case History; Bairoil, Wyoming

by R.J. Brokmeyer, SPE, D.C. Borling, SPE, and W.T. Pierson, SPE, Amoco Exploration and Production Sector

This paper was prepared for presentation at the Permian Basin Oil & Gas Recovery Conference held in Midland, Texas 27-29 March 1996

Abstract

The Lost Soldier Tensleep field tertiary performance has a noteworthy case history that demonstrates how an incremental 13% of original oil in place will be recovered from this sandstone reservoir using carbon dioxide. Located in south central Wyoming, the Lost Soldier Tensleep has been under carbon dioxide (CO_2) injection since 1989. From 1989 through 1995, a 61% hydrocarbon pore volume (HCPV) slug of CO_2 was injected into the reservoir to recover an incremental 13.6 million barrels of tertiary oil. Prior to CO_2 injection, the reservoir was producing 2,500 bopd at a 97% watercut. Within one year, oil production exceeded 10,000 bopd and is currently 6,000 bopd.

Since its start-up in 1989 with 47 producers and 40 injectors, the subject flood has used a water-alternating-gas (WAG) method and line drive pattern to process approximately 840 acres of reservoir. Illustrations of actual WAG injection cycles and resultant offset production response are included in this case history. The Lost Soldier flood has been so successful that one producing well had an incremental rate of 1700 bopd, carrying the unique distinction of having the highest peak incremental CO_2 response that the authors could find documented in U.S. literature.

Information contained in this paper provides observations and conclusions about field performance optimization as well as reservoir management philosophy. Several actions were completed during the early to mid-1990's to maximize recovery and profitability. These included maintaining miscible reservoir pressure, increasing natural gas liquid recovery, converting to sour reinjection, and exploiting downdip oil potential.

Introduction

The Lost Soldier Tensleep field is located in the Great Divide Basin of Wyoming approximately 40 miles northwest of the city of Rawlins (Figure 1). Lost Soldier is the larger of two fields located near the town of Bairoil, Wyoming, the other being Wertz field. Combined Lost Soldier and Wertz field production is 10,000 bopd and 1,500 bbl/day of natural gas liquids. The majority of current production comes from the Pennsylvanian age Tensleep and the Mississippian age Madison carbonate. The original oil in place (OOIP) in all horizons in both fields approaches 1 billion barrels.

Currently, besides the Lost Soldier Tensleep, CO_2 is being injected into the Wertz Tensleep[1] and the Lost Soldier Darwin/Madison reservoirs.

CO_2 for the Bairoil fields is supplied from Exxon's LaBarge project[2]. The CO_2 is transported via pipeline 120 miles from its Shute Creek plant in southwest Wyoming to a point 19 miles northwest of Bairoil, where it is transferred to an operator-owned spur line for final delivery.

Field History and Development

The Lost Soldier Field was discovered in 1916 when Bair Oil Company drilled a well 265 feet into the first Frontier formation. The well produced 200 bopd. The Dakota, Lakota, Morrison, and the Sundance sands were discovered prior to 1926[3]. In 1930, the Tensleep was discovered, and the initial well flowed 2,435 bopd. In 1947, the Darwin sandstone and Madison carbonate were discovered and produced at a combined rate of 1,045 bopd. In 1948, the Flathead sandstone was discovered immediately above the granitic basement. Amoco purchased the properties in 1975.

Development was slow for the Lost Soldier Tensleep until 1942. During the early 1940's, 16 wells were completed, producing between 2,500 and 8,000 bopd. Since its discovery, the Tensleep has been the most prolific of the nine productive horizons in the Lost Soldier Field. Primary production is attributed to a combination of fluid expansion, water influx, and gravity drainage. Peripheral water injection began in 1962, and pattern waterflood was initiated in 1976. The pattern development in the late 1970's resulted in a 16 acre north to south line drive pattern still used today.

Cumulative production from the Lost Soldier Tensleep through 1995 is 120 million barrels of oil (50% of OOIP).

Current production is 6,000 bopd, 1,000 bbl/day ngl, 75,000 bwpd, and 75,000 mcfd gas (Figure 2).

Geology and Reservoir Properties

The Tensleep sandstone at Lost Soldier is found at an average depth of 5,000 feet. The field structure, a faulted anticline (Figure 3) with the crest located at a depth of 3,800 had the original oil water contact at 7,000 feet (Figure 4). The structure dips 30 to 40 degrees on the west flank and up to 90 degrees on the east flank. Severe faulting exists on the east flank, and the north end is defined by a 2,500 foot thrust fault.

The Tensleep is predominantly an eolian sandstone deposit with an average gross thickness of 600 feet and a net thickness of 210 feet. Net pay porosity and permeability average 9.9% and 31 md, respectively. Initial water saturation was 10% in this oil wet reservoir.

A type log of the Tensleep sandstone is shown in Figure 5. The Tensleep is capped by an 80 foot tight, sandy dolomite zone. The best reservoir rock quality is found in Zone A, followed by Zones B and C. Some intervals in Zone A exceed 18% porosity and 500 md permeability. The Amsden shale separates the Tensleep from the Darwin formation below.

The reservoir oil is a conventional black oil with a stock tank gravity of 35° API and contained only 136 scf /bbl of gas initially (Table 1). The CO_2 minimum miscibility pressure was determined experimentally to be 2,350 psi. The original reservoir pressure was 2,520 psi.

Initial Project Implementation

The Bairoil fields were identified as excellent CO_2 flood candidates in 1978[1]. However, the lack of a CO_2 source delayed further evaluation. In 1984, Exxon developed its huge LaBarge Field, and in 1985, Amoco authorized CO_2 floods for the major horizons at Bairoil. First injection began in October 1986 into the Wertz Tensleep[1] and in July 1989 in Lost Soldier Tensleep.

Field Facilities. The Lost Soldier and Wertz fields share common production and injection facilities (Figure 6). Individual wells are produced into a well testing site where each well is tested at least weekly. The gas is separated at the tester and sent to the recycle facility, while the remaining oil and water flow to the central processing facility. From here, the oil is piped to a nearby refiner in Sinclair, Wyoming. The water is sent to the central water injection plant, and the remaining gas is sent to the CO_2 recycling facility.

The water injection facility repressures the produced water (along with 20,000 bbls of water makeup) to 2,600 psi. The water is distributed throughout the field via a radial system. Individual injection wells are pressure controlled from a central automation system.

The Bairoil CO_2 recycling facility receives approximately 110 MMCFD of inlet gas at 180 psi. The plant contains compression and natural gas liquid extraction. The CO_2 is repressured to 2,300 psi for reinjection into the field. Prior to reinjection, any CO_2 purchases are added to the total injection stream. The CO_2 is delivered into a lateral trunk system. Most injection wells are rate-controlled during their CO_2 half cycles.

Early Reservoir Pressurization. At the end of primary, reservoir pressure had decreased to 800 psi. Waterflooding had increased the pressure to only 1600 psi, 750 psi below the MMP. Since a miscible flood was necessary, the reservoir pressure had to be increased.

Figure 7 shows the reservoir pressure as measured in three wells during repressurization in 1988. The average pressure was increased by 1,200 psi during a period of 4 months by overinjecting 3 million barrels of water and by shutting in several high water oil ratio producing wells.

Flood Performance. Figures 8a and 8b show the production and injection performance curves for the Lost Soldier Tensleep Field. The relatively thick pay interval, good porosity, high permeability, presence of fractures, and close pattern spacing contributed to rapid tertiary response, as shown in figure 8a. In most cases, oil response to CO_2 injection occurred within six to eight months after start-up. Figure 8b shows CO_2 injection during mid-1989 with the initial response achieved later that same year. The response trend was briefly interrupted during late-1989 because of a major vessel malfunction. During this period, the operator determined that earlier reservoir repressurization rendered downhole electric submersible pumps ineffective. After facility problems were resolved, nearly all 47 producers were converted from artificial lift to flowing conditions. Just a few months later, production increased sharply from 5,000 to 10,000 BOPD. During this time, water injection averaged 75,000 BWIPD and CO_2 injection 65,000 MCFD. As expected, water production gradually declined while CO_2 production gradually increased.

A peak production of 11,000 BOPD was achieved when CO_2 injection was gradually increased from 65,000 to 90,000 MCFD. Two separate events are responsible for the oil drop in mid-1991. The first event was a scheduled field shut down because of yearly maintenance requirements at the CO_2 cycling facility. The second was a function of the reservoir. After the plant shut down work was completed, several large volume producing wells declined sharply in oil rate. This was caused by reduced sweep efficiency as a result of offset injection well conformance. Field performance was changed and signaled the first sign of a maturing field. High levels of flush oil production after a planned shut down period could no longer be achieved. The Wertz tertiary flood also demonstrated this performance.

Oil rate gradually improved during the next 12 months and eventually stabilized at the 9,500 BPD level. CO_2 injection averaged 90,000 MCFD and water 70,000 BPD. As a result of the 1992 plant maintenance shut down, oil production again dramatically declined from 9,500 to 8,000 BPD. By late-1992, oil rate began a steep decline from 8,000 to 6,500 BPD lasting six months. Water injection during this time was

75,000 BPD and CO_2 injection was 85,000 MCFD. By mid-1993, the Lost Soldier tertiary flood entered a mature state. Multiple patterns experienced high gas production volumes in conjunction with tertiary decline. A new reservoir management philosophy with conformance and reservoir pressure maintenance priorities would be required to flatten the decline.

The same variability in porosity, permeability, spacing, and fracture characteristics attributed to the rapid response are also responsible for steep tertiary oil decline rates. Rapid response and steep decline rates present a challenge for proper reservoir, pattern, and individual well management.

Individual Well Performance. Figure 9 shows the water and CO_2 injection rates for Lost Soldier 148 from 1989 and 1990. Daily injection volumes are shown on the plot. The CO_2 injection was rate controlled at 5.5 mmcfd, while the water injection was pressure controlled at 2,000 psi. The WAG ratio was 1:1 based on reservoir pressure compensated hydrocarbon pore volume.

During waterflood, the water injection rate averaged 3,000 bbl/d. After initiating CO_2 injection into this well, the water injection rate would begin at 4,000 to 5,000 bbl/d and decline rapidly to 2,000 bbl/d over the water half cycle. This phenomenon is attributed to a drop in reservoir pressure during the CO_2 half cycle, then a rebuilding of pressure during the water half cycle. While this behavior is observed in about 50% of the wells in Lost Soldier, only 20% of the Wertz injection wells exhibit this behavior. The remaining wells have both a constant water injection rate and pressure on the water half cycle. The constant water and CO_2 rates are attributable to fracturing characteristics.

The offset producing well to 148, Lost Soldier 30, increased in production from 100 to 1800 bopd within 9 months of CO_2 injection startup (Figure 10). Lost Soldier 30, the most prolific well in the field, had the largest single well incremental tertiary response (1700 bopd) known to the authors (Figure 11).

Current Project Management

The Lost Soldier Tensleep flood is being managed in conjunction with other reservoirs at Bairoil. Many of the original patterns that were on a 1:1 gas to water ratio (GWR), remain on a 1:1 GWR because they continue to produce significant tertiary oil. Each pattern is managed based on cash flow generation in relation to other patterns in the field. In general, the less mature or the better a pattern's cash flow, the higher it's GWR.

Each pattern is also evaluated for gas oil ratio (GOR) and water oil ratio (WOR) response. If the GOR in the producer rises too rapidly or gas breaks through soon after the CO_2 injection cycle is initiated, then the half-cycle slug size is reduced while maintaining the same GWR.

Some of the better patterns have GWR's higher than 1:1 to accelerate oil production, as long as the higher GWR does not cause uncontrollable increases in the GORs of offset production wells.

Mature Field Performance. The Lost Soldier Tensleep produces 6,000 bopd, 75,000 bwpd, and 75,000 mcfd from 52 producing wells (Figure 8a). CO_2 injection averages 80,000 mcfd, and water injection is 80,000 bpd into 60 wells.

Cumulative recovery through 1995 is 120 MMBO, representing 50% of the OOIP (Figure 12). Estimated ultimate recoveries are: 1) primary: 19.9%, 2) waterflood: 24.4%, 3) CO_2: 9.9%, and 4) ultimate: 54.2% of the OOIP. The current cumulative gross CO_2 utilization is 13.0 mcf/bo; net utilization is 4.6 mcf/bo. To date a 61% HCPV slug of CO_2 (177 BCF) has been injected into the reservoir (Table 2). Only four of the original CO_2 injectors have been put on chase water injection. Each of these wells has poor injection conformance.

In the Wertz Tensleep CO_2 flood, a 64% HCPV slug of CO_2 has been injected to recover 14 MMBO (9.4% of the floodable OOIP). Currently, Wertz is producing 2,000 bopd, 80,000 bwpd, and 25,000 mcfd (Figure 13). The higher recoveries expected for Lost Soldier are attributable to better reservoir quality.

The operator's reservoir management philosophy involves maintaining reservoir pressure above 2800 psi in order to keep producers flowing and maintain miscibility, to manage each pattern on profitability by allocating available CO_2 into areas with the best utilization, and to expand into new patterns as CO_2 becomes cost-effectively available.

Mature Individual Well Performance. Lost Soldier 148 remains on a 1:1 GWR. Average water injection rates have increased over time from 4000 to 6000 bpd. CO_2 injection rates have also increased from 5600 to 6800 mcfd (Figure 14). The performance of the offset producing well, Lost Soldier 30, has shown an increase in GOR from nil to 20,000 scf/bbl, indicating a high degree of maturity. Current production is 350 bopd, 5000 bwpd, and 7000 mcfd (figure 14).

Pattern Expansions. Since the initial flood installation, nine patterns have been expanded with CO_2 injection. The pay quality in these areas is not as good as in the initial flood area. However, average peak incremental tertiary response of 295 bopd has been achieved. Remaining expansion opportunities have been ranked, but they are much more risky as a result of geologic complexity. They are being pursued systematically so as to utilize surplus equipment from abandoned patterns in the Wertz field.

Downdip Oil. Reductions in operating costs and other improvements in lease profitability have allowed the operator to pursue drilling downdip producers. These wells are located at or near the oil water contact and produce oil in a portion of the reservoir very near the residual oil saturation to water. This residual oil saturation is a result of water influx during early field development. The performance of Lost Soldier 214 is shown in Figure 16. This well, the structurally lowest well in the field, is located within 200 vertical feet of the oil water contact. The well was drilled in 1987 prior to CO_2 injection and was completed at a non-commercial 2 bopd and 3,900 bwpd. CO_2 injection commenced in the offset injectors in 1989, and within one year, the oil rate had increased to 300 bopd. Ultimate recovery from this well is estimated to be 500 MBO.

Three wells were drilled in 1995 to further test the concept of downdip oil. Three additional wells are planned for 1996; contingent on the results of the wells drilled in 1995. With continued success, three more wells could be drilled in 1997.

Reservoir Repressurization. Between 1989 and 1994, reservoir pressure decreased in portions of the reservoir from 2,800 to less than 2,400 psi. As a result of changes that had occurred to the CO_2 injection stream, portions of the reservoir had lost miscibility (Table 3). The MMP for pure CO_2 is 2,350 psi; however, the MMP had increased to 2,544 psi due to contamination by various components[4] (nitrogen and methane) and the recent higher efficiency extraction of natural gas liquids from the injection stream. In addition to causing some wells to flow intermittently, the operator suspected that the loss of miscibility was causing a reduction in the flood's efficiency.

In conjunction with a plant shutdown in early 1995, 10 producing wells were either shut in or had back pressure applied while injection into offset wells continued. The goal was to overinject 1 million barrels of fluid over a two month period. During this period, bottom hole pressure was measured bi-weekly in 11 key wells. At the end of the two month period, pressure had increased by an average of more than 400 psi (Table 4). After resuming production following the plant shutdown, oil increased from 5,800 to 6,500 bopd. Unfortunately, a facility problem lowered water quality and plugged several injection wells, allowing reservoir pressure and production to decrease. The operator has taken additional actions to increase injection into areas of low pressure by improving water quality, adding injection wells via recompletion from other horizons, and implementing wellbore stimulations.

Injection Conformance. As the Lost Soldier tertiary flood continues to mature, modifying injection conformance will require use of the same work process that was implemented for Wertz[5]. The primary objective for injection well conformance work focuses on maximizing oil recovery by changing either vertical or areal injection fluid distribution. When conformance work is successful, individual injection-production well combinations can recover up to 140,000 barrels of additional oil and extend tertiary recovery life by as much as two years. Candidate selection and placement of a specific conformance control technique are the two processes that have repeatedly proven critical for implementing successful conformance work.

Essential data required for candidate selection includes evaluation of: 1) detailed geologic correlation of zones to locate bypassed reserves, 2) historic fluid injection conformance, 3) offset production well performance, 4) injection fluid travel time to offset producers and, 5) well history to determine cement bonding. Near wellbore techniques include use of sand plugbacks, mechanical plugs, cement and matrix cement squeezes, low molecular weight polyacrylamide gel, or a combination of the four to achieve effective zone isolation. More complicated areal treatments for deep formation penetration include alpha olefin surfactants that generate an in situ foam and high molecular weight polyacrylamide materials that form an in situ gel.

NGL Recovery. Increased NGL production from 38,000 to 63,000 gallons per day was achieved during late 1994 as a result of tripling the Reid Vapor Pressure (RVP) from 12 to 40. This was a significant profitability improvement for the Bairoil tertiary projects because an outside operator was capable of transporting and fractionating a 40 RVP raw mix product to C3, C4, and C5+ components for an attractive fee.

To avoid potential CO_2 corrosion, metallurgy in critical areas was upgraded with stainless steel components. Adequate refrigeration and plant process stability were also verified. Then, the 40 RVP product was achieved through lowering of the fractionator reboiler temperature from 390 to 290 degrees F. An amine unit was installed to ensure removal of hydrogen sulfide (H_2S) and CO_2 from the fractionated product.

Sour Reinjection. Since initiation of CO_2 injection during mid-1986, CO_2 recovered from tertiary flooded wells was sweetened using a LoCAT process. The 1700 ppm H_2S content of produced gas was reduced to less than 100 ppm before reinjection. However, dirty LoCAT service caused multiple downstream mechanical problems, as well as increased oil downtime and reduced processing plant throughput.

Mid-1995, the entire CO_2 recycling plant and associated facilities were successfully converted from sweet to sour service. Results included a significant 30% reduction to variable plant operating expenses (chemicals, maintenance labor costs, and electrical power), and improved major equipment runtime efficiency from 98.0 to 99.7%. After eight months of operation, CO_2 reinjection has averaged 950 ppm of H_2S and there have been no major problems operating a formerly sweet plant in a sour service environment.

Positive implications emerged from sour injection at Bairoil. There was no impact to the community, because the existing emergency injection-shut-down system was deemed acceptable by the town of Bairoil. Plant dehydration efficiency was improved so that piping systems now operate with zero free water and therefore no sulfide stress cracking potential. The original vent system was converted to a closed system, leading to the incineration unit and emergency flare stack. Finally, Sulfur Dioxide emissions were reduced through the use of additional compression and process modification.

Facility Consolidation. As the Bairoil tertiary recovery projects mature, reduced operating costs were and continue to be a critical issue for improving profitability. Consolidating surface production facilities offered one method of reducing these costs.

During 1994, the Wertz tertiary project had a significant increase in the number of marginally economic patterns. As a result, several patterns were either converted to chase water injection or shut-in. These changes caused fluid throughput into the Wertz production facilities to decline from 130,000 to 65,000 BFPD. Since the Lost Soldier main production battery was not at capacity, an opportunity to consolidate existed.

Consolidating Wertz and Lost Soldier production was completed during early 1995. The project included installing nearly one-half mile of 16 inch diameter flowline between the two fields and installing a low-volume vapor-recovery unit at the Lost Soldier main battery. As a result of this work, several Wertz production facilities were salvaged, including two primary production facilities, two major fluid transfer stations, one gas compressor, and a flare stack. In addition, five emergency discharge pits were reclaimed and chemical storage eliminated.

Conclusions

1. Ultimate incremental tertiary recovery of 13% of the original oil in place confirms the technical success of CO_2 miscible injection into the Lost Soldier Tensleep formation.
2. Reservoir management is critical from project inception throughout all stages of maturity to optimize product margins.
3. Evaluating all components of the field production system can result in dramatic improvements in field profitability, thereby allowing exploitation of growth opportunities.
4. The Tensleep reservoir could be repressured in a very short time period to attain miscibility pressure.
5. Sour reinjection can be a viable alternative to costly sweetening of injected gas streams.

Nomenclature

bopd = barrels of oil produced per day
bwpd = barrels of water produced per day
mcfd = thousands of standard cubic feet of gas per day
mmcfd = millions scf of gas per day
GOR = gas oil ratio, mcf/bo
GWR = gas to water ratio, res bbl/res bbl
HCPV = hydrocarbon pore volume
RVP = Reid Vapor Pressure
OOIP = original oil in place
NGL = natural gas liquids

Acknowledgments

The authors thank Amoco Exploration and Production Sector for permission to publish this paper; Mike DeHerrera and Ron Colwell for their assistance with the Bairoil plant and facilities; and acknowledge the efforts of countless other individuals who have been involved with the project since its inception.

References

1. Kleinsteiber, S.W.: "The Wertz Tensleep CO_2 Flood: A Review of the Engineering Design and Initial Performance," paper SPE 18067 presented at the 1988 SPE Annual Technical Conference and Exhibition, Houston, TX, Oct. 2-5.
2. Hunter, J.K., and Bryan, L.A.: "LaBarge Project: Availability of CO_2 for Tertiary Projects," *JPT*, (Nov., 1987) 1407-10.
3. Internal Amoco Studies.
4. Sebastian, R.S., Wenger, R.S. and Renner, T.A.: "Correlation of Minimum Miscibility Pressure for Impure CO_2 Streams," paper SPE 12648 presented at the 1984 SPE Symposium on Enhanced Oil Recovery, Tulsa, OK, April 15-18.
5. Borling, D.C.: "Injection Conformance Control Case Histories Using Gels at the Wertz Field CO2 Tertiary Flood in Wyoming", SPE/DOE paper 27825 presented at the Ninth Symposium on Improved Oil Recovery held in Tulsa, OK, USA, 17-20 April 1994.

Author Biographies

Ron Brokmeyer is a Staff Petroleum Engineer with Amoco Exploration and Production Sector in Denver, Colorado, specializing in the reservoir management of tertiary injection and waterflood projects. He joined Amoco in 1985 after earning a BS degree in Petroleum Engineering from Texas Tech University.

Dan Borling is a Staff Petroleum Engineer with Amoco Exploration and Production Sector in Cairo, Egypt, specializing in operations engineering. Prior to moving to Egypt, Dan worked on Bairoil properties for 5 years. He joined Amoco in 1981 after earning a BS degree in Geological Engineering from South Dakota School of Mines and Technology.

Weldon Pierson is a Senior Petroleum Engineer with Amoco Exploration and Production Sector in Denver specializing in the reservoir management of CO_2 injection and waterflood projects. He joined Amoco in 1989 after earning BS and MS degrees in Petroleum Engineering from Texas Tech University.

SI Metric Conversion Factors

cp x 1.0	E -03	= Pa s
ft x 3.048	E -01	= m
ft^2 x 9.290 304	E -02	= m^2
ft^3 x 2.831 685	E -02	= m^3
in. x 2.54	E+00	= cm
bbl x 1.59	E -01	= M^3
lbf x 4.448 222	E+00	= N
md x 9.869 233	E -04	= µm2
psi x 6.894 757	E+00	= kPa
mile x 1.609	E+00	= km
acre x 4.047	E+03	= m^2
gal x 3.785	E+00	= liter

Table 1 -- Lost Soldier Tensleep- Pertinent Data

Crude Gravity, °API	35	
Reservoir Temp, °F	158	
Original Pressure, psig	2,520	
Bubble Point Pressure, psig	875	
Current Reservoir Pressure, psig	2,800	
Original B_o, res bbl/stb	1.12	
Crude Viscosity, cp	1.38	
Recoveries	MMBO	% OOIP
Original Oil in Place	240	100
Ultimate Primary	47.8	19.9
Ultimate Waterflood	58.5	24.4
Current Incremental Tertiary	13.6	5.7
Ultimate CO_2 Tertiary	23.7	9.9
CO_2 Tertiary in flooded area only	23.7	13.2
Ultimate including Tertiary	130	54.2

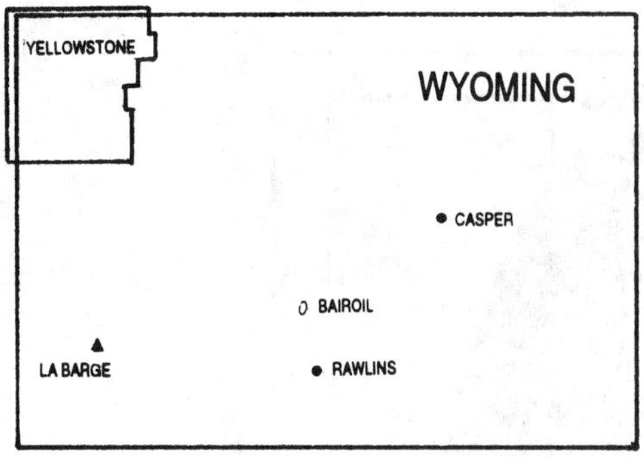

Fig. 1 - Location map for the Lost Soldier Tensleep Field

Table 2 -- CO_2 Utilization

CO_2 Injected (BCF)	177
CO_2 Produced (BCF)	114
Gross CO_2 Utilization (mcf/bo)	13.0
Net CO_2 Utilization (mcf/bo)	4.6
Slug size injected (% HCPV)	61%

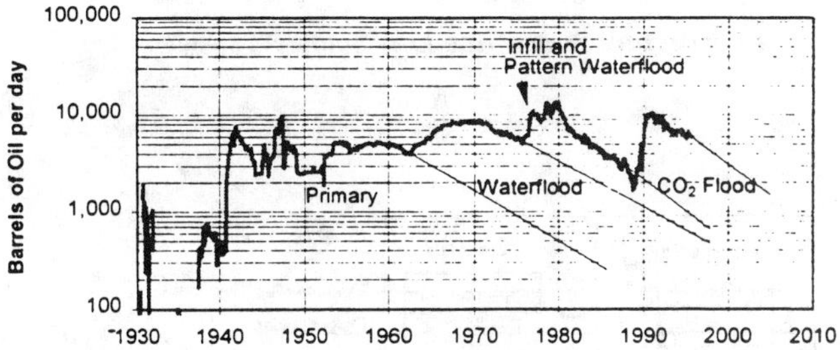

Fig. 2 - Historical Lost Soldier Tensleep oil production (1930 - 1995)

Table 3 -- Minimum Miscibility Pressure

Injectant	MMP (psig)
Pure CO_2	2,350
Pure purchases	2,420
Pure recycle (no NGL recovery)	2,399
Recycle (with 12 RVP product)	2,502
Recycle (with 40 RVP product)	2,563
Current mix (recycle and purchases)	2,544

Table 4 -- 1995 Repressurization Data

Well	Pre-pressure	Post-pressure	Increase
1	2,128	2,884	756
206	2,586	2,866	280
171	2,741	2,962	221
12	2,463	2,982	519
94	2,586	3,066	480
21	2,341	2,839	498
73	2,448	2,732	284
49	2,295	3,089	794
88	3,016	3,067	51
198	2,242	2,880	638
90	2,283	2,463	180

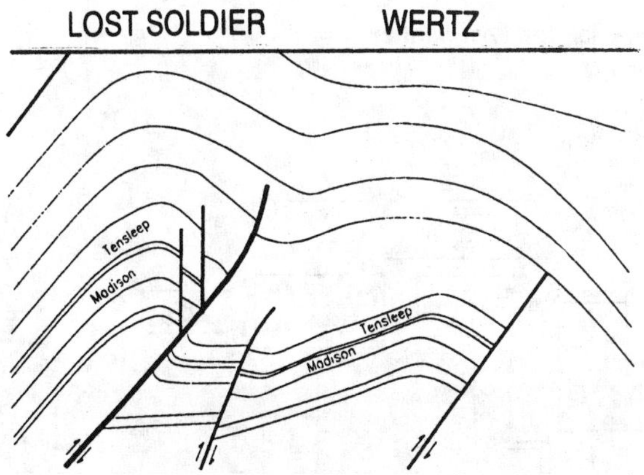

Fig. 3 - Structural cross section of the Bairoil properties

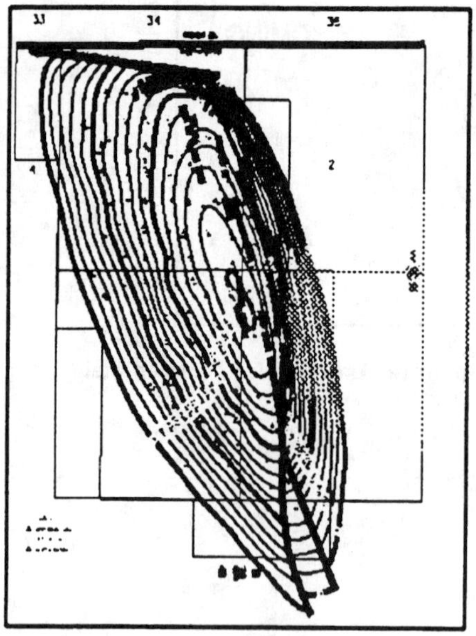

Fig. 4 - Structure map of the Lost Soldier Tensleep

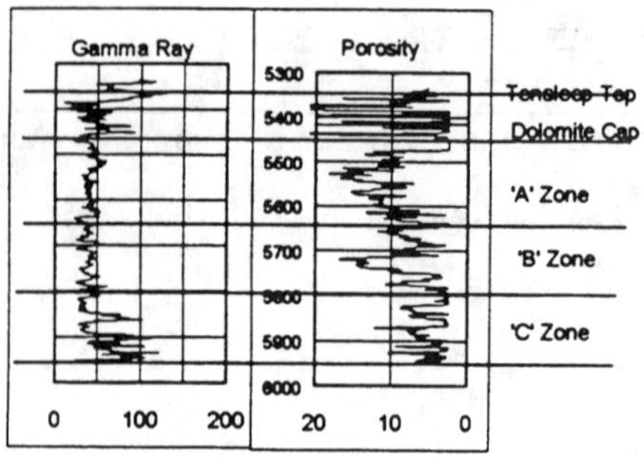

Fig. 5 - Type log for Lost Soldier

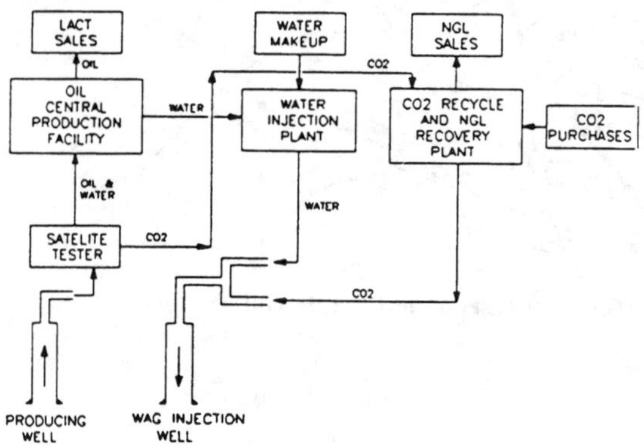

Fig. 6 - Bairoil field facilities

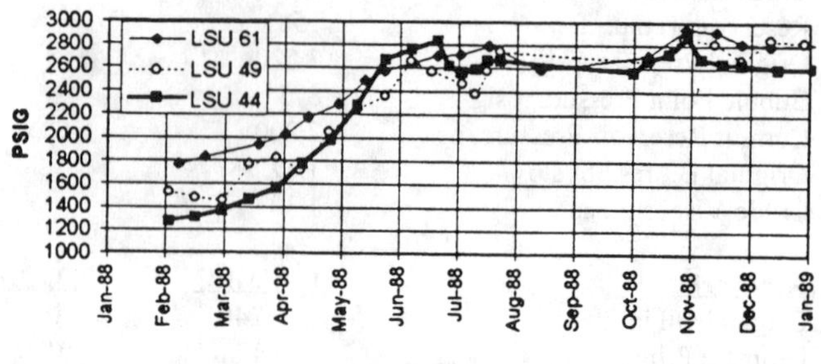

Fig. 7 - Results of 1988 reservoir pressurization in wells 61, 49, 44. Data shows a 1,200 psi increase over 4 months.

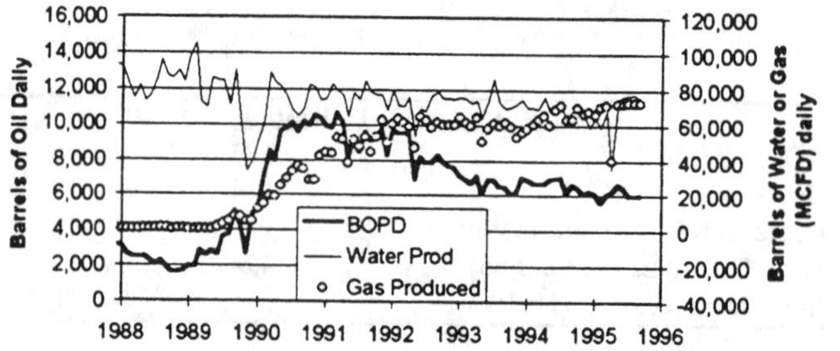

Fig. 8a - Oil, water and gas production from 1988 - 1995

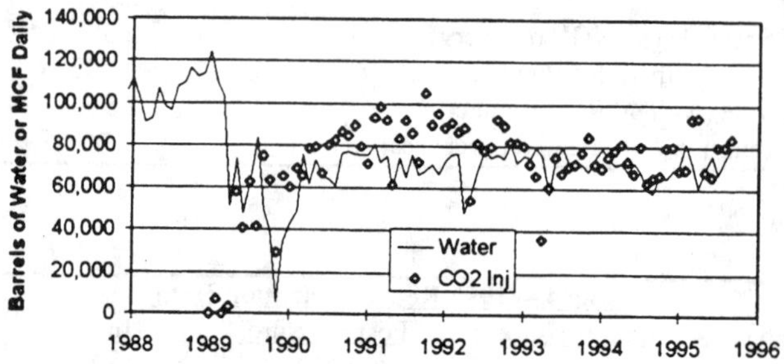

Fig. 8b - Water and CO_2 injection from 1988 - 1995

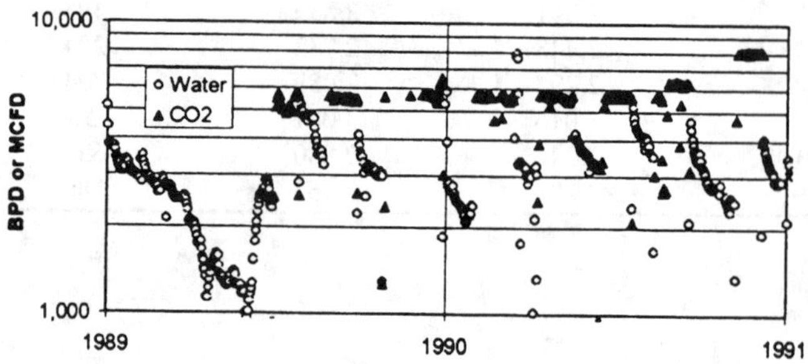

Fig. 9 - Injection performance curve for Lost Soldier 148 (1989 - 1990)

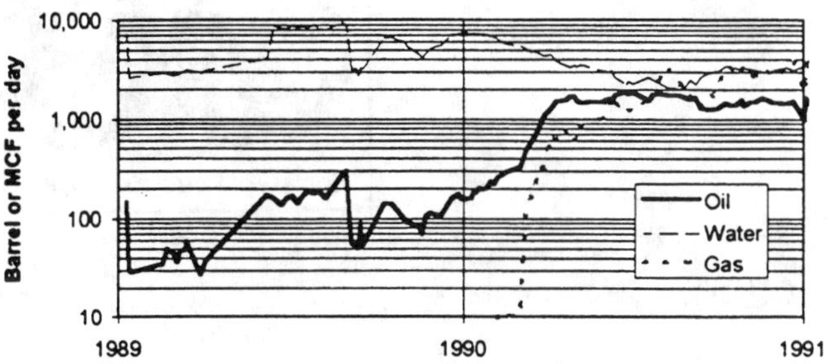

Fig. 10 - Production performance curve for Lost Soldier 30 (1989 - 1990).

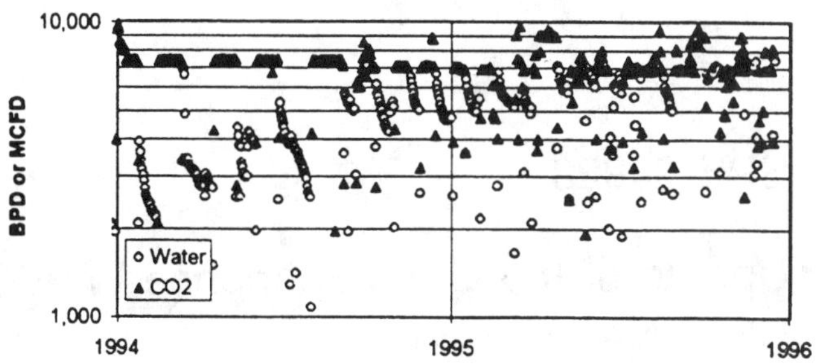

Fig. 14 - Injection performance curve for Lost Soldier 148 (1994 - 1995)

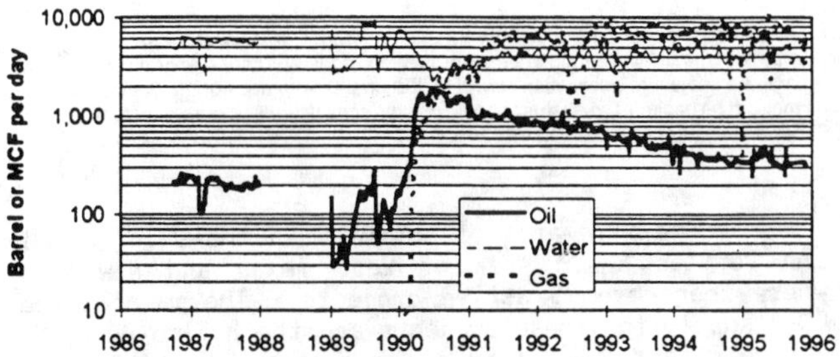

Fig. 11 - Production performance curve for Lost Soldier 30 (1986 - 1996). Best CO_2 response in U.S.

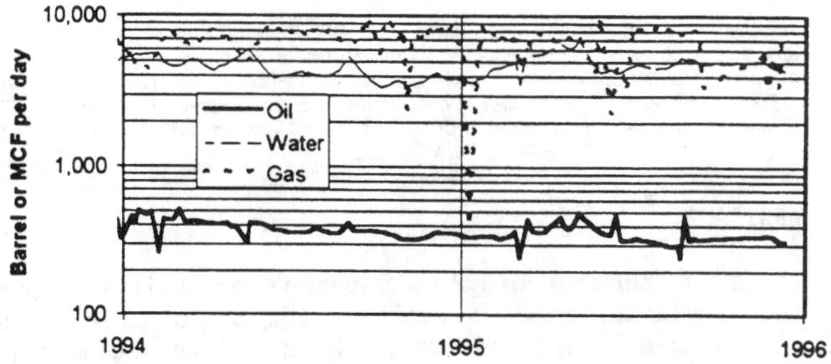

Fig. 15 - Production performance curve for Lost Soldier 30 (1994 - 1995)

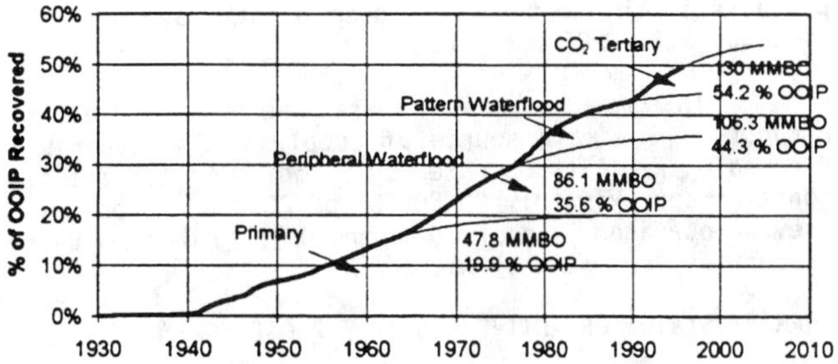

Fig. 12 - Lost Soldier Tensleep cumulative oil production since discovery (% OOIP recovered versus time)

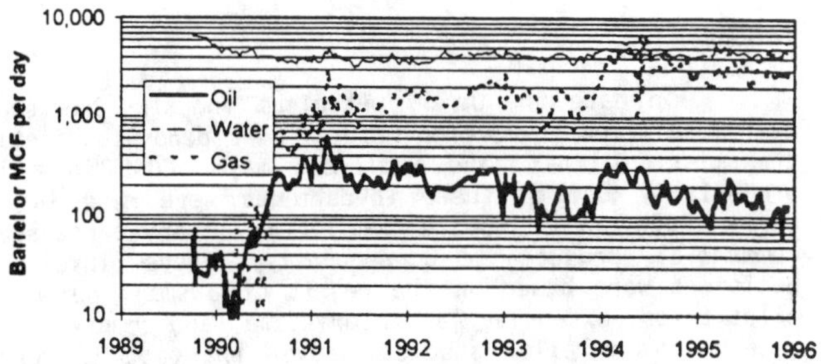

Fig. 16 - Production performance for Lost Soldier 214 (1989 - 1995)

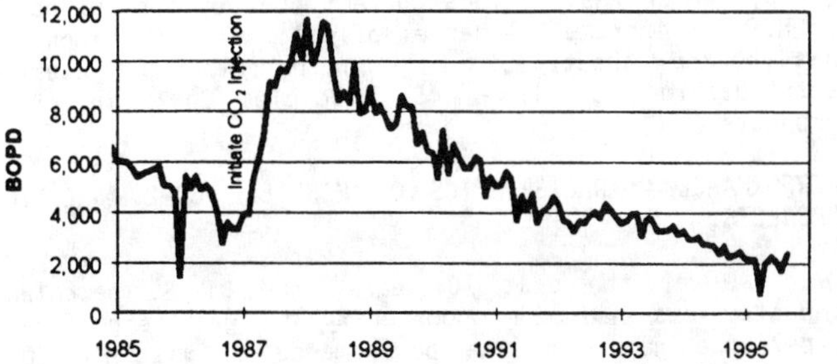

Fig. 13 - Oil production for the Wertz Tensleep CO_2 Flood

SPE 24928

Update of Industry Experience With CO_2 Injection

R.E. Hadlow, Exxon Co. USA

SPE Member

This paper was prepared for presentation at the 67th Annual Technical Conference and Exhibition of the Society of Petroleum Engineers held in Washington, DC, October 4–7, 1992.

ABSTRACT

The main purpose of this paper is to utilize current industry experience to evaluate the performance to date of miscible CO-2 injection projects. Also included is a summary of innovations being implemented by industry to improve recovery from existing CO-2 projects.

Experience to date has shown that on balance, CO-2 injection projects have performed better than anticipated and offer a significant opportunity for improving ultimate recovery from existing reserves.

INTRODUCTION

To date the U. S. petroleum industry has spent over $7 billion to develop natural CO-2 deposits, install transmission lines, and implement major CO-2 injection projects. Most of these investments were made in the early 1980s, prior to the completion of any large-scale commercial project. Consequently, these investment decisions were based on the result of a small number of pilot tests, extensive laboratory data, and comprehensive reservoir simulation studies. Over the past 10 years, industry has gained significant experience from implementing major CO-2 injection projects. It is the purpose of this paper to summarize some of the more important findings.

Currently there are 45 active miscible CO-2 projects in the U. S. The geographical location of these projects and the main source of CO-2 supply is shown on Fig. 1.

In the southeast part of the U. S., there are five active projects. The main source of CO-2 supply is the Jackson Dome, which contains a reserve on the order of 1 to 3 trillion cubic feet (TCF) of CO-2. A Shell-operated pipeline having a capacity of about 200 million cubic feet per day (MMCF/D) transports the CO-2 to projects in Mississippi and Louisiana.

There are two active projects in Oklahoma that are supplied CO-2 from the Enid Fertilizer plant.

The main area of activity in the U. S. has been the Permian Basin Area of West Texas and New Mexico, where there are 33 active projects. The main source of CO-2 supply for these projects are the McElmo Dome, Sheep Mountain, and Bravo Dome which combined contain between 20 and 40 TCF of CO-2. Three major trunk lines, which have an unpowered capacity of about 1.5 billion cubic feet per day (BCF/D) of CO-2, have been constructed to transport CO-2 from these natural deposits to the Permian Basin Area. Large gas processing plants in the Val Verde Basin are another important source of supply for CO-2 projects in the Southwest. These plants have a capacity of about 200 MMCF/D of CO-2.

In the Northwest, there are five active CO-2 projects. The main source of supply is the LaBarge Gas Processing Facility operated by Exxon. The LaBarge field contains approximately 20-60 TCF of CO-2 and supplies the Chevron-operated Rangely Webber Unit and the Bairoil properties, operated by Amoco.

CURRENT STATUS OF ACTIVE MISCIBLE PROJECTS

The status of the 45 miscible CO-2 injection projects is shown in Table I. In total, industry is injecting about 2.4 BCF/D of CO-2 into the 45 active projects. Of this amount, approximately 1 BCF/D or 40% is being produced and reinjected. The incremental production response is 142,000 barrels oil per day (BOPD), which results in a current CO-2 utility of about 17 MCF per incremental barrel of oil. To date, industry has booked about 1.9 billion barrels of reserves attributable to these 45 miscible CO-2 injection projects.

PERFORMANCE CHARACTERISTICS OF MISCIBLE CO-2 INJECTION PROJECTS

Over the past 10 years, the U. S. petroleum industry has had the opportunity to gain significant experience regarding the performance of miscible CO-2 injection projects. Listed below are some of the more important observations.

UPDATE OF INDUSTRY EXPERIENCE WITH CO-2 FLOODING

EARLY BENEFITS FROM IMPROVED RESERVOIR MANAGEMENT

Prior to injecting an expensive solvent, most operators implemented improved reservoir management activities to insure that conditions were optimum for CO-2 displacement. These activities included such things as infill drilling to improve pattern uniformity, reallocation of water injection to achieve optimum reservoir pressure, improvement in injection profiles, and a closer monitoring of producing wells. As a result of better reservoir management, a substantial increase in production from the existing waterflood operation was noted prior to CO-2 injection.

A case in point is the Means San Andres Unit operated by Exxon. Concurrent with the implementation of the CO-2 injection project, Exxon initiated an infill drilling program to reduce the well spacing from 20 to 10 acres. In addition, the injection pattern was changed from a 5-spot to an inverted 9-spot. The additional reserves associated with CO-2 injection helped to justify the economics of the infill drilling program. Conversely, the closer well spacing improved the sweep efficiency of the CO-2, thereby improving the economics of the CO-2 injection project. Thus the projects complemented one another.

The results are shown on Fig. 2. This figure shows a plot of actual production rate through 1990, and a forecast out to the year 1995. Also shown is the base production decline without the project. Notice the sharp rise in production in 1983, resulting from the infill drilling and improved secondary recovery operations. Also, notice that the CO-2 injection did not commence until November 1983. The production rate continued to rise as CO-2 response began to take effect and peaked at a rate of about 15,000 BOPD. The production has gradually declined to the current level of about 14,000 BOPD, which represents a 10-fold increase over the rate without the project. Based on reservoir studies, Exxon estimates that CO-2 injection will be responsible for about 65% of the increased recovery and the infill drilling and improved waterflood operations will account for the remaining 35%.

PRODUCTION RESPONSE ENCOURAGING

Because of gravity override, viscous fingering, reservoir heterogeneity, and lack of experience from field-wide commercial projects, there was an initial concern by industry that the production response from CO-2 injection might be disappointing. Experience to date, however, has shown that incremental production response to CO-2 injection has been outstanding. This is illustrated by Fig. 3, which shows the production history of 4 major CO-2 projects. These are typical response curves and are representative of industry's overall experience. The actual production from the unit is shown by the solid line and the estimated decline without the project is shown by the dashed line. The difference between the two represents the production response from the CO-2 injection project. It should be noted that the production from the total unit is shown even though CO-2 injection is not occurring throughout the entire unit area.

The graph in the upper left corner of Fig. 3, shows the production response in the Seminole-San Andres Unit, that is operated by Amerada Hess. CO-2 injection began in March 1983, using a water alternating gas (WAG) ratio of 1.4 to 1. To date, approximately 700 BCF of CO-2 has been injected, which represents about 25% of a hydrocarbon pore volume (HCPV). The current incremental production response due to CO-2 injection is over 30,000 BOPD, which accounts for 65% of the total unit production.

The graph in the upper right corner shows the production history of the Denver San Andres Unit, operated by Shell. CO-2 injection began in April 1984, using the continuous injection method. In 1988, Shell switched from continuous CO-2 injection to a 1 to 1 hybrid WAG. To date, about 1034 BCF of CO-2 has been injected, representing about 17% of a HCPV. Production response from CO-2 injection is currently in excess of 20,000 BOPD or about 50% of the total unit production.

The lower left graph shows the production history of the Conoco-operated Ford Geraldine Unit which produces from the Delaware Sandstone Reservoir. The response from this project is a textbook case. Without the project, the Field probably would have reached its economic limit and would have been abandoned. In February 1982, Conoco initiated CO-2 injection at a rate of about 20 MMCF/D, using the continuous CO-2 injection method. CO-2 injection during the next several years was erratic because of down-time associated with the plant supplying the CO-2. However, in 1986, Conoco obtained a more reliable supply of CO-2 and has since injected CO-2 on a steady rate. Note the sharp response in incremental oil production starting in 1986. The current production response is in excess of 1500 BOPD.

The graph in the lower right of this figure shows the production history of the Willard San Andres Unit operated by Arco. CO-2 injection was implemented in January 1985, using a 1 to 1 WAG ratio. To date, Arco has injected about 144 BCF of CO-2, representing 10.5% of a HCPV. Currently the production response is over 3,000 BOPD which represents about 60% of the total unit production.

In summary, it is obvious that the production response from the CO-2 injection has been outstanding and will continue to grow over the next few years as these projects mature.

CO-2 BREAKTHROUGH

Another concern of industry was the possibility of massive premature CO-2 breakthrough which could cause serious curtailment of production. Experience to date has shown that early CO-2 production has occurred in essentially all the projects, but at rates in line with predictions. The favorable CO-2 retention is illustrated by the top portion of Figure 4, which shows a plot of cumulative CO-2 retention versus cumulative CO-2 injection for 5 major projects. Of the 5 projects shown, Rangely has had the most severe CO-2 breakthrough problems, but it still has a satisfactory retention level in excess of 65%. Experience has shown that excessive CO-2 production can be alleviated by increasing the WAG ratio or using polymers and foams.

The bottom section of Figure 4 illustrates the effectiveness of CO-2 in mobilizing the trapped residual oil. Plotted is CO-2 utility which is defined as the amount of CO-2 required to recover an incremental barrel of oil. Three stages in the life of the project are shown. The first bar represents the cumulative CO-2 utility to date; the second bar, the current CO-2 utility; and the third bar, the estimated final CO-2 utility for the project. Notice that for every project the current CO-2 utility is significantly below the cumulative. This is a very favorable indication that a significant bank of oil has been mobilized and that the efficiency of the CO-2 is continuing to improve. Also notice that the current CO-2 utility is below the final utility in two of the projects. Overall, the CO-2 utility rate appears to be in line with the forecast and this is a very positive sign that the CO-2 is doing an efficient job of mobilizing the trapped residual oil.

LOSS OF INJECTIVITY

Another major concern of industry was the loss of injectivity in the WAG projects. Since most of the CO-2 injection projects are being conducted in reservoirs which have an average permeability of less than 10 millidarcies, loss of injectivity could have a major, adverse impact on the economics. As shown by Fig. 5, experience to date has indicated that most projects will experience some loss of injectivity during the water injection cycle of the WAG process. The first bar on Fig. 5 represents the injectivity of water under waterflood operations and is used as a benchmark to compare the injectivity under the WAG process. The second bar shows the injectivity of CO-2 compared to the waterflood injectivity. With the exception of Slaughter, the CO-2 injectivity exceeds the injectivity of water under waterflood operations. This is to be expected because the viscosity of CO-2 is much lower than the viscosity of water. The third bar shows the injectivity of water following CO-2 injection. Substantial loss of injectivity was experienced in Denver, Cordona Lake, and Slaughter. There was essentially no loss of injectivity at Means and Rangely and a gain of injectivity at SACROC.

In summary, it appears that an average CO-2 project will experience about a 20% loss of injectivity in the WAG process. In cases where the losses of injectivity has been severe, the problem has been minimized by decreasing the WAG ratio, increasing the injection pressure, or adding additional injection wells. In no case, has the loss of injectivity seriously impaired the economics of a major CO-2 injection project.

RESERVOIR SIMULATION CAPABILITY

The ability to predict CO-2 performance through the use of reservoir simulation is a critical tool in designing and implementing a successful CO-2 injection project. Experience has shown that industry has developed simulation capability that can accurately predict CO-2 performance. To date industry has obtained sufficient production history under CO-2 flooding to calibrate and evaluate the effectiveness of reservoir simulators. Numerous technical articles have been published which indicate the simulators do a good job of matching and predicting reservoir performance.

WATER BLOCKING

Because most CO-2 floods were initiated late in the life of the waterflood, there was concern that water blocking would be a severe problem. Experience to date suggests that water blocking has not had a significant impact on CO-2 flood performance. The main evidence which supports this conclusion is the excellent results obtained from the WAG projects. If water blocking was a severe problem, the injection of water in the WAG cycle would have had a detrimental effect on recovery. To the contrary, the WAG projects appear to be performing better than those using the continuous injection method.

CORROSION

Because CO-2 is a corrosive material when combined with water, there was uncertainty regarding the severity of corrosion. Experience indicates that corrosion has not been a severe problem mainly because of the preventative measures taken by industry. Methods to reduce corrosion include the use of corrosion inhibitors, separate CO-2 injection lines, stainless steel wellheads, and fiber glass gathering systems.

PRODUCTION PROBLEMS

The most severe production problem experienced to date has been the plugging of the tubing strings by asphaltines and gyp deposition. There also have been some problems with rod pumps gas locking and electrical submersible pump cycling when large amounts of CO-2 break through have occurred. Overall production problems have not been a major factor and have been minimized by an active well monitoring program.

ULTIMATE RECOVERY

Initial studies indicate that CO-2 flooding would increase the ultimate recovery by about 8 to 14% of OOIP. Since most of the major CO-2 projects are still in their early-to-mid life, the final amount of recovery is still speculative. However, performance to date has been encouraging and it appears that the incremental reserve estimates are conservative and that the final recovery will exceed initial expectations.

Although the major CO-2 floods are still a long way from being completed, there is one project which has been successfully flooded on a commercial scale basis. This project was conducted by Chevron in the south part of the SACROC unit. It consists of 4 inverted 9-spot patterns on 40-acre spacing, covering slightly over 600 acres. The results are shown on Figure 6. The upper left of the figure shows the location of the 4-pattern area. The upper right of the figure shows the production history of oil, water, and CO-2. Notice that production rose from 800 BOPD to about 3200 BOPD; a 4-fold increase. The lower left of the chart shows a plot of water-oil ratio versus cumulative oil production. Shortly after CO-2 injection began in June 1981, the water-oil ratio declined sharply from about 30 to around 8. This is evidence that the injected CO-2 has successfully mobilized a bank of trapped residual oil left after waterflooding.

The data graph on the lower right illustrates the final results of the CO-2 flood. The incremental oil recovery was about 10% of OOIP. The overall CO-2 utilization was about 10 MCF per incremental barrel and about 2/3 of the injected CO-2 was recovered. These results are very encouraging considering that project was conducted in the Canyon Reef formation on 40-acre spacing.

INNOVATIONS TO IMPROVE CO-2 RECOVERY

Although the CO-2 floods have been very effective in increasing ultimate oil recovery, there will still be between 50 to 60% of the oil left in the reservoir when these projects are completed. Consequently, there is significant incentives for industry to develop ways of improving the effectiveness of existing CO-2 projects. Following are some of the more successful innovations being implemented by industry to increase the recovery from existing CO-2 projects.

HYBRID WAG

One of the more successful innovations to improve recovery from CO-2 injection projects has been the use of the hybrid WAG process. Fig. 7 is a simple schematic which illustrates the hybrid process. The upper left shows the continuous injection or slug process in which a predetermined bank of CO-2, usually between 20-40% of a HCPV is injected on a continuous basis followed by the injection of chase water to drive the solvent bank through the reservoir. This technique has the advantages of early production response, better injectivity, and minimizing water blocking. On the other hand, it has the disadvantages of poor sweep efficiency and less efficient use of CO-2.

The upper right portion of this figure illustrates the WAG process. This method involves the alternate injection of CO-2 and water. Normally a volume of CO-2 equal to about 1 to 4% of a HCPV is injected on a continuous basis. Next a volume of water is injected to achieve the desired WAG ratio. The cycle is continued until the total solvent bank has been injected. The advantages of the WAG process are higher CO-2 utility, reduced CO-2 production, and greater ultimate recovery. The disadvantages are slower oil response, loss of injectivity, and greater susceptibility to water blocking.

The hybrid process shown at the bottom of the chart combines the best features of the WAG and continuous injection methods. In the hybrid process, a bank size of between 8 to 12% of a HCPV of CO-2 is injected on a continuous basis. Next the WAG process is implemented using the desired WAG ratio until the entire solvent bank has been injected. This is followed by chase water. The advantages of the hybrid process are (1) the early production response associated with continuous injection method and (2) greater ultimate recovery and CO-2 utilization characteristic of the WAG process. In essence, it results in the best of both of the other methods. The hybrid WAG has been successfully applied in the Denver Unit, operated by Shell, and the Dollarhide Unit, operated by Unocal.

The advantages of the hybrid WAG process referred to by Shell as the Denver Unit WAG is shown by Figure 8. This graph is based on reservoir simulation studies. Plotted is cumulative incremental CO-2 production in dimensionless terms versus time in years for three CO-2 injection methods. The recovery from the conventional 1 to 1 WAG is shown by the dashed line; the dotted line shows the recovery from the continuous method, and the solid line shows the 1 to 1 Denver Unit WAG. Notice that the 1 to 1 Denver Unit WAG has the same early production response as the continuous injection process. It also results in the same ultimate recovery as the 1 to 1 WAG. Based on the results of this simulation study and the favorable results of field tests, Shell has converted from continuous injection to the hybrid WAG process.

TAPERED WAG

Tapering the WAG ratio has also proved effective in reducing CO-2 production and increasing the effectiveness of CO-2 injection. A case in point is the Rangely Unit operated by Chevron. In order to reduce CO-2 production, Chevron is utilizing a tapered WAG process in which the WAG ratio is increased in step-wise fashion at predetermined solvent bank sizes. The general plan would be to use a 1 to 1 until 30% HCPV is injected. At this point, Chevron would switch to a 2 to 1 WAG until 40% of HCPV of CO-2 was injected. From 40 to 50% HCPV, a 3 to 1 WAG would be utilized after which chase water would be injected. An example of the favorable production response to the tapered WAG process is shown by Figure 9. This graph shows the cumulative incremental oil production in % OOIP versus HCPV of solvent injected for a single pattern. Notice the sharp increase in the slope of the incremental recovery line indicating a substantial improvement in CO-2 utility after shifting from a 1 to 1 WAG to a 2 to 1 WAG. In addition to improving solvent utility, the tapered WAG process will allow more wells to be on CO-2 injection for a given supply of CO-2.

INCREASING THE SOLVENT BANK SIZE

Another innovation implemented by industry is the use of a larger than planned solvent bank. It has been a recognized fact that the more solvent injected, the greater the incremental oil recovery. Unfortunately, it follows the law of diminishing returns. This is illustrated by Figure 10, which shows the results of a reservoir simulation study of the Means CO-2 project using varying solvent bank sizes. This unit, operated by Exxon, is employing a WAG ratio of 2 to 1. As illustrated by Fig. 10, the larger the bank size, the greater the ultimate recovery. But the increment gets smaller and smaller. The optimum bank size for a particular project will depend upon economics factors such as crude price, CO-2 cost, and the amount and timing of the incremental oil recovery. Fortunately, the ultimate solvent bank size for any project does not have to be finalized at the start of the project. This decision can be made later in the life of the project when the operator has a better handle on crude price and production response. Also, it should be mentioned that the optimum solvent bank size should be determined on an individual pattern basis rather than on a total project evaluation.

EXPAND THE PROJECT AREA

To reduce risk and optimize cost, the initial injection area in most projects was limited to the better sections of the reservoir. Expansion to other parts of the reservoir could be accomplished at a later date if the economics looked attractive. Since most of the major investments were made in implementing the initial project, the economics of project expansion are much easier to justify. Because of the successful performance of CO-2 projects to date, most projects have experienced significant expansions. Expansions can occur both laterally and vertically. Examples of lateral expansions have been experienced in the Denver Unit, the Seminole Unit, and the Rangely Unit. Examples of vertical expansion include the South Wasson project, operated by Shell and the Lower San Andres project, operated by Exxon. The Transition Zone in the Wasson and Seminole Fields represent future vertical expansion potential.

FOAMS AND POLYMERS

Foams and polymers are being tested by industry as a means of improving the vertical and areal sweep efficiency of CO-2 injection projects. Polymers and gels have been used primarily to improve vertical sweep in such projects as Rangely, Means, Maljamar, and Little Creek. Overall, the success rate has been above average.

Foams have been used to improve the mobility ratio between the CO-2 and reservoir crude. Success to date has been limited. Foam tests have been conducted at SACROC, Rangely, Wertz and Seminole. Recently, a joint industry/DOE sponsored foam test was implemented in the East Vacuum Unit operated by Phillips. Considering the magnitude of the potential for increasing recovery, it is likely that industry will continue to devote a great deal of effort to develop effective foams for use in CO-2 injection projects.

NEAR MISCIBLE DISPLACEMENT

One of the most significant opportunities for expanding the use of CO-2 could be the application of CO-2 flooding in reservoirs in which a near miscible condition exists. To date, industry has focused on conducting projects in reservoirs in which the CO-2 miscibly displaces the oil. However, laboratory and field data indicate that substantial incremental recovery over waterflooding can be achieved if the CO-2 displacement process is conducted in a near-miscible condition. For example, Figure 11 shows a typical plot of a slim tube test to determine miscibility. As shown, the displacement efficiency of CO-2 is plotted against reservoir pressure. At pressures above 1300 psig, the displacement efficiency exceeds 90% and the displacement process is considered to be miscible. However, at 1200 psig, the displacement would be near the miscible level and the recovery efficiency would be 80%, substantially above the recovery level of waterflooding. Field experience also supports this data. For example, when CO-2 injection was initiated at North Cross, the reservoir pressure was slightly below miscibility. Shell noted good reservoir response to CO-2 injection both below and above the minimum miscibility pressure. A similar response was noted at Dollarhide by Unocal as the pressure was being raised by CO-2 injection to achieve miscibility.

CONCLUSIONS

Experience to date from CO-2 injection projects have indicated the following.

(1) Significant production response has been noted from improved reservoir management activities initiated prior to CO-2 injection.
(2) Production response from CO-2 injection has been encouraging and in most projects has exceeded expectations.
(3) CO-2 retention has been excellent and the amount of CO-2 breakthrough has been in line with predictions.
(4) Original estimates of CO-2 utility appear to be on target.
(5) Loss of injectivity in the WAG process averages about 20% but has been manageable.
(6) Industry has developed satisfactory simulation capability.
(7) Water blocking does not appear to be a significant problem.
(8) Corrosion problems have been kept under control by an active preventative maintenance program.
(9) Production problems have been manageable.
(10) Ultimate recovery appears reasonable and will probably be exceeded.
(11) The future for CO-2 injection is bright, considering the success of existing projects and the innovations being implementing by industry to improve the performance of CO-2.

ACKNOWLEDGEMENT

The author wishes to thank Exxon Company, U.S.A. for permission to publish this paper, and also appreciates the cooperation and assistance from technical representatives of other companies who supplied data on the performance of their CO-2 projects.

REFERENCES

1. Poole, E.S.:"Evaluation and Implementation of CO-2 Injection at the Dollarhide Devonian Unit," Paper SPE 17277, presented at the SPE Permian Basin Oil and Gas Recovery Conference,Midland, Texas, March 10-11, 1988.

2. Brownlee, M.H. and Sugg, L.A.:"East Vacuum Grayburg San Andres Unit CO-2 Injection Project: Development and Results to Date," Paper SPE 16721, presented at the 62nd Annual Technical Conference and Exhibition, Dallas, Texas, September 27-30, 1987.

3. Pittaway, K.R. and Runyan, E.E.:"The Ford Geraldine Unit CO-2 Flood: Operating History," Paper SPE 17278, presented at the SPE Permian Basin Oil and Gas Recovery Conference, Midland, Texas, March 10-11, 1988.

4. Langston, M.V., Hoadley, S.F. and Young, D.N.:"Definitive CO-2 Flooding Response in the SACROC Unit," Paper SPE/DOE 17321 presented at the SPE/DOE Enhanced Oil Recovery Symposium, Tulsa, Oklahoma, April 17-20, 1988.

5. Magruder, J.B., Stiles, L.H. and Yelverton, T.D.:"A Review of the Means, San Andres Full-Scale CO-2 Tertiary Project," Paper SPE/DOE 17349, presented at the SPE/DOE Enhanced Oil Recovery Symposium, Tulsa, Oklahoma, April 17-20, 1988.

6. Larsen, W.K.:"Rangely Weber Sand CO-2 Project: A Case History," presented at the Denver SPE MiniSymposium, November 18, 1987.

7. Kirkpatrick, R.K., Flanders, W.A. and DePauw, R.M.:"Performance of the Twofreds CO-2 Injection Project," Paper SPE 14439, presented at the 60th Annual Technical Conference and Exhibition of the SPE, Las Vegas, Nevada, September 22-25, 1985.

8. Rowe, H.G., York, S.D. and Ader, J.C.:"Slaughter Estate Unit Tertiary Pilot Performance," Paper SPE 9796, presented at the 1981 SPE/DOE Enhanced Oil Recovery Symposium, Tulsa, Oklahoma, April 5-8, 1981.

9. Flanders, W.A., Stanberry, W.A. and Martinez, M.:"Review of CO-2 Performance of the Hansford Marmaton Unit," Paper SPE/DOE 17327, presented at the SPE/DOE Enhanced Oil Recovery Symposium, Tulsa, Oklahoma, April 17-20, 1988.

10. Brink, J.L., Davenport, C.B., Patel, K.S., Patel, Y.R., Thakur, G.C., and Winzinger, R.,:"Design of a Major CO-2 Flood, North Ward Estes Field, Ward County, Texas," Paper SPE 19654, presented at the 64th SPE Annual Technical Conference and Exhibition, San Antonio, Texas, October 8-11, 1989.

11. Lin, E.C. and Poole, E.S.,:"Numerical Evaluation of Single-Slug, WAG, and Hybrid CO-2 Injection Processes, Dollarhide Devonian Unit, Andrews County, Texas," SPE Reservoir Engineering, November 1991.

12. Frey, D.D., Stein, M.H., Pariani, G.J., and Walker, R.D.:"The Slaughter Estate Unit CO-2 Flood: A Comparison Between Pilot and Field-Scale Performance," SPE 19375, presented at the SPE Production Technology Symposium, Lubbock, Texas, November 13-14, 1989.

13. Kleinsteiber, S.W.,:"The Wertz Tensleep CO-2 Flood: A Review of the Engineering Design and Initial Performance," SPE 18067, presented at the 63rd SPE Annual Technical Conference and Exhibition, Houston, Texas, October 2-5, 1988.

14. Baxley, P.T., Crump III, J.G., and Miller, W.C.:L"Production Performance of the Wasson Denver Unit CO-2 Flood," SPE/DOE 24156, presented at the SPE/DOE 8th Symposium on Enhanced Oil Recovery, Tulsa, Oklahoma, April 22-24, 1992.

15. Mizenko, G.J.:"North Cross (Devonian) Unit CO-2 Flood: Status Report," SPE/DOE 24210, presented at the SPE/DOE 8th Symposium on Enhanced Oil Recovery, Tulsa, Oklahoma, April 22-24, 1992.

TABLE 1 - CURRENT STATUS OF ACTIVE MISCIBLE CO-2 PROJECTS

GEOGRAPHICAL AREA	NUMBER OF PROJECTS	CURRENT CO-2 INJ RATE MMcf / D	CURRENT CO-2 PROD MMcf / D	INCREMENTAL TERTIARY OIL PROD MB / D	INCREMENTAL TERTIARY RECOVERY MM Bbls
Southwest (Tex. - N. Mex.)	33	1650	550	110	1,500
Northwest (Utah, Colo., Wyo.)	5	450	250	25	250
Oklahoma	2	70	40	4	40
Southeast (La., Miss.)	5	240	200	3	80
	45	2410	1040	142	1,870

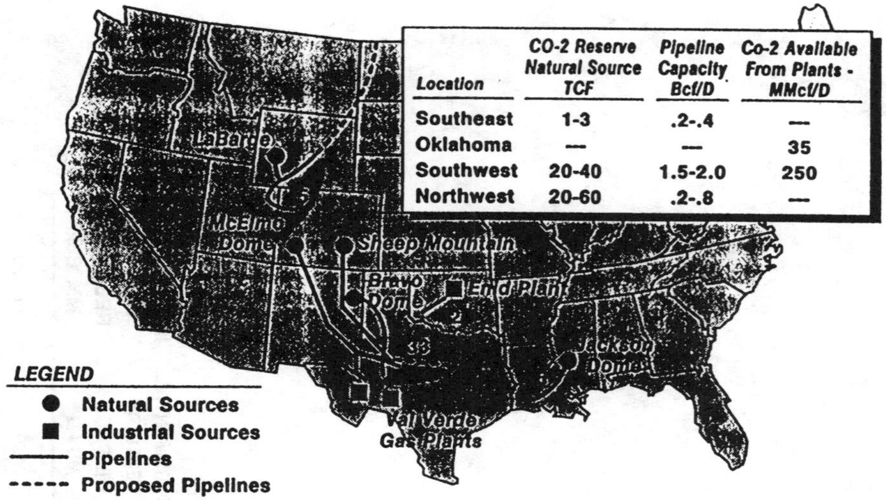

FIGURE 1 - LOCATION OF ACTIVE MISCIBLE CO-2 PROJECTS & MAJOR CO-2 SUPPLY SOURCES

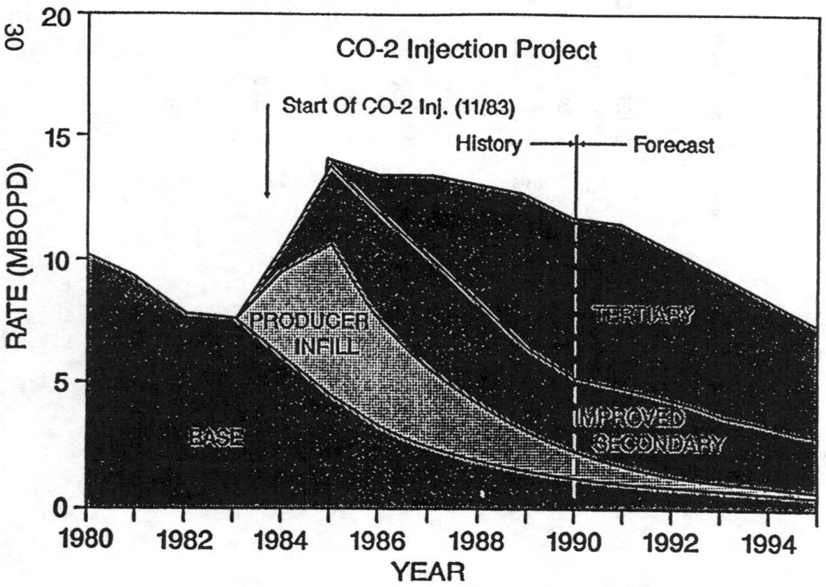

FIGURE 2 - RESPONSE COMPONENTS OF THE MSAU PROJECT AREA

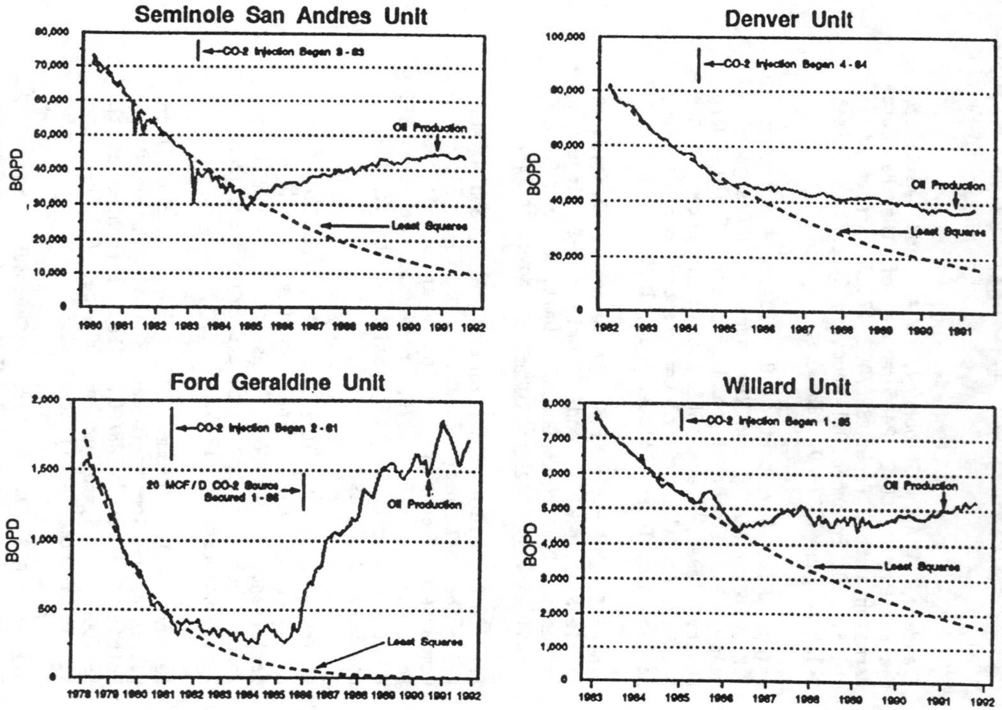

FIGURE 3 - PRODUCTION RESPONSE OF MAJOR MISCIBLE CO-2 PROJECTS

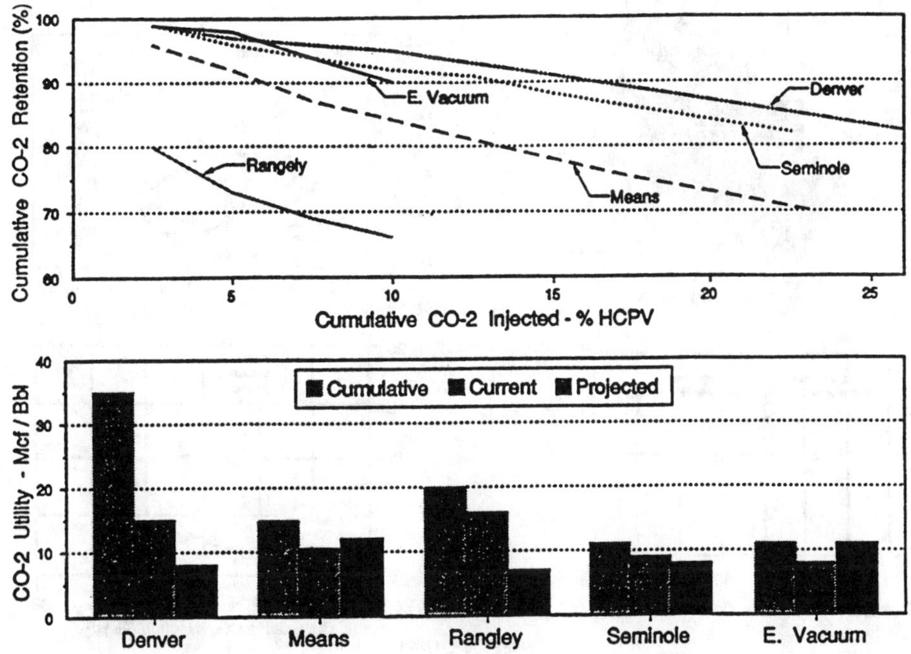

FIGURE 4 - CO-2 RETENTION AND UTILITY OF 5 MAJOR CO-2 INJECTION PROJECTS

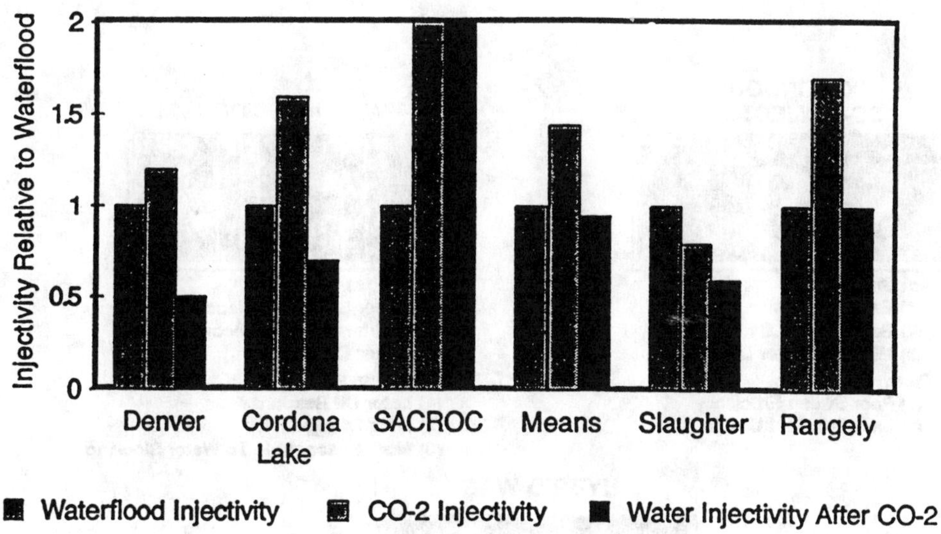

FIGURE 5 - W A G INJECTIVITY DATA

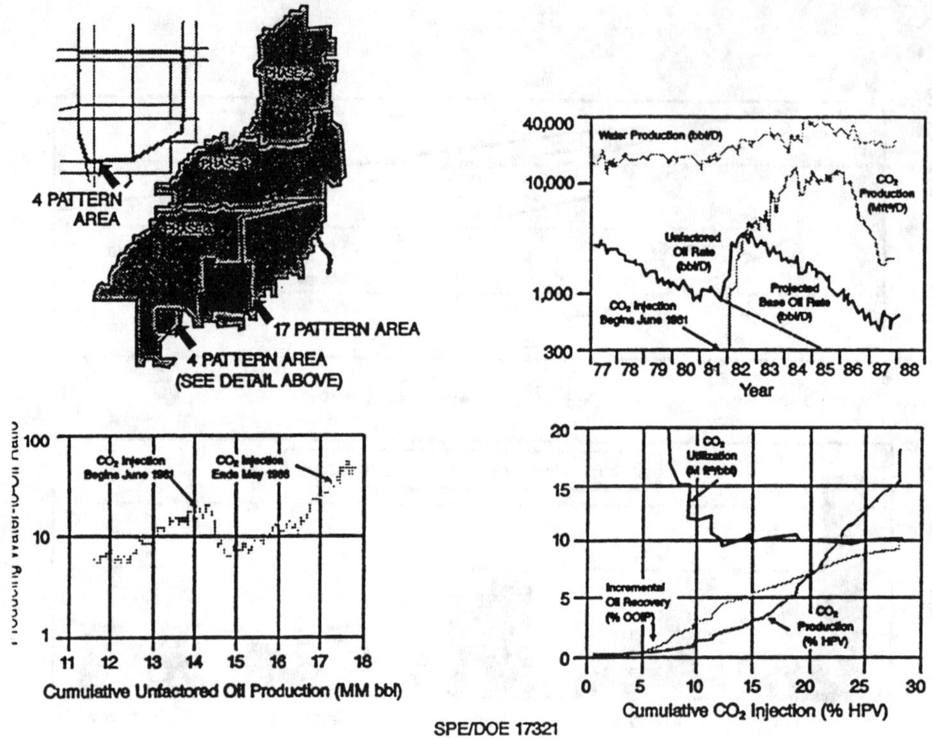

FIGURE 6 - DEFINITIVE CO-2 FLOODING RESPONSE OF PATTERN AREA - SACROC UNIT

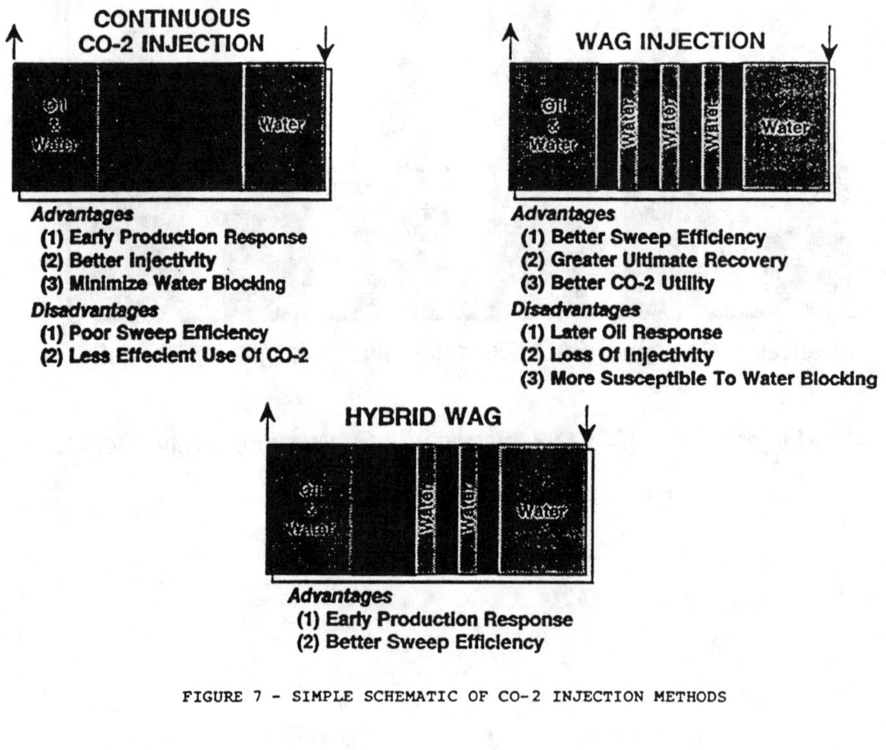

FIGURE 7 - SIMPLE SCHEMATIC OF CO-2 INJECTION METHODS

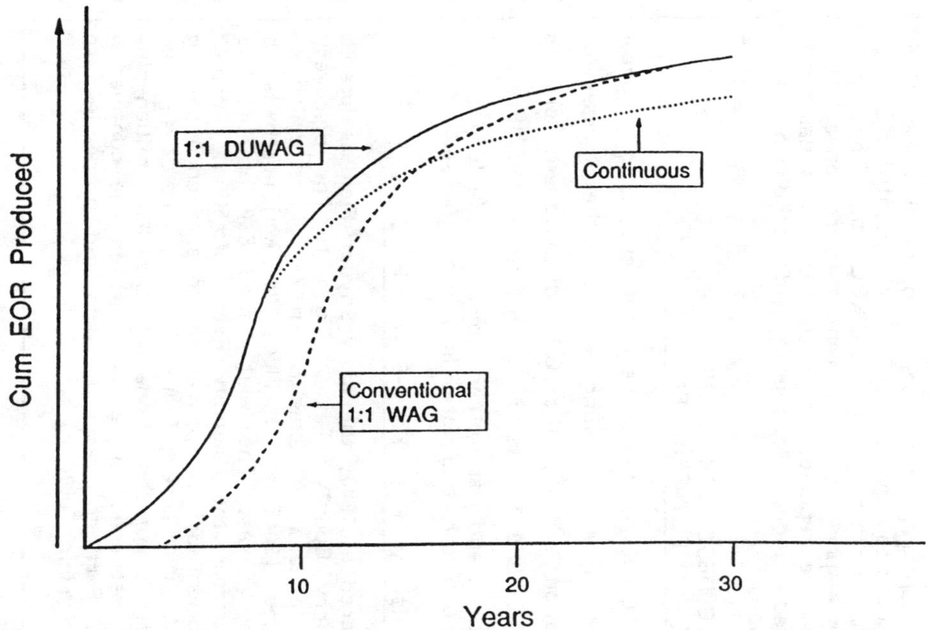

SPE/DOE 24156

FIGURE 8 - COMPARISON OF INCREMENTAL EOR PRODUCED

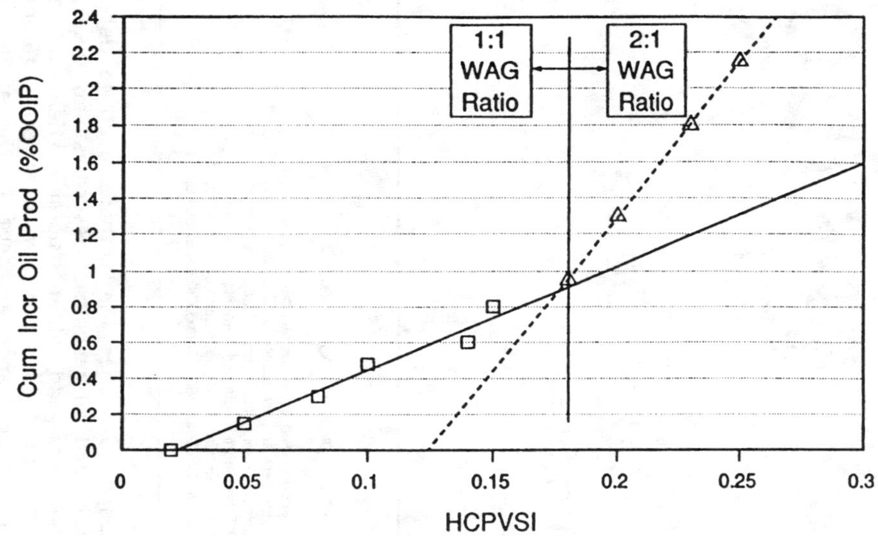

FIGURE 9 - EFFECT OF CHANGING WAG RATIO ON INCREMENTAL OIL RECOVERY

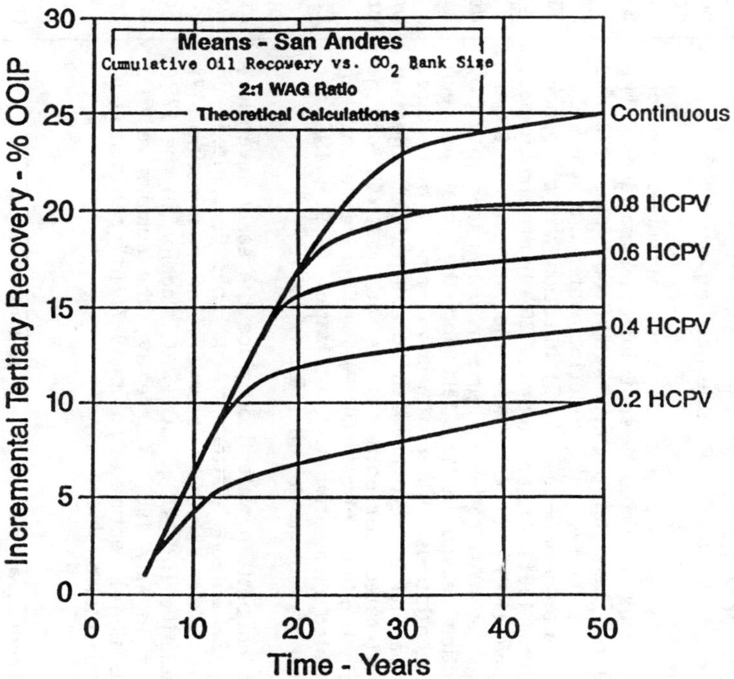

FIGURE 10 - EFFECT OF SOLVENT BANK SIZE ON OIL RECOVERY

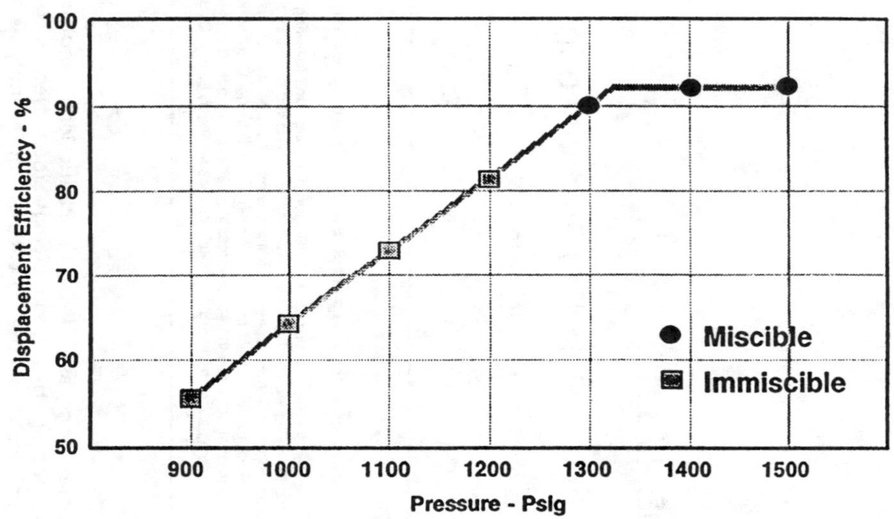

FIGURE 11 - TYPICAL SLIM TUBE MISCIBILITY TEST

SPE/DOE
Society of Petroleum Engineers
U.S. Department of Energy

SPE/DOE 17321

Definitive CO_2 Flooding Response in the SACROC Unit

by M.V. Langston, S.F. Hoadley, and D.N. Young, Chevron U.S.A. Inc.

SPE Members

This paper was prepared for presentation at the SPE/DOE Enhanced Oil Recovery Symposium held in Tulsa, Oklahoma, April 17–20, 1988.

ABSTRACT

A substantial increase in oil production resulting from CO_2 flooding has been clearly identified in two multi-pattern areas of the SACROC Unit. Analysis of the two areas permitted the identification of oil response to CO_2 injection with greater accuracy than has previously been possible at SACROC. The areas include the 600 acre [2.43×10^6 m^2] Four Pattern Area (4PA) and the 2700 acre [10.93×10^6 m^2] Seventeen Pattern Area (17PA). Located in the Kelly-Snyder Field of Scurry County, Texas, the 50,000 acre [202.3×10^6 m^2] SACROC Unit is the world's largest CO_2 miscible flooding project.

The 4PA encompasses 24 wells arranged in four contiguous inverted 9-spot injection patterns. The area has been on pattern waterflood since 1972 and was at a 95 percent producing water cut when CO_2 water-alternating-gas (WAG) injection was commenced in June 1981. An approximate 30% hydrocarbon pore volume (HPV) of CO_2 was injected over a 5-year period at WAG ratios ranging from two to eight. CO_2 injection ceased in May 1986 and the area has been on continuous water injection since that time. Incremental oil recovery attributable to CO_2 injection is estimated currently to be at least 9% of the original oil in place (OOIP). This represents an estimated cumulative CO_2 utilization of 9.5 Mft^3 per barrel of incremental oil [1692 m^3/m^3].

Also on pattern waterflood since the early seventies, the Seventeen Pattern Area has exhibited an approximate 5% OOIP recovery after injecting 17% cumulative HPV CO_2. CO_2-WAG flooding in the 17PA began in May 1981. Currently, the cumulative CO_2 utilization is estimated to be 9.7 Mft^3 per barrel of incremental oil [1728 m^3/m^3].

This paper examines the methods used to determine CO_2 mobilized oil response, describes how the effects of workovers and other "normal" field operations were accounted for, and evaluates the influence of activities in patterns adjacent to the study areas.

References and illustrations at end of paper.

INTRODUCTION

A substantial increase in oil production resulting from CO_2 flooding has been clearly identified in two multi-pattern areas of the SACROC Unit. The intent of this paper is to document that response. CO_2 performance reported herein is that which has been observed under "normal" field conditions and operations.

SACROC DESCRIPTION AND EARLY PROJECT PERFORMANCE

The SACROC Unit has been the subject of a great many papers dealing with the reservoir description[1,2,4,5,7], the CO_2 displacement process [5,8,9], CO_2 transmission [3], performance of the CO_2 project[6,10], and many other topics. The history provided below, therefore, is only a synopsis.

Early History of the Kelly-Snyder Field

Discovered in 1948, the Kelly-Snyder Field is located in Scurry County, Texas (Fig. 1). The discovery well, Standard of Texas Brown 2-#1, was drilled to 6,700 feet [2042 m], 9 miles [14.5 km] northwest of Snyder, Texas. The well flowed 530 bbl/D [84.3 m^3/d] from the Canyon Reef formation. Further development drilling proved up an area encompassing some 84,000 acres [340×10^6 m^2]. To date this discovery represents one of the last billion-plus barrel reservoirs to be found within the continental U.S.A. Pertinent reservoir data and properties are summarized in Table 1.

Description of the Canyon Reef Reservoir

The Kelly-Snyder Field produces from a thick carbonate buildup known as the Horseshoe Atoll. The chain of limestone mounds that make-up the atoll extend for a distance of 175 miles [281.6 km] and occupies most of the northern portion of the Midland Basin[1]. The eastern side of the atoll represents an area where major oil accumulations of the Pennsylvanian age occur.

The reef complex developed in the early Permian time when sediment filled the previously undisturbed Midland Basin. Active mound building ceased and sediment began covering the younger parts of the atoll until even the tallest reef pinnacles were buried under Dean siltstone. After completion of the burial process, subsidence occurred west of the Horseshoe atoll resulting in a westward tilting of the entire reef complex. Due to good lateral reservoir communication, updip hydrocarbon migration filled the reef mounds along the atoll's eastern edge. As a result over forty fields exist today in the Horseshoe Atoll. The most prominent are the Kelly-Snyder, the Lion-Diamond "M", the Sharon Ridge, and the Cogdell Fields.

The SACROC Unit constitutes the major portion of the Kelly-Snyder Field. Within the Unit, the reservoir thickness varies from 800 feet [244 m] in the crest of the reef to approximately 10 feet [3 m] on the flanks and covers an estimated 50,000 acres [202.3×10^6 m^2]. In the study areas the reef mound thickness varies from 300 feet [91 m] to less than 50 feet [15 m].

Formation of the SACROC Unit

The field was developed rapidly with over 1600 wells on production by 1951. The individual operators within the field soon realized that the primary recovery efficiency would be very low as the average reservoir pressure dropped by 50% and only 4.5% of the original oil in place (OOIP) was recovered. From the production behavior, the primary producing mechanism was determined to be solution gas drive. The ultimate recovery using the reservoir's natural energy was estimated to be 19% of OOIP.

In March 1953 the SACROC (Scurry Area Canyon Reef Operators Committee) Unit was formed to improve oil recovery. A unique center-line injection pattern was implemented in 1954 when 53 wells located along the crest of the reef, which runs lengthwise through the center of the field, were converted to injection service (Fig. 2). This injection scheme allowed large volumes of water to be injected into the thickest portion of the reef, achieving quick reservoir re-pressurization and an increase in oil production.

Design of the CO_2-WAG Project

Over the next 15 years it became apparent that the center-line injection system would be incapable of supporting production on the flanks of the reef. An intercompany engineering planning committee was formed in 1968 and was charged with investigating the feasibility of modifying or replacing the center-line system. Pattern flooding with CO_2 was determined to be the most economic alternative.[2] The plan eventually implemented called for inverted 9-spot injection patterns to be flooded in a water-alternating-gas (WAG) process.

Due to limited CO_2 supplies, the field was divided into three areas (or "phases") of similar hydrocarbon pore volume (Fig. 2) which were to be flooded sequentially. The original center-line area was not incorporated into the pattern injection scheme, but was expanded with additional injectors to encompass most of the area behind the leading edge of the center-line flood front.

Early Performance of the CO_2-WAG Project

Although Kane[10] and Cameron [6] discuss early phase performance of the project in great detail, some review is necessary to emphasize that project performance due to CO_2 injection was masked by response to 9-spot pattern waterflooding.

CO_2 injection commenced in the Phase 1 area in January of 1972 following two months of pattern water injection. As the majority of the injection patterns in Phase 1 were initially below 1,600 psi [11 MPa], a pre-CO_2 water slug was injected into low pressure patterns to ensure that minimum miscibility pressures (about 2,300 psi [16 MPa]) were attained. The pre-CO_2 water was followed by intentional overinjection during the ensuing CO_2-WAG cycles. Consequently, average pattern pressures in Phase 1 were in excess of 2,400 psi [16.6 MPa] by April 1973, an 800 psi [5.5 MPa] increase in 15 months' time. Phase 1 production response is evident in Fig. 3. The oil rate increased over three-fold from 30,000 bbl/D [4.77×10^3 m^3/d] in January 1972 to greater than 100,000 bbl/D [15.9×10^3 m^3/d] within 18 months.

Phase 2 pattern waterflooding began in June 1972. Phase 2 response to pattern waterflooding (Fig. 3) is not as noticeable as in Phase 1. Because of the long, narrow geometry of Phase 2 about the center-line pattern (Fig. 2), this phase received good waterflood support prior to pattern flooding. The sharp rise in oil rate that occurs in late 1971 is due to expansion of the Phase 2 centerline and installation of artificial lift equipment. A rising oil rate is apparent several months after the beginning of CO_2 injection in March 1974.

Phase 3 oil production response to 9-spot waterflooding, which began in April 1973, is very evident in late 1973 (Fig. 3) and coincides with rising water cuts. Although CO_2 injection began in Phase 3 in November 1976, only small amounts were injected until 1980.

In summary, determination of individual Phase response to CO_2 injection is difficult for two major reasons:

1) the rapid increase in pattern-area reservoir pressure had a profound effect on production capacity;

2) unlike most modern CO_2 miscible injection projects, the SACROC CO_2 project was implemented as a secondary project and the

bulk of the early response must be attributed to the waterflood component of the WAG process.

Given this situation, one would expect the best opportunities for isolating oil production response to CO_2 injection would be found in Phase 3 due to the long period of pattern waterflooding in that Phase prior to CO_2 injection. In the following sections we examine two examples of production response clearly attributable to CO_2 injection. Both of the areas examined are in Phase 3.

FOUR PATTERN AREA PERFORMANCE

After 1981, CO_2 processing efficiency became evident in a tertiary application as CO_2 injection was expanded into near watered-out areas in Phase 3. Production increases were observed for oil, natural gas liquids, and hydrocarbon gas following CO_2 injection. In particular, the Four Pattern Area (4PA) and the Seventeen Pattern Area (17PA) showed a definite and quantifiable impact from CO_2 injection. Because neither of these areas were unduly affected by workovers, temporarily abandoned wells that were returned to production, infill wells, or changing well counts, they had good potential for serving as examples of CO_2 processing performance at SACROC.

4PA Description

The 4PA is located on the south flank of Phase 3, south and east of the center-line injection system (Fig. 2). It is important to emphasize that the 4PA was never designated as a pilot area but was given special emphasis as a study area only after clearly discernible response to CO_2 injection was observed. This situation was advantageous in that it created an opportunity to study and document performance response to CO_2 injection under "normal" field conditions. As configured for pattern injection, the area is comprised of three, 160 acre [647.5×10^3 m^2], inverted 9-spot patterns and one smaller, and somewhat irregular, pattern. Total area is 603.1 acres [2441×10^3 m^2]. Four of the 24 original wells in the area have been recompleted as pattern injectors.

Determination of Pattern Factors

Some of the production reported below is factored. The term "factored production" refers to production allocated to a pattern, or pattern group, from a particular well. Two methods of factoring were used in the following analysis. Under one scheme, well factors were determined based on pattern geometry. Under the other, the results of streamline modelling were used to determine well factors.

Under the scheme based on geometry, a well located completely within a pattern would credit 100% of its producing stream to that pattern. A pattern side well straddling two pattern boundaries, however, would allocate 50% of its producing stream to each of the two adjacent patterns. Similarly, 25% would be allocated to a pattern from a well located on a pattern corner.

Chevron's Generalized Streamtube Model[11] was used to generate streamline contours of the 4PA during the periods of interest, 1978-86, based on average annual flow rates. Total reservoir rates for each well in the 4PA and in the immediately adjacent patterns were accounted for. Annual pattern balancing allocation factors were calculated for each 4PA edge well for each year, 1981-86. Average factors representing the period 1978-80 were applied during 1972-80 as production-to-injection ratios were fairly constant during this time. The streamline factors were developed as an alternative to geometric factors for the purpose of allocating the producing stream of the 4PA boundary wells. Although they did not differ greatly from geometric factors, they were slightly higher.

Examination of streamline-model developed contour maps also resulted in an estimate of the areal influence of CO_2 injection. This area was assumed to be the area contained within the "no-flow" boundaries implied by the streamline contours (Fig. 4).

4PA Pore Volume Calculations

The HPV of the 4PA was determined using data from the digital geologic database established by the SACROC Unit Reservoir Description Project[14]. In that study, net HPV included zones above the oil-water contact (OWC) with porosity values greater than 4%. In cases where the wellbore did not extend to the OWC, "missing-interval" porosity values were assigned using contour map values generated from deeper wells in the area.

For the purposes of our study, the wellbore footage between the top of the reef and the lowest open interval was included in the HPV calculation. The "lowest open interval" refers to the lowest point in the wellbore open to the formation, whether that point be defined by perforations or the bottom of an open hole completion. Using this technique and 4PA pattern boundaries, the HPV is determined to be 42,110 Mbbl [6.7×10^6 m^3]. The same technique was also used to estimate the HPV contained within streamline generated no-flow boundaries as discussed above. The HPV within the streamline boundaries was determined to be 43,904 Mbbl [6.98×10^6 m^3]. Using an oil formation volume factor from the 1973 simulation study[7] (1.57 reservoir volumes per standard volume), we determined that corresponding 4PA OOIP is 26,764 Mbbl [4.255×10^6 m^3] and 27,904 Mbbl [4.436×10^6 m^3], respectively.

4PA Primary Performance

The 24 wells in the area were placed on flowing production after completion in 1950 (Fig. 5). All production in the area was shut-in by September 1962 as the reservoir pressure in the vicinity was declining and the unit allowable was considerably below unit capacity. The area was returned to production in 1969 and a substantial but short-lived flush response is evident. Most of the wells were again shut-in over the next 3 years due to rising GORs and increasing water cuts. Cumulative factored (geometric) production to the start of pattern waterflooding (April 1973) is assumed to represent the end of primary at 19% of the area's OOIP. This is consistent with earlier studies indicating primary recovery would be 19% of OOIP, field-wide.

Review of bottom-hole pressure data from 1954 to 1970 reveals that the 4PA was apparently isolated from the influence of the nearby center-line injection pattern. We believe a permeability barrier may exist on the west and northwest sides of the 4PA. Additional evidence is provided by the fact that the water produced in the 4PA prior to pattern waterflooding appears to have migrated from the southeastern OWC and not from the center-line pattern (west and north). Furthermore, we can identify fill-up occurring after the area went on pattern waterflood. If the 4PA received strong center-line support prior to pattern flooding, we would not have expected a free gas saturation to exist in the area, nor would we have expected the rising GORs.

4PA Secondary Performance and Determination of 4PA Incremental Response to CO_2 Injection

The 4PA was placed on pattern waterflood in April of 1973. Approximately 1 HPV of water was injected between waterflood commencement and the start of CO_2 WAG injection in June of 1981. CO_2 was injected at an average rate of 13,600 Mft^3/D [385.1×10^3 m^3] until May of 1986. Cumulative CO_2 injection is 24.8 Bft^3 [702.3×10^6 m^3]. This volume represents 29.2% of the pattern-area HPV.

It was necessary to consider both streamline-factored and unfactored 4PA production to determine secondary and tertiary recovery as the waterflood mobilized production came from inside and outside the 4PA and the CO_2 mobilized production came from inside the 4PA only. Streamline generated allocation factors were applied to the 4PA boundary wells to separate and identify the secondary oil associated with the 4PA. For our determination of CO_2 mobilized production we established, as a base case, a production forecast for continued waterflood operations using unfactored production volumes. A production decline rate was determined from the pre-CO_2 injection period and extrapolated into the future. This projection was necessary because the 4PA was not at waterflood residual oil saturation at the time of CO_2 injection. The difference between the projected and actual production represents incremental oil production due to CO_2 injection (Fig. 6). Ultimate incremental oil production due to CO_2 injection is estimated to be 2.74 MMbbl [435.6×10^3 m^3].

Although the decline curve analysis was performed on the aggregate production of the 4PA, we also examined individual well performance by forecasting WOR ratio against cumulative oil production. Estimated ultimate incremental oil production for the individual wells was approximately 2.63 MMbbl [418.1×10^3 m^3].

Unlike other CO_2 floods in West Texas carbonate fields[12,13], the Canyon Reef reservoir injectivity is strongly stimulated by CO_2 injection. Increased injectivity typically occurs following the first CO_2 cycle and this behavior is observed in the 4PA. The water injection rates in the 4PA doubled following CO_2 injection but the injection pressure remained constant. Corresponding higher water production rates were observed as a result of increased injectivity.

Although the 4PA was producing at a surface oil cut of 5% prior to CO_2 injection, we do not credit any of our estimated incremental oil to accelerated waterflood production. If there was any acceleration of waterflood reserves, it represents a small volume as a 5% oil cut is a near-uneconomic WOR of 19:1. Also, any acceleration that was achieved was possible only because of stimulation of the injection wells by CO_2. The WOR actually decreased as the water production rate doubled, reflecting the formation of an oil bank mobilized by the CO_2 (Fig. 7).

The secondary and tertiary ultimate recoveries are estimated to be 21.7.% and 10.2% of the geometric pattern-area OOIP respectively. Streamline-area OOIP recoveries are 20.8% and 9.8%. At the end of 1987, the combined primary, secondary and tertiary recovery is approximately 49% with a projected total ultimate recovery of 50% of the OOIP for the pattern-area.

The cumulative CO_2 utilization, which is defined as the ratio of cumulative CO_2 injection to cumulative incremental oil production, is 9.5 Mft^3/bbl [1692 m^3/m^3] (Fig. 8). This figure also illustrates cumulative CO_2 production expressed as a function of cumulative CO_2 injection. Note that approximately two-thirds of the CO_2 injected has been produced on an HPV basis.

The effect of WAG ratio (or slug size) on CO_2 performance at SACROC has not been quantified. Because the timing of individual pattern cycles do not coincide with each other, total pattern area WAG ratios are difficult to determine. The cumulative WAG ratio, however, is approximately 3:1 for the 4PA.

Influence of Wellwork in the 4PA

Some well-work was performed during the period that CO_2 was being injected into the 4PA. Although most of the work was maintenance, there were three workovers (Fig. 2) which resulted in noticeable improvement in oil production rates and corresponding higher water production rates.

SACROC Unit Well #323-4 was recompleted in November 1981. Oil production increased to a stabilized rate of 229 bbl/D [36.4 m^3/d] from a pre-workover oil rate of 6 bbl/D [0.95 m^3/d]. Water production increased to 2,752 bbl/D [438 m^3/d] from 134 bbl/D [21.3 m^3/d]. Over the next 18 months, the oil rate continued to rise and the water rate held constant, indicating response to the on-going CO_2 injection.

SACROC Unit Well #323-6 was recompleted in May 1982. Oil production increased from 24 bbl/D [3.8 m^3/d] to 165 bbl/D [26.2 m^3/d] and water production increased from 272 bbl/D [43.2 m^3/d] to 2246 bbl/D [357.1 m^3/d]. This well also showed distinct signs of CO_2 response within several months.

SACROC Unit Well #322-2 was recompleted in June

1983. Oil production increased from 75 bbl/D [11.9 m^3/d] to 230 bbl/D [36.6 m^3/d] and water production increased from 172 bbl/D [27.4 m^3/d] to 953 bbl/D [151.5 m^3/d]. This well did not exhibit the typical CO_2 performance response that was apparent in the other two recompleted wells.

We estimate these workovers represent, at most, 411 Mbbl [65.34x10^3 m^3] in incremental oil reserves. While not all of those reserves are CO_2 mobilized, it is doubtful that the well-work would have been performed had there not been an effort to maximize our CO_2 recovery efficiency. These workovers were proposed with the expressed intent of capturing CO_2 mobilized oil in previously excluded zones.

SEVENTEEN PATTERN AREA PERFORMANCE

The purpose of the detailed analysis performed on the 4PA was to provide a benchmark for future estimates of CO_2 performance at SACROC. We have also examined the much larger Seventeen Pattern area. The size of this area precludes the in-depth type of study performed on the 4PA.

Seventeen Pattern Area CO_2 Performance

The 17PA area covers approximately 2,700 acres [10.93x10^6 m^2] and contains over 100 wells. This area is northeast of the 4PA, also in Phase 3, and is adjacent to the center-line pattern (Fig. 2). Unlike the 4PA, primary and secondary recovery in the 17PA was influenced by center-line injection. The center-line floodfront is believed to have moved through the area prior to pattern injection and swept additional oil into the 17PA. This increased the cumulative oil recovery and results in an optimistic waterflood processing efficiency for this area.

Identification of the swept volume using streamlines for the 17PA was not possible as the area in question is too large. The HPV was calculated using the lowest open interval as a base (as in the 4PA analysis) and using the existing geometric pattern boundaries. The HPV determined this way is 124,456 Mbbl [19.79x10^3 m^3]. The corresponding OOIP is 79,100 Mbbl [12.58x10^3 m^3].

Geometrically factored production volumes were used for our analysis of CO_2 performance. The assumption was that the peripheral wells should be factored to exclude oil produced from outside; however, the degree of 17PA CO_2 processing is greater than that of the adjacent patterns and the use of geometrically factored production has probably resulted in a conservative analysis.

The decline curve analysis (Fig. 9) was similar to the procedure used for the 4PA. The only difficulty that arose was making the projection for the incremental oil after 1987. It was apparent that tertiary recovery had not peaked and forecasting the ultimate recovery was questionable. The trend established by a brief decline in production which occurred from early 1985 to late 1986 was used to forecast production beyond 1987.

The 17PA has been on active CO_2-WAG flooding since May 1981. Cumulative processing through December 1987 is approximately 17.5% HPV. The cumulative WAG ratio is approximately 5:1. Incremental oil recovery due to CO_2 injection is estimated to be 5.5% OOIP. Ultimate incremental recovery for the 17PA should reach 7.5% of OOIP, but will depend on the amount of CO_2 eventually injected. The cumulative utilization ratio is 9.7 Mft^3/bbl [1728 m^3/m^3] (Fig. 10). This figure also illustrates cumulative CO_2 production expressed as a function of cumulative CO_2 injection. Note that approximately one-third of the CO_2 injected has been produced on an HPV basis.

CONCLUSION

Review of these areas in the Phase 3 section of the SACROC Unit has resulted in the definition of a quantitative response to CO_2 injection.

Injection of CO_2 in the Four Pattern Area will ultimately be responsible for an estimated 2.74 MMbbl [435.6x10^3 m^3] of incremental oil production. Cumulative utilization of the injected CO_2 will be at least 9.5 Mft^3/bbl [1692 m^3/m^3].

Injection of CO_2 in the Seventeen Pattern Area has been responsible for an estimated 5.93 MMbbl [942.8x10^3 m^3] of incremental oil production through the end of 1987. Cumulative utilization of the injected CO_2 is 9.7 Mft^3/bbl [1728 m^3/m^3].

ACKNOWLEDGMENT

We wish to thank Chevron U.S.A. Inc., and the working interest owners of the SACROC Unit, for permission to publish this paper. We also thank the production engineering staff at the SACROC Unit office in Snyder, TX. Without their cooperation this work would not have been possible.

REFERENCES

1. Vest, E. L., Jr.: "Oil Fields of the Pennsylvanian-Permian Horseshoe Atoll, West Texas," Geology of Giant Petroleum Fields, Memoir No. 14, The American Association of Petroleum Geologists, (1970) 185-203.

2. Hull, Paul: "SACROC: An Engineering Conservation Triumph," Oil and Gas J. (Aug. 17, 1970) 57-62.

3. West, James: "Line Will Move 240 MMCFD of CO_2," Oil and Gas J. (Nov. 8, 1971) 53-56.

4. Cramer, T. D. and Massey, J. A.: "Carbon Dioxide Injection Project, SACROC Unit, Scurry County, Texas," paper 361-D presented at the API Second Annual Meeting, Div. of Production, March 6-8, 1972.

5. Dicharry, R. M., Perryman, T. L., and Ronquille, J. D.: "Evaluation and Design of a CO_2 Miscible Flood Project -- SACROC Unit, Kelly-Snyder Field," J. Pet. Tech. (Nov. 1973) 1309-1318; Trans., AIME, 255.

6. Cameron, J. T.: "SACROC Carbon Dioxide

Injection: A Progress Report," paper 365-B presented at the API Annual Meeting, Production Department, Los Angeles, Apr. 5-7, 1976.

7. Brummet, W. M., Jr., Emanuel, A. S., and Ronquille, J. D.: "Reservoir Description by Simulation at SACROC--A Case History," J. Pet. Tech. (Oct. 1976) 1241-1255.

8. Rosman, A. and Zana, E.: "Experimental Studies of Low IFT Displacement by CO_2 Injection," paper SPE 6723 presented at the SPE-AIME 52nd Annual Fall Technical Conference and Exhibition, Denver, Oct. 9-12, 1977.

9. Graue D. J.: "SACROC Tertiary CO_2 Pilot Project," paper SPE 7090 presented at the 5th Symposium on Improved Methods for Oil Recovery, Tulsa, Oklahoma, April 16-19, 1978.

10. Kane, A. V.: "Performance Review of a Large-Scale CO_2-WAG Enhanced Recovery Project, SACROC Unit, Kelly-Snyder Field," J. Pet. Tech. (Feb. 1979) 217-231.

11. Martin, J. C., and Wegner, R. E.: "Numerical Solution of Multiphase, Two-Dimensional Incompressible Flow Using Streamtube Relationships," Soc. Pet. Eng. T. (Oct. 1979) 313-323.

12. Patel, P. D., Christman, P. G., and Gardner, J. W.: "An Investigation of Unexpectedly Low Field-Observed Fluid Mobilities During Some CO_2 Tertiary Floods," paper SPE 14308 presented at the SPE 60th Annual Technical Conference and Exhibition, Las Vegas, Sept. 22-25, 1985.

13. Pittway, K. R., Albright, J. C., and Hoover, J. W.: "The Maljamar Carbon Dioxide Pilot: Review and Results," paper SPE/DOE 14940 presented at the SPE/DOE 5th Symposium on Enhanced Oil Recovery, Tulsa, Apr. 20-23, 1986.

14. Swulius, T. M.: "Porosity Calibration of Neutron Logs, SACROC Unit," J. Pet. Tech. (Apr. 1986) 468-476.

CANYON REEF RESERVOIR AND FLUID PROPERTIES

Average Porosity (%)	9.41
Average Permeability (mD)	3.03
Oil Formation Volume Factor (bbl/bbl)	1.51
Initial Reservoir Pressure (psi)	3122
Reservoir Temperature (°F)	130
Bubble Point Pressure (psi)	1850
Oil Viscosity at 3122 Psi (cp)	0.35
Solution GOR (ft^3/bbl)	1000
Oil Gravity (°API)	41
Water - Oil Mobility Ratio	0.3
CO_2 - Oil Mobility Ratio	8.0

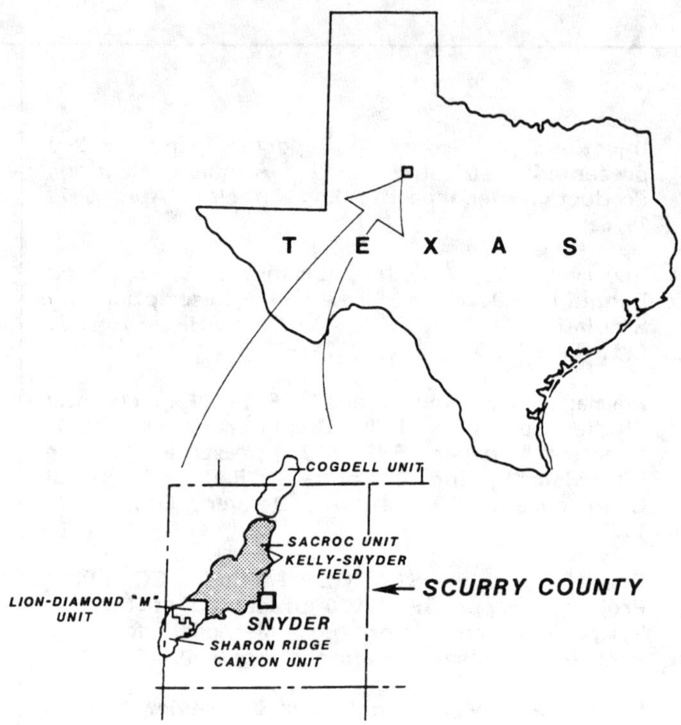

Fig. 1—Canyon Reef fields in Scurry County.

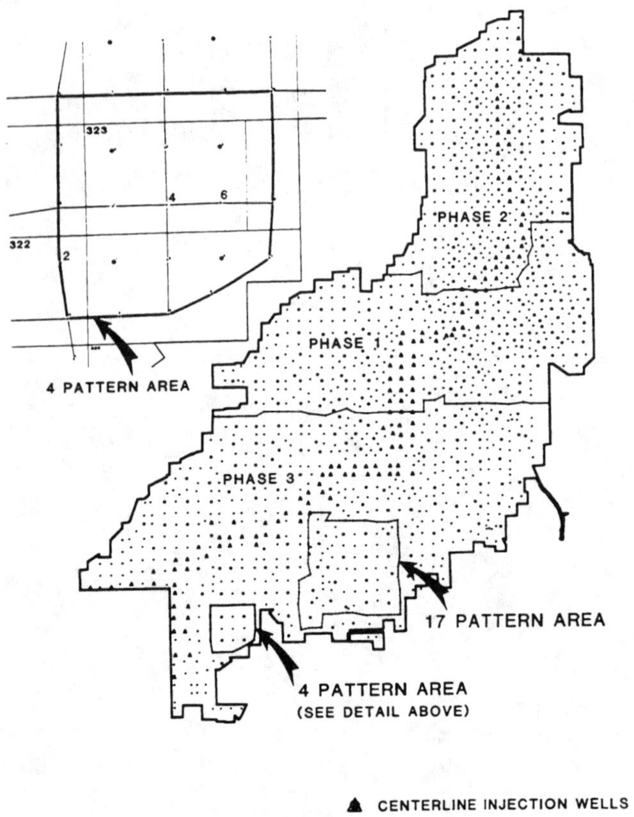

Fig. 2—Map of SACROC Unit.

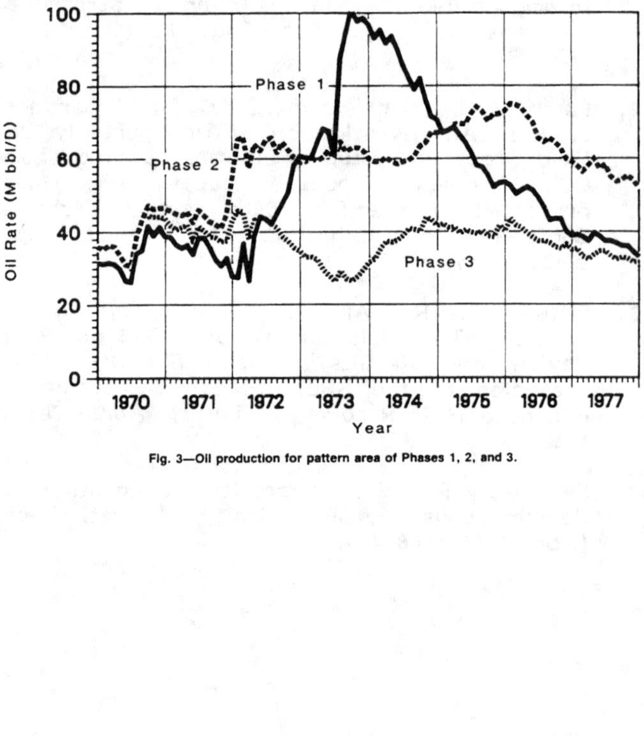

Fig. 3—Oil production for pattern area of Phases 1, 2, and 3.

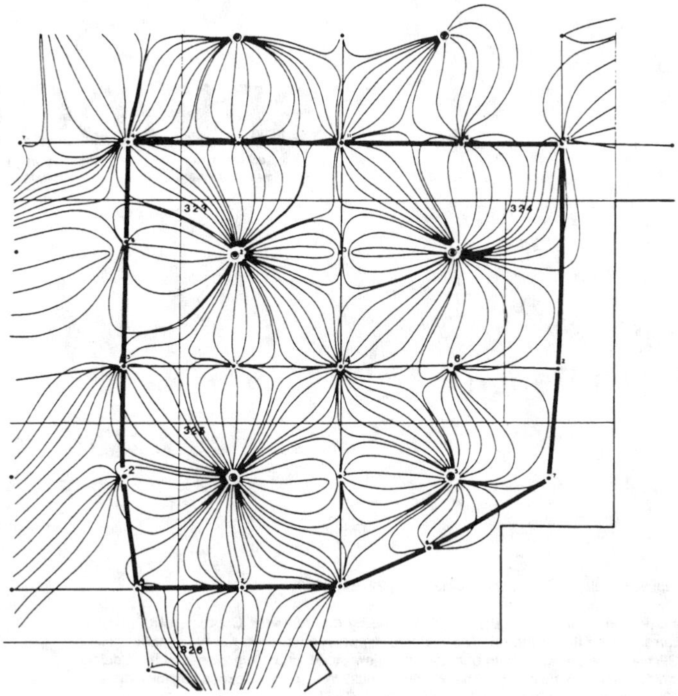

Fig. 4—Streamline map of the Four Pattern Area.

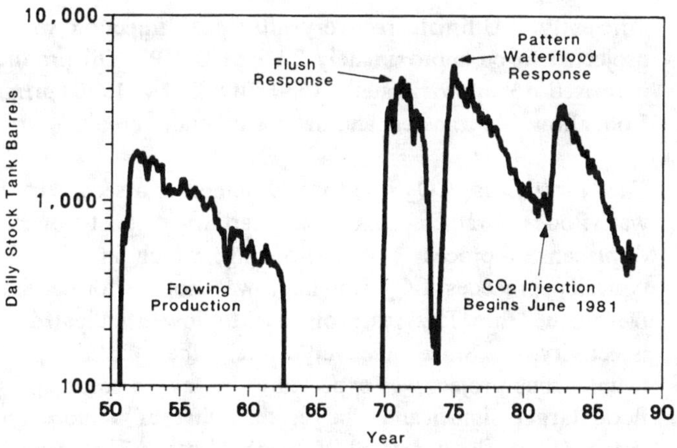

Fig. 5—Oil production history of the Four Pattern Area.

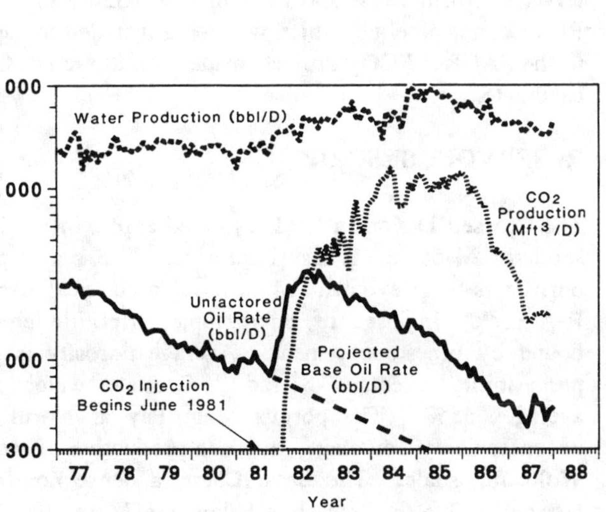

Fig. 6—Four Pattern Area performance summary.

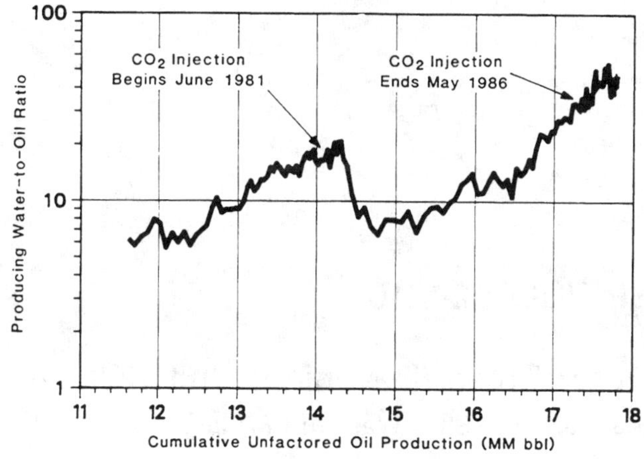

Fig. 7—Four Pattern Area WOR performance.

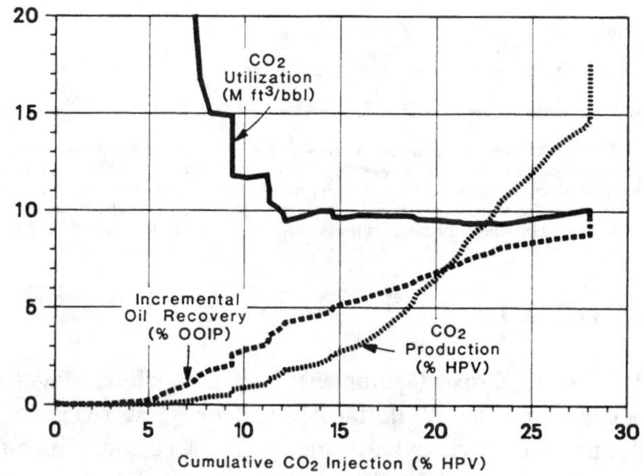

Fig. 8—Four Pattern Area processing efficiency.

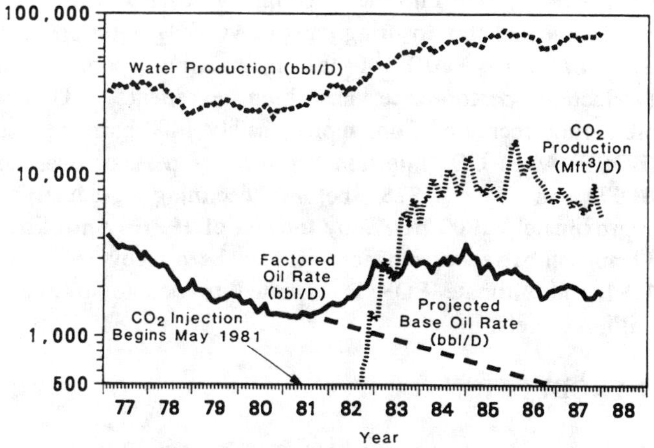

Fig. 9—Seventeen Pattern Area performance summary.

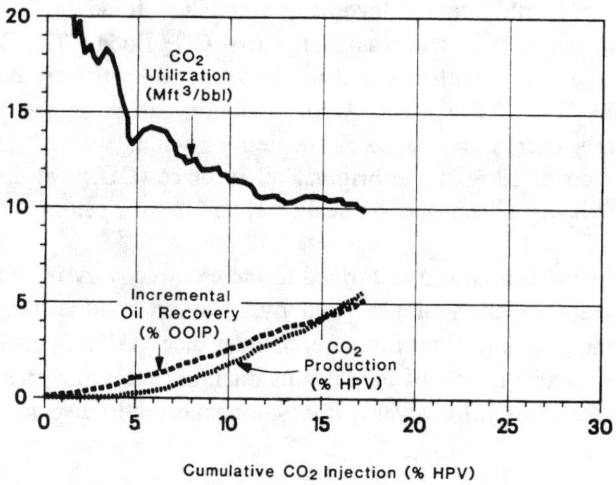

Fig. 10—Seventeen Pattern Area processing efficiency.

SPE/DOE
Society of Petroleum Engineers
U.S. Department of Energy

SPE/DOE 24210

North Cross (Devonian) Unit CO_2 Flood: Status Report

G.J. Mizenko, Shell Western E&P Inc.

SPE Member

This paper was prepared for presentation at the SPE/DOE Eighth Symposium on Enhanced Oil Recovery held in Tulsa, Oklahoma, April 22–24, 1992.

ABSTRACT

The North Cross (Devonian) Unit CO_2 flood, Shell's first, began in 1972. While fairly small by West Texas standards, the project provides an example of a successful, mature CO_2 flood. Seventy-eight billion cubic feet (BCF) of pipeline CO_2 have been injected into the reservoir. Recycled produced gas has augmented this to bring cumulative CO_2 injection to 128 BCF or nearly 70% of the hydrocarbon pore volume. Production performance has been excellent. Unit oil production increased from approximately 1400 barrels per day (BOPD) when CO_2 injection began to its peak of over 2600 BOPD in late 1978 before declining gradually to approximately 1900 BOPD by the end of 1991. An estimated 11 million barrels of enhanced oil have been recovered through 1991 and ultimate EOR is expected to double to over 22 million barrels.

INTRODUCTION

The purpose of this paper is to summarize the performance of the North Cross (Devonian) Unit CO_2 flood. This project began in 1972 and was Shell's first CO_2 flood. The Devonian reservoir was discovered in 1944 and competitively developed by Shell, Texaco and Arco. Solution gas expansion provided the energy for primary depletion which was estimated to recover 13% of the original oil in place (OOIP) at discovery. Primary oil production peaked at 637 barrels per day in 1960.

In 1964, a Unit was formed to increase recovery from the field via reinjection of gas plant hydrocarbon residue gas. While the gas injection pressure maintenance (GIPM) project was seen as a means of partially maintaining reservoir pressure and increasing oil recovery, it was not expected to displace oil very efficiently. Ultimate recovery after gas injection was only projected to be approximately 20% of OOIP. Oil production increased to approximately 1650 BOPD by 1970 primarily from allowable transfers and net gas-oil- ratio credits.

The North Cross CO_2 flood was designed as a secondary (pre-waterflood) rather than a tertiary (post-waterflood) displacement process. Waterflooding, which in West Texas typically precedes CO_2 flooding, was not performed in the Devonian Main Pay reservoir due to low anticipated water injectivity. Since recovery after the partial pressure maintenance project was expected to be low, a secondary CO_2 flood target significantly larger than that of a more typical tertiary CO_2 flood existed at North Cross. The large EOR target at North Cross and the Unit's proximity to the Canyon Reef Carrier pipeline, which was being installed to supply CO_2 to the SACROC CO_2 Project, made it an attractive CO_2 flood candidate.

RESERVOIR DESCRIPTION

The Crossett Devonian field is located approximately 60 miles south of Midland, Texas (Figure 1). The major producing horizon is the Devonian "Main Pay," a classical Archie Type II-A rock composed of microscopic, tripolitic chert grains bound by limestone cement. Average porosity is 22% and permeability averages 5 md. Connate water saturation averages 35%. The porous Main Pay at North Cross is unconformably overlain by non-productive Permian-aged Wolfcamp shale. The Basal Chert, a dense non-productive horizon at North Cross lies below the Main Pay. The oil accumulation is defined by a stratigraphic trap bounded on the north, south and west flanks by porosity deterioration and

References and figures at end of paper.

erosional thinning and on the east by a producing water level at -3040 ft subsea (Figure 2). The reservoir oil was initially saturated and a gas-oil contact existed at -2860 ft subsea. Net to gross ratios range from 70-100% in individual wells and pay continuity between wells is very good.

ORIGINAL CO_2 FLOOD DESIGN

The CO_2 flood was initially designed to flood the oil column within the North Cross Unit with four 160 acre nine-spot patterns (40 acre well spacing) in the interior of the Unit. Numerical reservoir simulation showed that this pattern would maintain miscible displacement over a larger area than a down-dip line-drive[1]. Incremental CO_2 flood recovery was expected to be approximately 20 million barrels or 38% OOIP. To achieve this, an estimated 73 billion cubic feet (BCF) of pipeline CO_2 were to be injected into four CO_2 injection wells. This represented about 40% of the oil-column hydrocarbon pore volume (HCPV). As the flood matured and produced gas became contaminated with CO_2 (over a "background" level of 2.5 mole percent) and could no longer be processed at an existing natural gas liquids (NGL) recovery plant, CO_2 contaminated gas was to be reinjected into existing injection wells. With recycled CO_2, total CO_2 injection was expected to exceed 200 billion cubic feet or over 100% HCPV.

To maintain high displacement efficiency, recycled CO_2-contaminated gas was to be segregated based on its CO_2 content. Produced gas exceeding 60 percent CO_2 was to be injected into wells which initially injected pipeline quality CO_2. Laboratory coreflood experiments conducted with North Cross crude suggested that high displacement efficiencies could still be achieved with produced gas at this CO_2 concentration. Produced gas with CO_2 content below 60 mole percent was to be injected into former updip hydrocarbon gas injectors (wells 25, 33 and 42). Injection of low CO_2 content produced gas in that region was expected to prevent excessive up-dip movement of CO_2 banks rather than to miscibly displace oil.

CO_2 FLOOD FACILITIES

All production is collected at a central battery where a production header and seven separators enable gas-liquid separation and segregation of gas on a well-by-well basis. Historically, produced gas with less than 2.5% CO_2 was delivered to an existing gas plant for NGL recovery (plant capacity has recently enabled some gas with over 2.5% CO_2 to be processed). Produced CO_2-contaminated gas with CO_2 content between 2.5% and 60% has been historically classified "lean", and gas with CO_2 content above 60% CO_2 has been classified "rich".

CO_2-contaminated produced gas is compressed with four 1350 horsepower Cooper-Bessemer GMV-10 STF integral compressors; two of the machines handle lean CO_2-contaminated gas and two serve the rich system. These machines were designed to compress CO_2-contaminated gas from 70 psig to 2200 psig in three stages. Dehydration occurs at the first interstage (250 psig). In 1985, a distillation column designed to recover heavier hydrocarbon components of CO_2-contaminated produced gas was installed downstream of the dehydration unit. This unit has produced 300 to 400 barrels of natural gas liquids per day (approximately 20% propane and 80% natural gasoline) since that time. To reduce compression costs, the two compressors handling rich gas were recylindered in 1986 as two stage units. Inlet rich gas now flows directly to the dehydration unit at 250 psig.

Pipeline CO_2 (approximately 93% CO_2 and 7% CH_4) is delivered via the Canyon Reef Carrier (CRC) pipeline from the Terrell Plant located approximately 50 miles southeast of the North Cross Unit. CO_2 is taken from the pipeline at approximately 1400 psi and compressed with a 375 horsepower compressor to sufficient pressure to enable blending with rich CO_2-contaminated produced gas prior to injection.

PROJECT PERFORMANCE

CO_2 Injection

Through December, 1991, approximately 128 BCF (nearly 70% HCPV) of CO_2 had been injected into the Devonian Main Pay. Of this, approximately 78 BCF was delivered from the Terrell Plant and the remainder was recycled and reinjected CO_2. Unit injection performance is shown in Figure 3. Since CO_2 injection began in April, 1972, four additional wells (in addition to the four initial CO_2 injectors) have been converted to CO_2 injection to improve CO_2 injection efficiency. Infill and edge development drilling has also reduced average spacing in the Unit to 31 acres per well (from a nominal 40 acres per well). Updated reservoir modelling indicates that with existing wells, ultimate CO_2 flood recovery is now expected to exceed 22 million barrels.

Recycled injection volumes have grown steadily since the two recycle systems have been brought on-line in 1975 as shown in Figure 4. Total produced gas reinjection averaged 26.6 MMCFPD during the fourth quarter of 1991. As anticipated in the initial project design, the CO_2 flood is nearing a "recycle-only" phase when no pipeline CO_2 will be injected. Due to discharge pressure limitations, recycle compressors are currently handling produced gas at their peak capacity. Injection well inflow and compressor discharge pressures are in equilibrium near maximum compressor working pressure, i.e. additional injected volumes would cause compressor discharge pressure to exceed the maximum working pressure. Due to this situation, pipeline CO_2 injection was temporarily ceased in May, 1991 to maintain compressor discharge pressure below maximum working pressure and prevent production curtailments.

Lean CO_2-contaminated recycle injection (produced CO_2 concentration between 2.5 and 60 percent) began in former hydrocarbon residue gas injectors (Wells 25, 33 and 42) in 1975. Through 1991, an estimated 16 BCF of produced CO_2 has been injected into this system at an average CO_2 concentration of 35%. Hydrocarbon residue gas injection ceased in 1978. Since that time, only CO_2-contaminated produced gas has been injected in the lean system. During the fourth quarter of 1991, lean recycled gas injection rates averaged 8.3 MMCFPD (approximately 40% CO_2).

The rich injection system began recycle operations in 1978 when highly CO_2-contaminated produced gas was available in sufficient quantity to load recycle compressors. In this system of eight injection wells, CO_2-contaminated produced gas injection followed pipeline CO_2 injection. An estimated 33 BCF of produced CO_2 has been injected in this system at an average CO_2 concentration of 66%. With pipeline CO_2, a total of 111 BCF have been injected into the eight CO_2 injectors. Recycle rates in this system averaged 18.2 MMCFPD (approximately 60% CO_2) during the fourth quarter of 1991.

Until recently, pipeline CO_2 had been delivered via the Canyon Reef Carrier pipeline from the Terrell Plant in southwest Texas where CO_2 is separated from hydrocarbon gas produced from deep wells in the JM and Brown Bassett fields. CO_2 delivery curtailments during the early to mid-1980's were the result of weak demand for natural gas. CO_2 deliveries to North Cross have been shut-in since May, 1991 due to recycle facilities processing constraints. Although additional pipeline CO_2 injection may result in additional oil recovery, this will likely require additional recycle compression capacity. A study to optimize purchased CO_2 slug size and balance the value of additional recovery with costs for additional recycle compression is ongoing.

Displacement Efficiency

In 1975, a Devonian Main Pay pressure core (North Cross Core Test #1) was acquired in a CO_2 swept region and confirmed that injected CO_2 was efficiently displacing oil at North Cross. The core penetrated the Devonian Main Pay reservoir approximately 155 feet from NCDU Well 64 (Figure 6) that at the time, had injected approximately 3.4 BCF CO_2. Prior to CO_2 injection, it is estimated that the reservoir pressure at this location was approximately 1400 psi. Evaluation of pressure core data, pressure tests, and wireline logs indicated excellent vertical sweep of the flood interval and a CO_2-residual oil saturation within the swept zone of approximately 3% (Figure 5). This level of residual saturation would only expected to be achieved after several pore volumes of CO_2 throughput and therefore not realized throughout the entire reservoir.

The extremely high displacement efficiency is attributed to a "developed-miscible" process[2], In this process, it is hypothesized that CO_2 is initially immiscible with the oil that it contacts. Light to mid-range hydrocarbons are extracted from the oil until the mixture becomes miscible with newly contacted oil at the displacement front. Heavier fractions of initially-bypassed oil are thought to be stripped by continued CO_2 injection until oil saturations are eventually reduced to levels observed in Core Test #1. Compositional analysis of produced gas and modelling of pressure test results are consistent with this "developed-miscible" process. While the laboratory measured "minimum miscibility pressure" (MMP) was 1650 psi[1], these data suggest that miscible-like displacement of oil (low residual oil saturation) is possible even where reservoir pressure at the time CO_2 injection began was significantly lower than MMP.

Injectivity

Although North Cross CO_2 injectivity has been reported to be lower than initially expected[1,2,3], refinements in reservoir modelling have significantly improved our ability to understand CO_2 injectivity at North Cross. To develop the initial numerical model[1,2] used to predict flood performance, GIPM gas injection and oil production rates were matched. To obtain the history match, absolute permeability was increased and gas-oil relative permeabilities were adjusted (no gas-oil relative permeability data existed). As reported[1,2,3], actual CO_2 injectivity was significantly lower than CO_2 injectivity predicted by the history-matched model. More recent modelling studies have focused on gas-oil relative permeability adjustments and honored absolute permeabilities. Using this approach, GIPM rates have been matched, but more importantly, CO_2 injectivity was also matched. In summary, these refinements have resolved previously-documented "problems" of lower than expected CO_2 injectivity.

Production

Field wide response to CO_2 injection began in 1974. By the end of 1991, an estimated 11 million barrels of oil above projected production from continued hydrocarbon gas injection have been produced. Cumulative oil production since discovery totals approximately 21 million barrels or 41% of OOIP. Unit production performance is shown in Figure 6.

After approximately 14 months of CO_2 injection, production response to CO_2 injection was first observed in Well 63 in mid-1973 This well exhibited increasing oil production and decreasing gas oil ratios (down to solution gas oil ratio) for approximately one year prior to substantial CO_2 breakthrough (Figure 7). As oil production from this well increased in response to offset CO_2 injection, it underwent a flow-pump-flow cycle as a result of reduced gas-oil ratio followed by increasing CO_2 production after breakthrough. This behavior was also observed in several other responding wells (at the end of 1991, four of the unit's 28 producers are beam pumped; the remaining producers flow). Oil production from Well 63 has been sustained above its pre-response level for nearly 19 years and through 1991 totals over 700,000 barrels since initial response. This well has produced over 800,000 barrels since the Unit was formed in 1964. The average cumulative for initial Unit producers (still producing, i.e. not converted to injection) is approximately 890,000 barrels since the Unit was formed.

Several other wells offsetting CO_2 injectors have also shown dramatic sustained response. For example, production performance for Well 36 is shown in Figure 8. Since its response to CO_2 injection in 1979, this well has produced nearly 1.2 million barrels of oil. While producers offsetting former hydrocarbon residue gas injectors (now lean CO_2 injectors) have not shown as dramatic response to CO_2 injection due to dilution of injected CO_2 with hydrocarbon gas, their production profiles have been impressive nonetheless. Figure 9 shows performance for Well 34 which offsets both lean and rich injectors. Since initial response in 1973, this well has produced over 600,000 barrels of oil.

Infill/Edge Drilling

Since the CO_2 flood began in 1972, twelve new producers have been drilled. Average spacing has been reduced from 40 acres to 31 acres per well. Most drilling has been concentrated in the eastern (downdip) part of the Unit (8 wells) and has served to extend development to the oil column's downdip edge. While not all of these wells have met with overwhelming success, two have produced more than 500,000 barrels of oil in their relatively short life and their average cumulative oil production is over 200,000 barrels. Four 20 acre infill wells have also been drilled to evaluate the potential to accelerate the CO_2 flood process by further reducing well spacing. One early infill, Well 28, has produced over 600,000 barrels of oil since it was drilled in 1978. Results of 1989-90 infill drilling are currently being evaluated to determine if additional 20 acre infill drilling is warranted. A summary of post-CO_2 flood startup drilling is summarized in Table 1.

Production Problems

Following oil production response and CO_2 breakthrough, producing wells have exhibited some problems associated with hydrate formation that occasionally result in brief shut-ins. Hydrates tend to form in flowlines downstream of wellhead chokes as a result of gas expansion cooling combined with free water in the production stream. Hydrate problems occur most often in wells with high GOR's and high CO_2 cuts. Several techniques have been utilized to manage this problem. In 1985, the two recycle compressors serving the rich injection system were recylindered from three-cycle to two-cycle units increasing suction pressure from 70 psig to 250 psig. The primary justification for this modification was to reduce compression costs but it also served to reduce the pressure drop from the wellhead to the facility inlet. This change has resulted in reduced hydrate formation tendency. Oil production rates were not adversely impacted.

In 1989, an automatic flowing well control (valve/choke) system was installed in a flowing well to regulate production rates and to prevent hydrate formation. This has resulted in decreased hydrate-related downtime and increased oil production. Additional installations are pending. To combat paraffin and hydrates together, a Paratrol™ system was installed in another North Cross producer in 1989. This system electrically heats wellbore fluids to prevent formation of paraffin which typically occurs in production tubing near the surface due to gas expansion cooling. This has also resulted in reduced downtime and increased oil production.

OUTLOOK/FUTURE PLANS

Oil production has shown a very gradual decline over recent years. This is the result of both an efficient displacement process and increasing injection rates. CO_2-contaminated recycle facilities are operating at capacity and injection rates have decreased through reduced CO_2 purchases in order to prevent compressor discharge pressure from exceeding its working limit. As this could have a detrimental impact on oil production rates, several options to increase recycle capacity and enable sustained injection rates are currently under evaluation. Additional infill drilling and CO_2 flood project expansion to drain unswept portions of the reservoir are also under evaluation. Follow-up or water injection will also be evaluated as a means of moderating gas production in patterns with high gas-oil ratios. If viable, water injection may reduce ultimate recycle facility requirements. Regardless of the outcome of these studies, the CO_2 project at North Cross is expected to continue well into the next century. Ultimate recovery from the North Cross Unit may approach 63% of OOIP with CO_2 flood operations alone contributing as much as 43% OOIP.

CONCLUSIONS

The North Cross (Devonian) Unit CO_2 flood began in 1972 and represents a very successful, mature CO_2 flood project. A total of 128 BCF of CO_2 had been injected into eight CO_2 injection wells and three former hydrocarbon gas injectors through December, 1991. Thirty-nine percent of the injected CO_2 has been provided by produced, recycled gas. Oil production response to CO_2 injection is evident throughout the Unit and by the end of 1991 has resulted in 11 million barrels of enhanced oil recovery.

A pressure core was acquired in 1975 in a CO_2 swept portion of the reservoir and showed that injected CO_2 was effectively displacing oil in the reservoir to an estimated residual saturation of approximately 3% pore volume. Pressure and production data suggest that in areas of the reservoir where reservoir pressure at the time CO_2 injection began was significantly below the laboratory-measured MMP, a "developed-miscible" process is responsible for high displacement efficiency. Previously-documented CO_2 injectivity "problems" have been resolved through improvements in reservoir modelling parameters.

Oil production increases as large as six-fold have been observed in individual producing wells. As a result of dramatic changes in pressure and gas-oil ratios, a flow-pump-flow cycle has been observed in a number of producers responding to CO_2 injection. Reducing the pressure drop between producers and compression facilities has reduced compression costs and has also alleviated hydrate problems. Other well control techniques to combat hydrates and/or paraffin have been employed.

It is estimated that ultimately, the existing North Cross Unit CO_2 flood may recover 43% of Unit OOIP and bring total recovery to 63% OOIP. Additional infill drilling could increase this total. Optimization studies to most efficiently operate the flood for the remainder of its life are ongoing.

REFERENCES

1. Henderson, L. E.: "The Use of Numerical Simulation to Design a Carbon Dioxide Miscible Displacement Project," *Journal of Petroleum Technology* (Dec. 1974) 1327-1334

2. Pontious, S. B. , and Tham, M. J.: "North Cross (Devonian) Unit CO_2 Flood - Review of Flood Performance and Numerical Smulation Model," *Journal of Petroleum Technology* (Dec. 1978) 1706-1714.

3. Frey, R.P.: "Operating Practices in the North Cross CO_2 Flood," paper presented at the Southwest Petroleum Short Course Seminar, Lubbock, Texas (April 1975).

TABLE 1

North Cross (Devonian) Unit
Drilling Since CO2 Injection Began

Well	Type	Date	Initial Rate Oil B/D	Initial Rate Water B/D	Initial Rate Gas MCF/D	Cumulative Production through 11/91 Oil MB	Cumulative Production through 11/91 Water MB	Cumulative Production through 11/91 Gas MMCF	Cumulative Production through 11/91 Gas % CO2
15	Edge	12/89	44	0	600	31	1	113	2
38	Edge	8/80	37	6	180	73	19	483	57
48	Edge	4/77	30	56	30	151	175	751	30
49	Edge	8/80	70	7	2100	67	19	1950	78
50	Edge	2/90	32	1	2200	18	1	945	5
57	Edge	3/77	454	0	765	500	117	5678	65
60	Edge	9/80	116	2	210	513	3	3357	26
66	Edge	2/78	104	150	340	288	104	1817	52
28	Infill	2/78	430	4	830	630	30	9254	80
80	Infill	12/77	32	26	520	40	7	513	75
144	Infill	11/89	56	14	3290	26	4	823	66
154	Infill	2/90	140	50	240	61	60	465	13

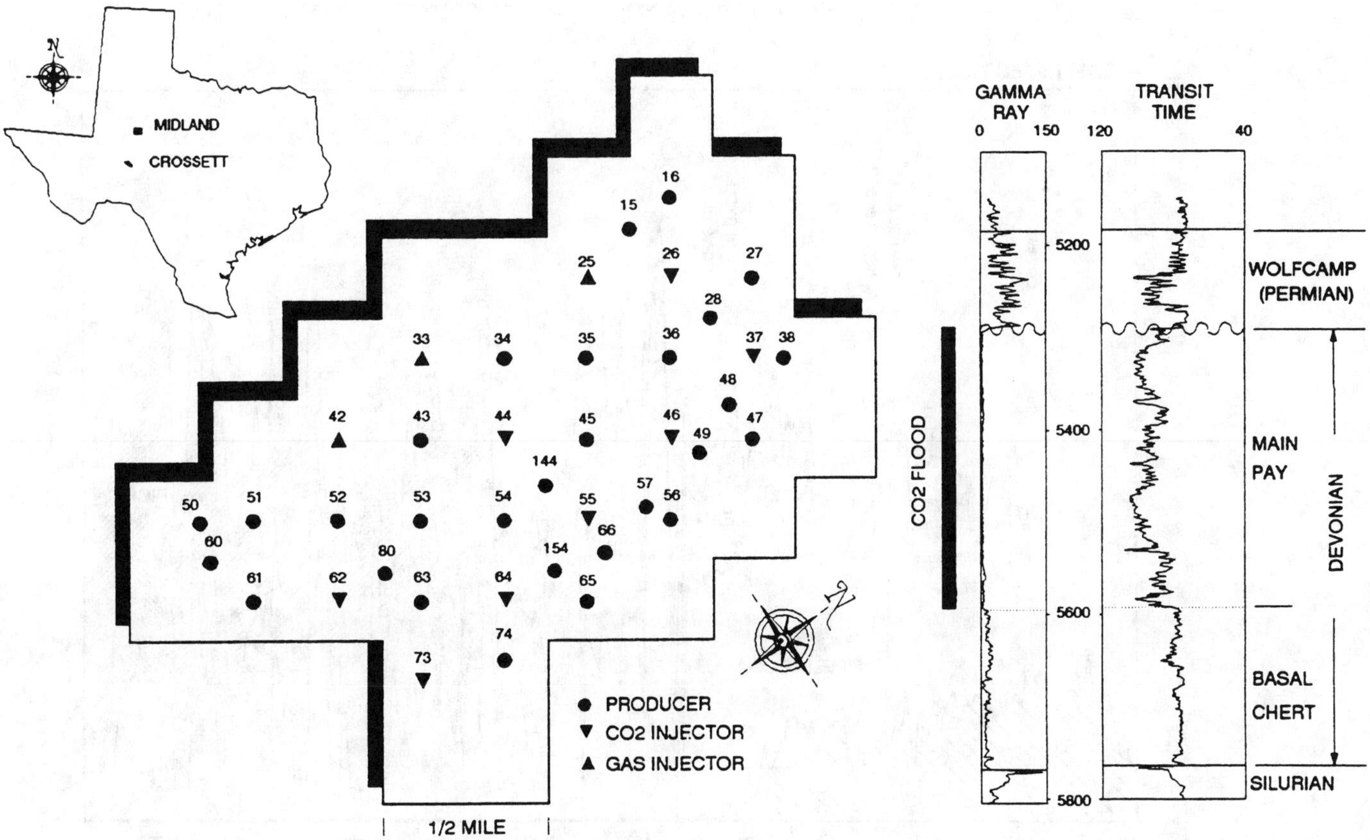

Figure 1 - North Cross (Devonian) Unit
Crossett Field, Crane and Upton Counties, Texas

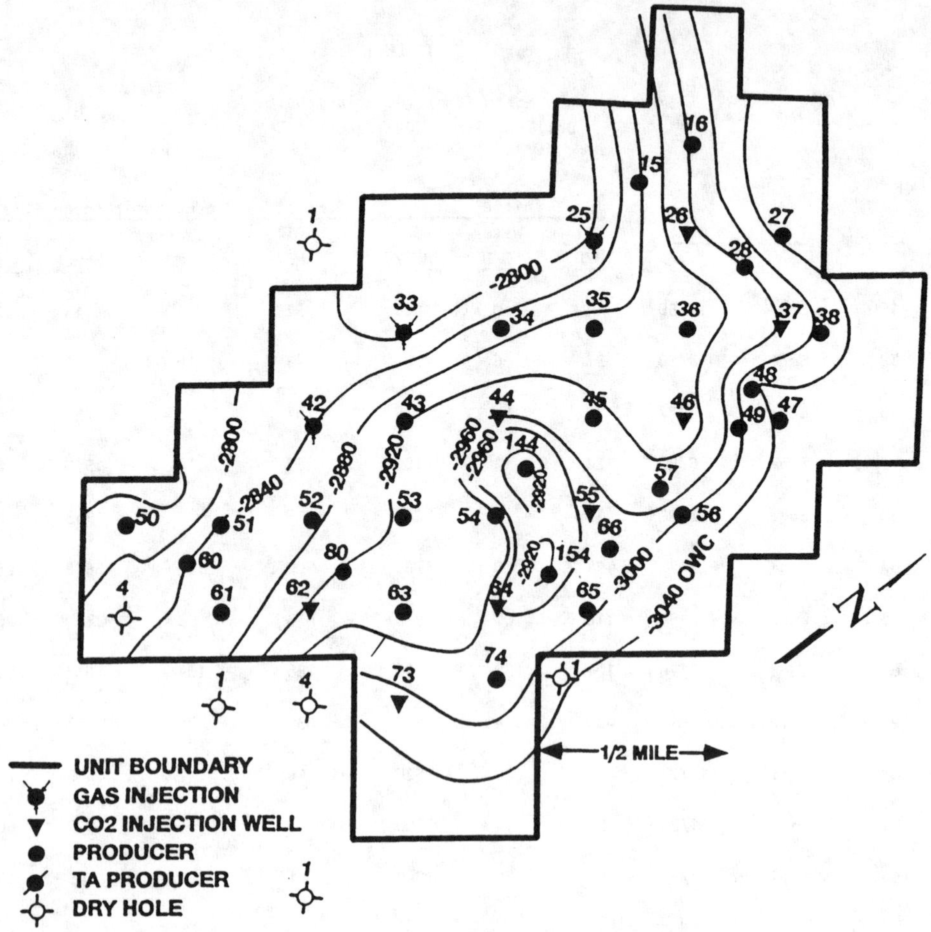

Figure 2 - Structure on Eroded Devonian (20' contour interval)

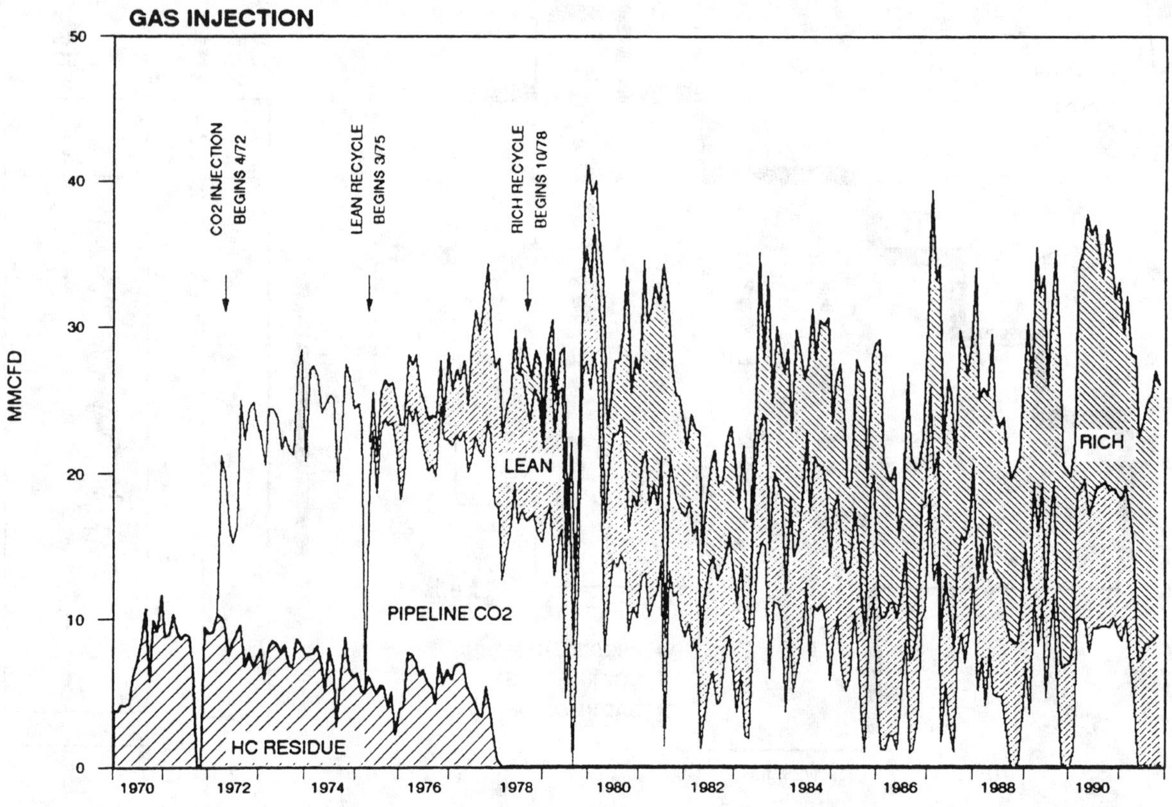

Figure 3 - Total Gas Injection by Injection System

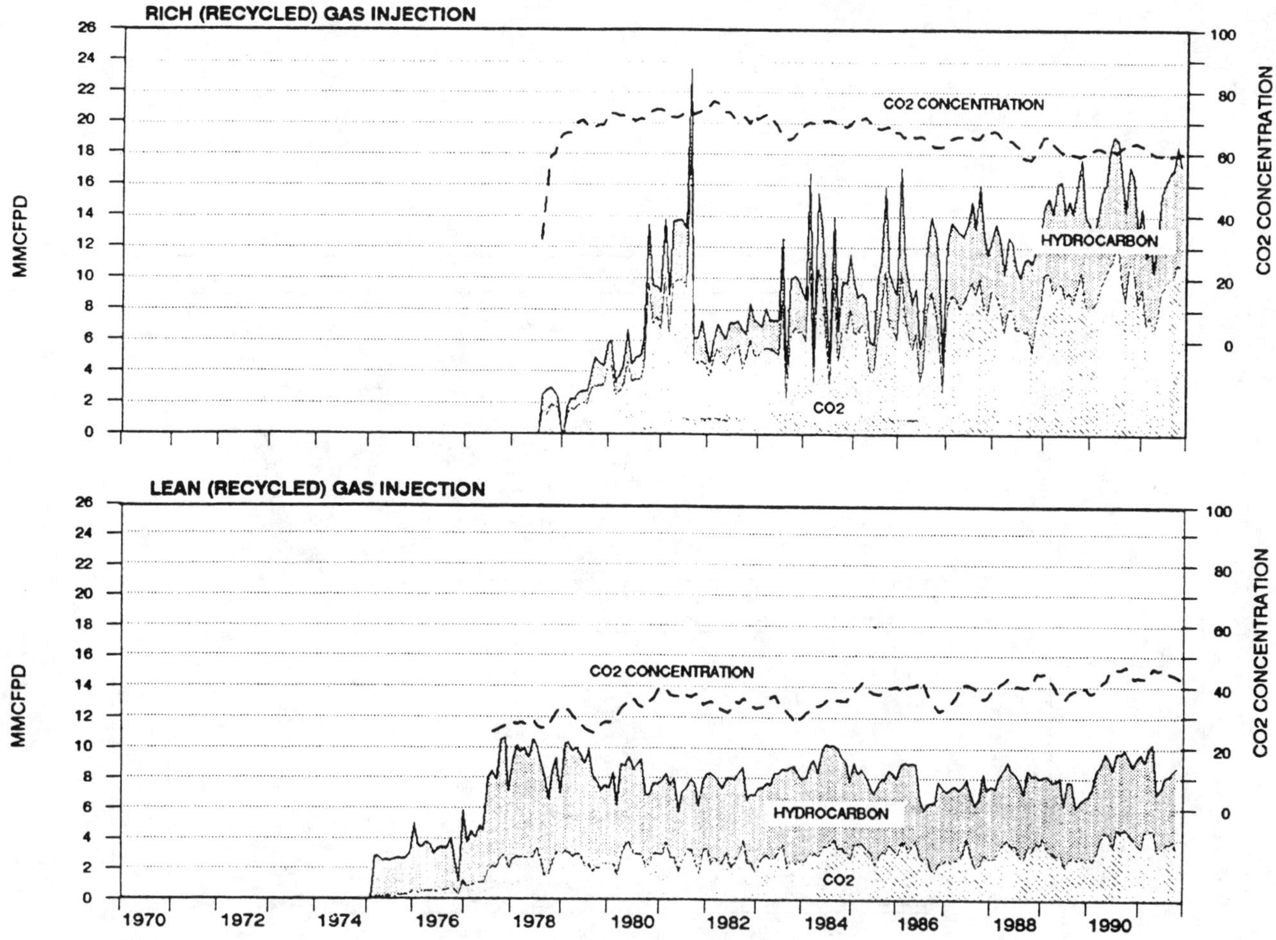

Figure 4 - CO2 Contaminated (Recycled) Gas Injection

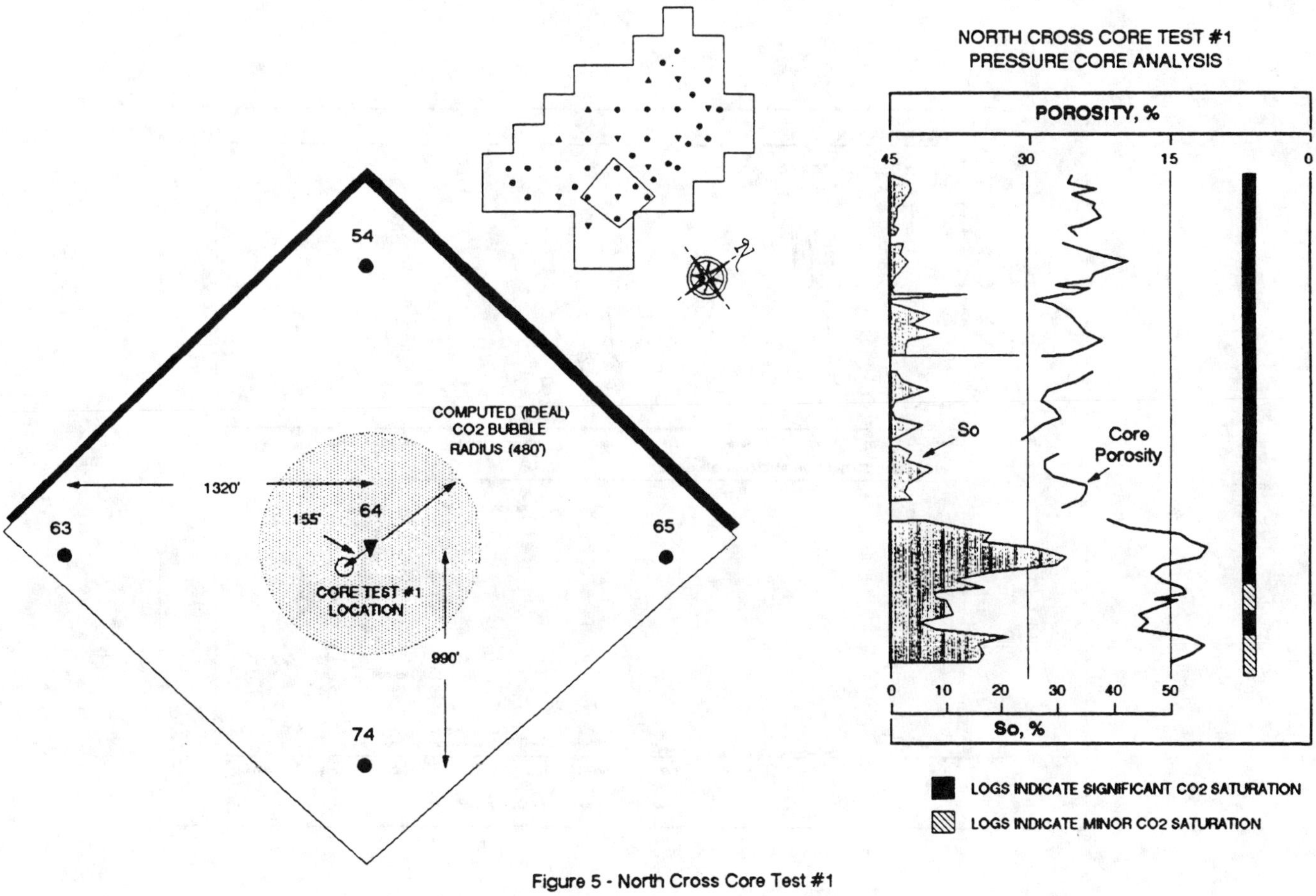

Figure 5 - North Cross Core Test #1

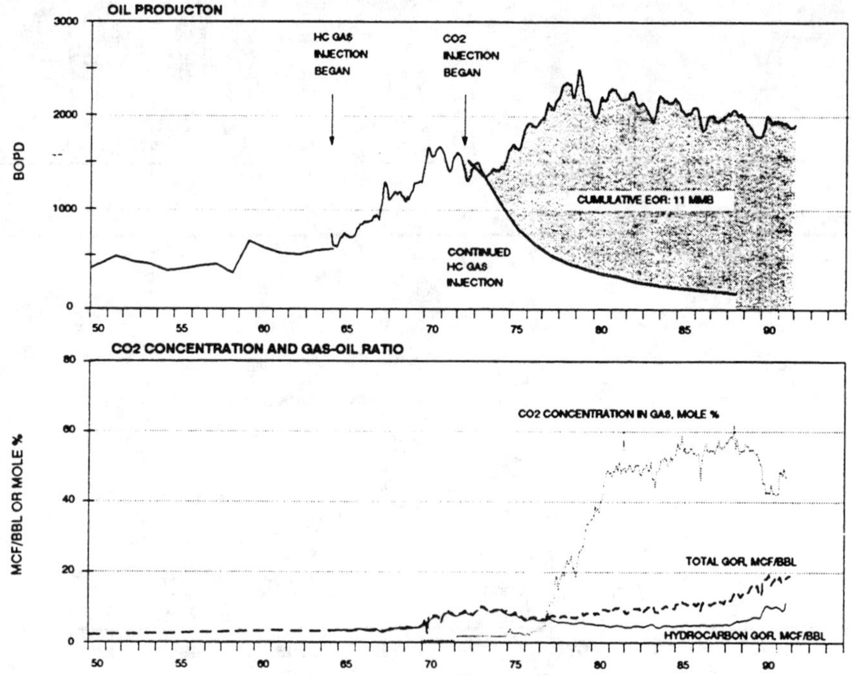

Figure 6 - North Cross Unit Production Performance and Predicted Performance with Continued Hydrocarbon Residue Gas Injection

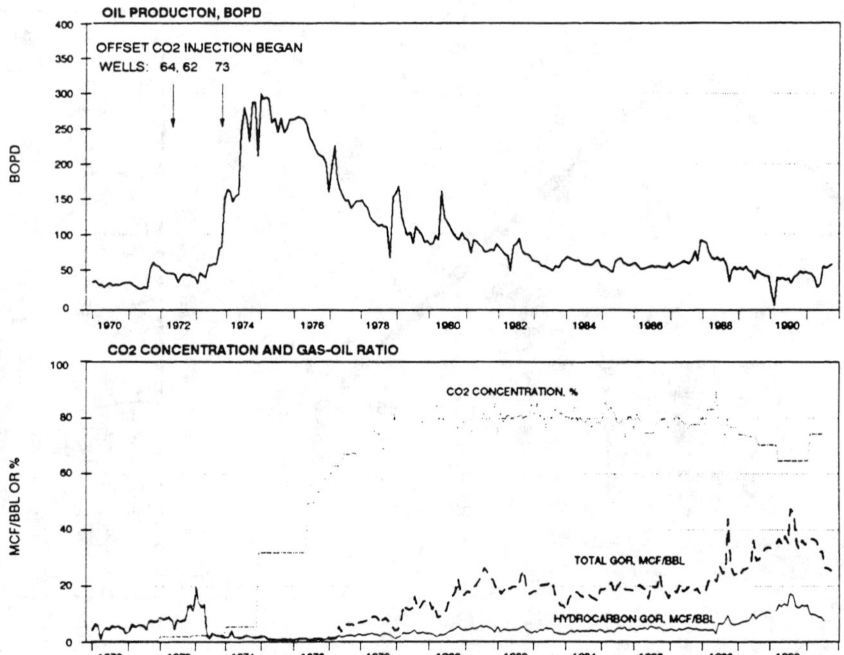

Figure 7 - Well 63: Response To CO2 Injection

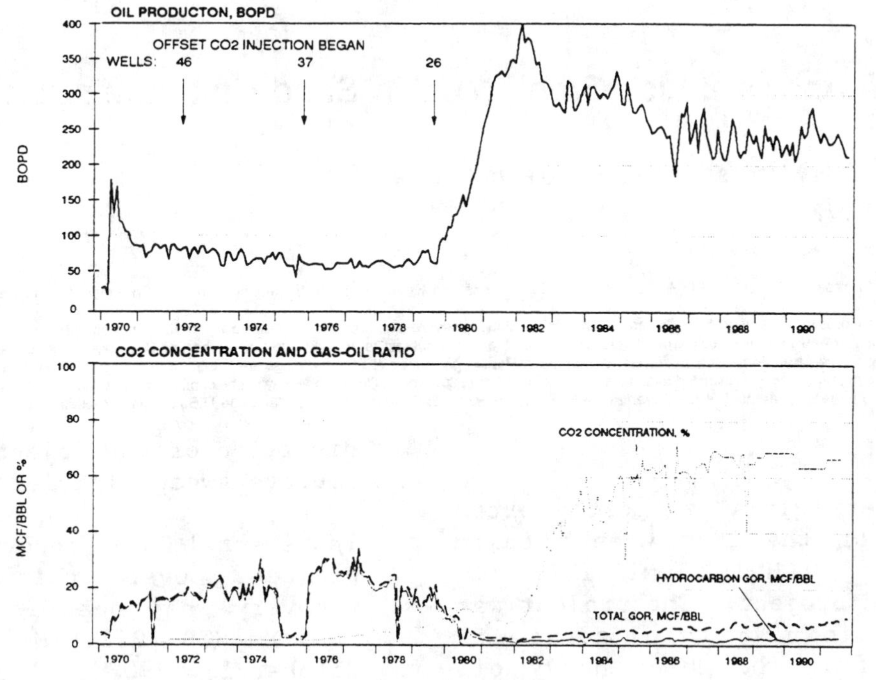

Figure 8 - Well 36: Response To CO2 Injection

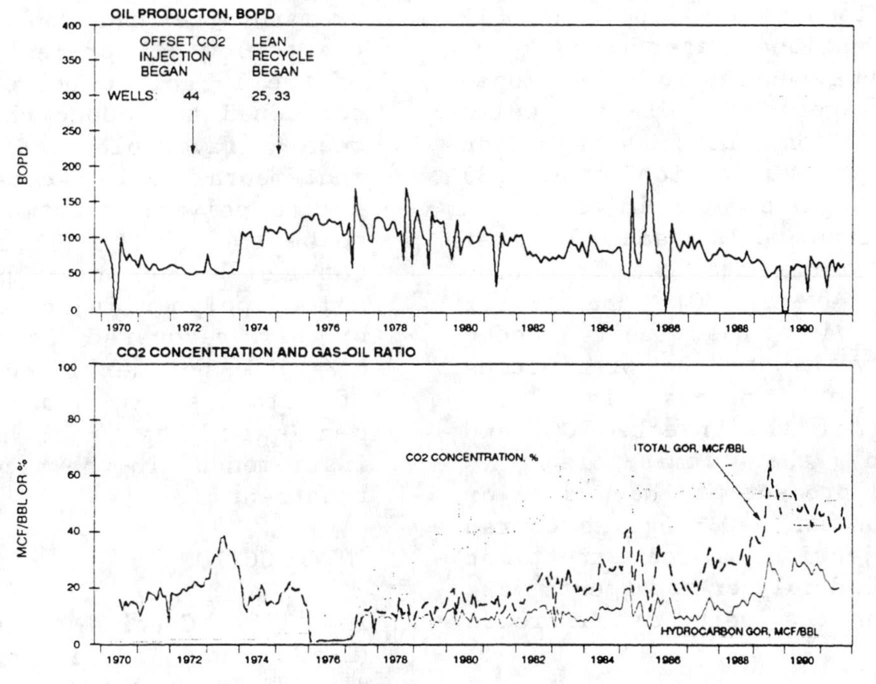

Figure 9 - Well 34: Response To CO2 Injection

SPE 24933

Society of Petroleum Engineers

Postmortem Analysis: Lick Creek Meakin Sand Unit Immiscible CO_2 Waterflood Project

P.D. Moffitt and D.R. Zornes, Phillips Petroleum Co.

SPE Members

This paper was prepared for presentation at the 67th Annual Technical Conference and Exhibition of the Society of Petroleum Engineers held in Washington, DC, October 4–7, 1992.

ABSTRACT

This paper gives a post mortem analysis of the Lick Creek Meakin Sand Unit (LCMSU) immiscible CO_2 waterflood project. The project was initiated in 1976 to increase oil recovery from the thin, heavy oil sand. A reservoir simulation study run prior to the project indicated an immiscible CO_2 waterflood would recover twice as much oil as a conventional waterflood (oil recovery of 16 percent OOIP for an immiscible CO_2 waterflood versus 8 percent OOIP for a conventional waterflood). The project was scheduled to be developed in four stages: 1) cycle all wells with CO_2, 2) continuous CO_2 injection into the unit injection wells, 3) alternate CO_2 and water injection in the injection wells, and 4) water injection only into the injection wells; however, CO_2 and water injection in the unit ended prematurely due to the precipitous drop in oil prices in 1986. Channeling of the injected CO_2 and water in the immiscible CO_2 waterflood project has been a major area of concern. During the course of CO_2 injection, a foam treatment and 15 gelled polymer treatments were applied in the unit to increase diversion of the injected fluids and improve sweep efficiency.

Incremental oil recovery from the project, over that of primary recovery, has been estimated to be 1.75 MM STB [0.278 MM m^3] of oil to October 1, 1990 (11.1 percent OOIP), compared to an initial projection of 16 percent OOIP incremental oil recovery. The lower than projected oil recovery was due to channeling of the injected CO_2 and water in the reservoir and also due to the premature termination of the project due to low oil prices. The results of the 15 gelled polymer treatments conducted to reduce channeling have been favorable overall with incremental oil recovery from the gelled polymer treatments estimated to be 114,700 STB [18,240 m^3] from a total of 76,800 active pounds [34,850 kg] of polymer injected or 1.49 STB of oil recovered per pound [0.52 m^3/kg] of polymer injected. The use of tracers in identifying high permeability channels was instrumental in the success of these treatments.

References and illustrations at end of paper.

INTRODUCTION

The Lick Creek Meakin Sand Unit (LCMSU) was an immiscible carbon dioxide waterflood project operated by Phillips in Bradley and Union Counties, Arkansas. The field was discovered in 1957 and unitized in 1975 for the purpose of secondary recovery. The immiscible CO_2 waterflood project was developed in four stages: 1) cycle all producing wells with CO_2, 2) large volume CO_2 injection into unit injection wells, 3) alternate CO_2 and water injection (WAG process) into unit injection wells, and 4) water injection only into unit injection wells (1,2). The purpose of the immiscible CO_2/WAG process in the thin, heavy oil sand was to reduce oil viscosity and also to increase reservoir oil volume due to oil swelling. The minimum miscibility pressure (MMP) for this reservoir could not be attained due to the relatively shallow reservoir depth.

Mobility control in CO_2 floods has long been an area of concern (3) and the Lick Creek Meakin Sand Unit project was no exception. During the course of CO_2 injection into LCMSU, multiple profile control treatments were conducted to improve the sweep efficiency of the immiscible CO_2 waterflood project. CO_2 injection into the unit injection wells started at injection wells 4G-1 and 27G-2 in February, 1976. Two weeks after initial injection of CO_2, breakthrough of CO_2 occurred at producing wells 5-1 and 6-1, approximately 940 feet [287 m] away from injection well 4G-1 (see Figure 1). Channeling of CO_2 in the LCMSU reservoir continued to be a problem even after water/alternate gas injection was initiated, and a foam treatment and two different types of gelled polymer treatments were performed in the unit in an attempt to minimize CO_2 channeling. This paper summarizes the results of the immiscible CO_2 waterflood project and also the treatments to improve sweep efficiency which were carried out in LCMSU; the results of this project provide a case history for immiscible CO_2 waterflooding and also provide guidance for mobility control in other reservoirs in which CO_2 is being injected.

RESERVOIR CHARACTERISTICS

The Lick Creek Meakin Sand Unit produces from the Meakin sandstone at an average depth of 2550 feet [777 m] and has an average net thickness of 8.4 feet [2.6 m]. The reservoir is truncated by a fault to the southeast and is limited on the north and west by low permeability and a water-oil contact. The Meakin Sand is a fine grained, well sorted, clean, unconsolidated sand which exhibits high permeability and high porosity. The permeability averages 1200 md. [1.18 μm^2] with an average porosity of 30.3 percent. The original oil-in-place was 15.8 MM STB [2.51 MM m^3] with approximately 4.5 MM STB [0.715 MM m^3] (28.3 percent OOIP) being produced during primary production prior to unitization. The primary producing mechanism was pressure depletion combined with a weak water drive. The oil produced from the reservoir is heavy, 17° API, and has a viscosity of 160 cp. [0.16 Pa·s] at the reservoir temperature of 118° F [47.8° C]. More complete reservoir and fluid characteristics for the LCMSU are given in Table 1. A plat of the unit is shown in Figure 1 giving the location of the 39 producing and 24 injection wells.

TOTAL UNIT PRODUCTION PERFORMANCE

The production performance of the total unit is shown in Figure 2 which gives oil production rate and producing water-oil ratio versus time. The starting dates for the four phases of the immiscible CO_2 waterflood project are shown as well as the dates of the five polymer/chromium treatments and the ten K-TROL® VC treatments. As can be seen from Figure 2, oil production from the unit has shown a very positive response to the immiscible CO_2 waterflood project with the greatest response evidenced after initiation of phase 2 of the project, the injection of large volumes of CO_2 into the permanent injection wells. Unit oil production increased from 8000 BOPM [1270 m^3/mo] prior to the project to a peak rate of 28,000 BOPM [4450 m^3/mo] after initiation of CO_2 injection. Incremental oil recovery from the immiscible CO_2 waterflood project was estimated to be 1,754,000 STB [278,900 m^3] of oil to October 1, 1990 (11.1 percent OOIP). Prior to the project, a reservoir modeling study (4) indicated an immiscible CO_2 waterflood project in the Lick Creek Meakin Sand Unit would recover twice as much oil as a conventional waterflood (oil recovery of 16 percent OOIP for an immiscible CO_2 waterflood versus 8 percent OOIP for a conventional waterflood). Actual incremental oil recovery, 11.1 percent OOIP, was less than projected incremental oil recovery, 16 percent OOIP, due to two reasons: 1) poor sweep efficiency of the injected CO_2 due to excessive channeling and 2) premature cessation of CO_2 and water injection in the unit due to the rapid decline in oil prices which occurred in 1986.

Figure 3 gives the injection history of the Lick Creek Meakin Sand Unit. The highest CO_2 injection rates occurred during Phase 2 and at the beginning of Phase 3. Cumulative CO_2 injected into the unit was 21.6 BCF [612 MM m^3] (242 % HCPV - this includes approximately 39 percent purchased and 61 percent recycled CO_2); this gives a utilization factor of 12.3 MCF CO_2 injected per barrel [2.19 M m^3/m^3] of incremental oil produced. The use of water/alternate gas injection did increase the sweep efficiency of the injected CO_2. This can be seen from Figure 4 which gives cumulative oil production versus cumulative CO_2 injection into the unit. The slope of the curve began to increase as larger volumes of water were injected into the unit; this indicates the water was effective in diverting the injected CO_2 into unswept areas of the reservoir and in increasing the unit oil recovery efficiency. Operating procedures were also changed to increase the areal sweep efficiency of the injected CO_2. This involved changing points of CO_2 injection and fluid withdrawal and also shutting in injectors or producers (2).

Figure 5 gives the variation of cumulative oil production across the unit. The northern portion of the unit has the largest net sand thickness and also exhibited the highest cumulative oil production during the immiscible CO_2 waterflood project. An areal map of cumulative gas production is shown in Figure 6. Cumulative gas production is highest in the northern portion of the unit as was the case for cumulative oil production. The majority of gelled polymer treatments to control channeling were conducted in the northern portion of the reservoir.

FOAM TREATMENT

As was mentioned previously, breakthrough of CO_2 from injection well 4G-1 to producing wells 5-1 and 6-1 occurred in two weeks. During phase 2 of the immiscible CO_2 waterflood project (CO_2 injection only at the permanent injection wells), large volumes of CO_2 began to be cycled between the injection and producing wells. This indicated the buildup of a free gas saturation and

® - registered trademark of Halliburton Services

that much of the oil in the reservoir was being bypassed by the injected CO_2. In an attempt to correct this situation, water injection was alternated with CO_2 injection to lower the mobility of the injected fluids (phase 3 of the project - see Figure 3).

A further attempt to control CO_2 channeling was made in February, 1978 when a CO_2 foam treatment was conducted on injection well 4G-1 to reduce CO_2 channeling to producing well 5-1. During the treatment, 58 barrels [9.2 m^3] of COR 180 foaming solution varying in concentration from 17 to 50 weight percent was injected in eight separate slugs at an injection rate of approximately 1 BPM [0.16 m^3/min]. The foaming solution was injected simultaneously with CO_2 at a rate of 1.4 MMCFPD [39.6 M m^3/day]. The wellhead pressure of injection well 4G-1 increased 15 to 20 psi [103 to 138 kPa] during the treatment and producing wells 5-1 and 6-1 experienced a reduction in CO_2 production rate, from approximately 1.5 MMCFPD [42.5 M m^3/day] before the treatment to 1.1 MMCFPD [31.1 M m^3/day] after the treatment. The reduction in CO_2 production rate was temporary, however, lasting only five hours. The failure of the CO_2 foam test was attributed to the very permeable streak located between injection well 4G-1 and producing well 5-1 (5). The volume of the channel was determined to be small (88 barrels [14 m^3]) based on measurement of response times during the foam test, but the permeability was calculated to range from 96 to 3025 Darcies [94.7 to 2985 μm^2] (depending on the fluid phase assumed to be flowing). The final report on the CO_2 foam test stated "future foam applications in the field should be limited to channels with permeabilities less than about 100 Darcies [98.7 μm^2]" (5).

GELLED POLYMER TREATMENTS

After the failure of the CO_2 foam test in February of 1978, a gelled polymer treatment was conducted on injection well 4G-1 in April of 1978. The gelled polymer treatment utilized 263 barrels [41.8 m^3] of polymer/chromium gel in addition to 422 barrels [67.1 m^3] of polymer solution injected for in situ crosslinking with aluminum citrate. The polymer used was Corco WD 160, a cationic polyacrylamide in solid powder form, and the polymer concentration varied from 1 to 2 lb/bbl [2.85 to 5.7 kg/m^3]. A preliminary assessment of the treatment indicated it produced 700 to 1400 barrels [110 to 220 m^3] of incremental oil.

Four further near-well gelled polymer treatments were performed in February of 1980 on injection wells 4G-1, 22G-2, 27G-2, and 34G-1. The aluminum citrate crosslinking process was not utilized for these treatments as it had been concluded from the previous gelled polymer treatment on injection well 4G-1 that aluminum gels were too weak for the very high permeability channels observed at LCMSU (6). The two polymers used in these treatments were N-Hance® 210, a cationic polyacrylamide powder, and Prod® polymer; both types of polymers were gelled using the chromium reduction-oxidation process (7). The volume of polymer gel required was determined by estimating the volume and permeability of the channels between injection and production wells (6). Table 2 gives the chemical quantities used for the treatment of the four wells and a typical injection

® - registered trademark of Hercules Chemical Company
® - registered trademark of Drilling Specialties

sequence. Preliminary results after the treatments indicated the four polymer/chromium treatments failed to completely plug the known channels from injection to producing wells. It was concluded the anionic polyacrylamide did not have the stability desired in the low pH CO_2 environment and that the gelation rate of the chromium gel was uncertain at low pH ranges.

Channeling of CO_2 continued to be a problem after the February, 1980 gelled polymer treatments. Additional CO_2 (270 MMCF [7.65 MM m^3] above contract quantities) was purchased and injected between April and August of 1983 in an attempt to increase oil production from the unit. The oil production rate and producing water-oil ratio of the unit, however, did not respond to the additional CO_2 injected. This lack of response was primarily attributed to major channeling problems between injection and producing wells; the injected CO_2 was flowing through channels without contacting additional oil (8). As a result of the continued CO_2 channeling problems, the utilization of further profile control treatments was investigated. Laboratory studies and the results of the previous gelled polymer treatments had raised some questions regarding the stability of the polymer/chromium gel in a low pH environment, so this type of gel was ruled out for further application at LCMSU.

Based on laboratory data supplied by Halliburton, the polymer gel selected for further profile control treatments was the Halliburton K-TROL VC process. This process involves the injection of a low viscosity acrylamide monomer which undergoes in situ polymerization and simultaneous crosslinking with an organic crosslinking agent to create an extremely strong and rigid gel; the non-ionic nature of this polymer gives it a stability advantage in the low pH CO_2 environment (9) and the rigid nature of the gel may reduce CO_2 diffusion into the gel. Laboratory tests confirmed that low pH would not interfere with the polymerization and crosslinking mechanism and that the in situ polymer gel process would be stable when subjected to a CO_2 WAG process (9). The low viscosity of the injected monomer ensures proper placement of the gel in the high permeability channel or fracture. A disadvantage of the K-TROL VC process was that relatively large quantities of polymer are required per barrel of gel formed (5 percent polymer or approximately 17.5 pounds of polymer per barrel [50 kg/m^3] of gel); this limits the volume of gel that can be economically injected in a polymer gel treatment.

Ten treatments were performed at LCMSU using the K-TROL VC process to form polymer gels. A typical injection sequence for the K-TROL VC treatments is given in Table 3. A chronological list of the 15 gelled polymer treatments performed at LCMSU is shown in Table 4. The next two sections discuss the response of the total unit and individual offset producing wells to the gelled polymer treatments on CO_2 injection wells.

TOTAL UNIT RESPONSE TO GELLED POLYMER TREATMENTS

Total unit response to the five polymer/chromium gel treatments and the 10 K-Trol VC treatments are shown in Figures 2, 3, 4, and 7. As shown in Figure 2, the oil production rate and producing water-oil ratio from the unit evidenced a significant response to the profile control treatments performed on nine of the 24 unit injection wells. The response of the producing water-oil ratio curve is especially pronounced after the K-TROL VC treatments were performed. Figure 7 is a plot of producing water-oil ratio versus cumulative oil produced. The decrease and stabilization of the producing water-oil ratio after the start of the K-TROL VC treatments is easily discernible from this plot; the producing water-oil ratio also stabilized after the polymer/chromium treatments, but the reduction in producing water-oil ratio was not as pronounced as for the K-TROL VC treatments.

A plot of the total unit CO_2 and water injection rates is shown in Figure 3. A reduction in the total unit CO_2 injection rate is evident after the start of the K-TROL VC treatments; a similar reduction was not observed after the five polymer/chromium treatments. Figure 4 shows a plot of cumulative oil produced versus cumulative CO_2 injected. An increase in the slope of the curve is evident after the start of the K-TROL gelled polymer treatments, indicating an increase in the efficiency of oil recovery due to the K-TROL treatments. A similar increase in slope is not as evident after the five polymer/chromium treatments. In summary, the response of total unit production to the K-TROL treatments was more pronounced than the response of total unit production to the polymer/chromium treatments. This response indicates gelled polymer treatments will be effective in increasing sweep efficiency in CO_2 floods if the gelled polymer system can tolerate a low pH environment. The next section discusses the response of individual offset producing wells to the gelled polymer treatments on CO_2 injection wells.

INDIVIDUAL PRODUCING WELL RESPONSE TO GELLED POLYMER TREATMENTS

The response of individual producing wells to gelled polymer treatments of offset injection wells was determined by estimating the incremental oil recovery due to the gelled polymer treatments. Decline curve analysis of the pre-treatment oil production rate was used to establish the oil recovery which would have occurred if the gelled polymer treatment had not been performed. This was compared with the actual oil recovery occurring as a result of the gelled polymer treatment and incremental oil recovery was calculated as the difference in these two oil recovery volumes. The production response to gelled polymer treatments on offset injection wells was also determined by examining the producing water-oil ratio and gas-oil ratio to ascertain if the profile control treatments had effectively diverted injection fluids. Figure 8 is a plat of the unit giving the location of the major channels as determined from tracer tests using fluorescein dye. As shown in Table 3, the injection of four barrels [0.64 m^3] of formation brine containing 1000 ppm fluorescein dye was incorporated into the K-TROL VC process and was used to determine the volume of K-TROL VC polymer gel to be pumped before shutting in those producing wells which produced fluorescein dye.

The series of four profile control treatments on injection well 4G-1 will be discussed first. The first two treatments on injection well 4G-1 were polymer/chromium gel treatments, while the later two treatments were K-Trol VC treatments. Figures 9 and 10 show the response of producing wells 3-1, 6-1, and 8-1 (these wells produced fluorescein dye when it was injected into well 4G-1) to the gelled polymer treatments conducted on injection well 4G-1. As can be seen, the oil producing rate, producing water-oil ratio and producing gas-oil ratio of the three offset producing wells responded very markedly to treatment number 3, the first K-TROL VC treatment on well 4G-1. As given in Table 4, the incremental oil resulting from treatment no. 3 on injection well 4G-1 was estimated to be 18,680 STB [2970 m^3] of oil. Lesser volumes of incremental oil were attributed to the other three treatments as is also shown in Table 4; the second polymer/chromium treatment evidenced the largest response of the remaining three treatments, producing 4710 STB [749 m^3] of incremental oil. Figure 11 gives the injection rates of CO_2 and water into injection well 4G-1; a definite reduction in both CO_2 and water injection rates was noted after the third treatment (first K-TROL VC treatment), but was not noted after the other three treatments. Wellhead injection pressure during the water cycle increased from a vacuum prior to the third treatment on injection well 4G-1 to 250 psi [1720 kPa] after the treatment.

Figures 12 and 13 show the response of the nine surrounding producing wells which produced fluorescein dye to gelled polymer treatments on injection wells 27G-2, 27G-3, and 27G-4. The first polymer/chromium treatment on injection well 27G-2 stabilized the producing water-oil ratio and gas-oil ratio of the offset producing wells and produced an estimated 13,860 STB [2204 m^3] of incremental oil (see Table 4). The producing water-oil ratio also stabilized following the first K-TROL VC treatment on injection well 27G-3, but the most pronounced production response from the nine offset producing wells was exhibited following simultaneous K-TROL VC treatments on injection wells 27G-2, 27G-3, and 27G-4. The producing water-oil ratio displayed a substantial decrease and the producing gas-oil ratio also decreased; incremental oil resulting from the three simultaneous treatments was calculated to be 25,470 STB [4050 m^3]. As can be seen from Figures 12 and 13, gelled polymer treatments on CO_2 injection wells can be very effective, especially if the polymer gel is tolerant to a low pH environment.

Figure 14 shows the oil production response of five offset producing wells to a K-TROL VC polymer gel treatment on injection well 15G-1. No breakthrough of fluorescein dye from injection well 15G-1 was detected in the offset producing wells prior to the treatment and as shown in Figure 14, the offset producing wells did not respond to the K-TROL VC treatment on the injection well. This indicates the high polymer concentration, low volume K-TROL VC treatments, while effective in treating fractures or high permeability channels, are not effective in matrix type treatments.

SUMMARY OF PRODUCTION RESPONSE

A summary of the responses of offset producing wells to the 15 gelled polymer treatments conducted in the Lick Creek Meakin Sand Unit is given in Tables 4 and 5. As can be seen, the response to the treatments has varied widely, but total incremental oil recovery resulting from the 15 treatments was estimated to be 114,700 STB [18,240 m^3] or 7640 STB [1215 m^3] of oil per treatment. This indicates profile control treatments can be effective in controlling high permeability channels and in increasing oil recovery in CO_2 floods. Table 5 summarizes incremental oil recovery values and treatment parameters for the two different types of gelled polymer treatments conducted in LCMSU. Incremental oil recovery per pound of polymer injected is higher for the polymer/chromium gel treatments than for the K-TROL VC treatments. Incremental oil recovery due to the polymer/chromium gel treatments was more difficult to evaluate, however, as the production response to the polymer/chromium gel treatments was not as clearly defined as the production response to the K-TROL VC treatments. The K-TROL VC treatments provided more clear cut, positive production responses in offset producing wells, even though the polymer/chromium gel treatments were considerably less expensive than the K-TROL VC treatments. Both types of treatments, however, were shown to be effective for increasing sweep efficiency in the LCMSU immiscible CO_2 waterflood project.

The average cost for the K-TROL VC treatments was approximately $25,400 per treatment. The treating cost per barrel of incremental oil recovered was consequently $3.73/STB [$23.46/m^3]. No cost information was available on the polymer/chromium gel treatments.

CONCLUSIONS

1. The application of an immiscible CO_2 waterflood project to the Lick Creek Meakin Sand Unit has increased oil recovery by 1.75 MM STB [0.278 MM m^3] (11.1 percent OOIP) to October 1, 1990. The primary mechanisms for increased oil recovery are viscosity reduction and swelling of the reservoir oil.

2. The water/alternate gas (WAG) injection process was effective in increasing the sweep efficiency of the injected CO_2 in the reservoir.

3. Two different types of gelled polymer treatments were used in the LCMSU profile control treatments. The Halliburton K-TROL VC process gave more clear cut, positive production responses than the polymer/chromium gel treatments and gave evidence of being more stable in the low pH CO_2 environment.

4. Incremental oil recovery resulting from 15 gelled polymer treatments applied to nine injection wells in the LCMSU immiscible CO_2 waterflood project is estimated to be 114,700 STB [18,240 m^3] from a total of 76,840 pounds [34,850 kg] of polymer injected or 1.49 STB of oil per pound [0.52 m^3/kg] of polymer injected. The treating cost for the 10 K-TROL VC treatments was approximately $3.73/STB [$23.46/m^3] of incremental oil.

5. The use of tracers prior to the gelled polymer treatments proved successful in predicting which producing wells would respond favorably to the gelled polymer treatments.

ACKNOWLEDGEMENTS

Acknowledgement is made to Phillips Petroleum Company and to Berg Enterprises Operator, Inc. for their permission to publish this paper. Acknowledgement is also made to

K. R. Schramko, T. B. Reid, P. L. Woods, and H. J. Robinson for their work in this area.

REFERENCES

1. Reid, T. B. and Robinson, H. J., "Lick Creek Meakin Sand Unit Immiscible CO_2-Waterflood Project," SPE/DOE 9795 presented at the SPE/DOE Second Symposium on Enhanced Oil Recovery held in Tulsa, Oklahoma (April, 1981).

2. Reid, T. B. and Buckner, R. G., "Performance Review: Lick Creek Meakin Sand Unit Immiscible CO_2 Waterflood Project," presented at the 1984 World Oil Conference held in Dallas, Texas (June, 1984).

3. Holm, L. W., "Evolution of the Carbon Dioxide Flooding Processes," JPT (November, 1987) 1337-1342.

4. Lumpkin, W. B., "Lick Creek Field, Meakin Heavy Oil Reservoir, Reservoir Simulation, Immiscible CO_2-Waterflood," Phillips Petroleum Company Engineering Report (September, 1975).

5. Chen, H. K., Thomas, C. P., and Pollin, J. S., "Evaluation of Foaming Agents for Restricting and Diverting the Flow of Carbon Dioxide at the Lick Creek Field," Phillips Petroleum Company Interim Engineering Report (March 20, 1980).

6. Boneau, D. F., "Lick Creek Meakin Sand Unit - Polymer Treatment of CO_2 Injectors," letter to H. J. Robinson (February 7, 1980).

7. Clampitt, R. L. and Hessert, J. E., "Method for Controlling Formation Permeability," U. S. Patent 3,785,437 (January 15, 1974).

8. Schramko, K. R., "Lick Creek - Meakin Sand Unit, Additional CO_2 Purchase Evaluation," Phillips Petroleum Company Engineering Report (July 20, 1984).

9. Woods, P., Schramko, K., Turner, D., Dalrymple, D., and Vinson, E., "In-Situ Polymerization Controls CO_2/Water Channeling at Lick Creek," SPE/DOE 14958 presented at the SPE/DOE Fifth Symposium on Enhanced Oil Recovery held in Tulsa, Oklahoma (April, 1986).

TABLE 1

LICK CREEK MEAKIN SAND UNIT
RESERVOIR AND FLUID CHARACTERISTICS

Producing Zone	Meakin sandstone
Productive Area	1407 acres
Depth	2550 ft.
Porosity	30.3 percent
Permeability	1200 md.
Net Thickness	8.4 ft.
Temperature	118°F
Original Pressure	1200 psi
Oil Gravity	17 °API
Oil Viscosity at T_{res}	160 cp.
Oil Formation Volume Factor	1.05 res bbl/STB
Initial Oil Saturation	60.3 percent
Water Salinity	71,550 ppm TDS
Original Oil-in-Place	15,845,000 STB
Oil Produced to 2-1-76	4,486,000 STB (28.3% OOIP)
Oil Produced to 10-1-90	6,815,000 STB (43.0% OOIP)
Primary Oil Produced to 10-1-90	5,061,000 STB (31.9% OOIP)
Secondary Oil Produced to 10-1-90	1,754,000 STB (11.1% OOIP)
Number of Producing Wells	39 maximum
Number of Injection Wells	24 maximum

TABLE 3

K-TROL VC GELLED POLYMER TREATMENT ON INJECTION WELL 22G-2

Injection Sequence

1. Inject 4 barrels formation brine containing 1½ lb. of fluorescein (1000 ppm). Displace with 100 barrels formation brine. Monitor offset producing wells by sampling for the presence of fluorescein dye.
2. Inject 2000 gal. 6% KCl water at 1 to 2 BPM.
3. Inject 16,800 gal. (400 barrels) of K-TROL VC gelled polymer solution at 1 to 2 BPM.
4. Displace with 23 barrels 6% KCl water at 1 to 2 BPM.
5. Shut in injection well 22G-2 at master valve to avoid back-surging. Leave well shut in for 5 days.
6. Shut in offset producing wells overnight. Shut in offset producing wells that produced fluorescein dye for five days.
7. After 5 days, resume normal injection in well 22G-2 and resume production in all offset producing wells.

TABLE 2

GELLED POLYMER TREATMENTS ON INJECTION WELLS
4G-1, 22G-2, 27G-2, AND 34G-1
POLYMER/CHROMIUM GELS

Chemical Quantities Used Per Well

Material	Type of Chemical	Wells 4G-1 and 34G-1 Quantity	Wells 22G-2 and 27G-2 Quantity
N-Hance 210	cationic polyacrylamide	750 lb.	450 lb.
Prod	polymer	300 lb.	300 lb.
N-Hance Cr	chromium (gelling agent)	120 lb.	80 lb.
N-Hance RA	reducing agent	35 lb.	35 lb.
N-Hance RAD	delayed reducing agent	265 lb.	160 lb.
Tretolite XC-102	biocide	10 gal.	10 gal.

Injection Sequence (Wells 4G-1 and 34G-1)

1. Inject 50 barrels fresh water containing 10 gal. biocide.
2. Inject 500 barrels of N-Hance 210 polymer solution prepared by dissolving 750 lb. N-Hance 210 in 500 barrels fresh water. Gel polymer solution by adding 500 ppm chromium and 1500 ppm delayed reducing agent "on the fly."
3. Inject 100 barrels of Prod polymer solution prepared by adding 250 lb. Prod polymer to 100 barrels fresh water. Gel polymer solution by adding 750 ppm chromium and 1000 ppm reducing agent "on the fly."
4. Flush gelled polymer solution to formation with one tubing volume of ungelled Prod polymer solution and shutin well for 72 hours. Then resume injection cycle.

TABLE 4

RESPONSE TO POLYMER TREATMENTS OF INJECTION WELLS

Injection Well	Date Treated	Type of Treatment	Treatment Number	Pounds of Polymer	Producing Wells Responding to Polymer Treatment	Incre-mental Oil, STB	STB / lb. polymer	Lifetime, months
4G-1	4-78	P/Cr	1	948	6-1	970	1.02	6
4G-1	2-80	P/Cr	2	1,000	3-1	4,710	4.71	14
22G-2	2-80	P/Cr	1	700	9-1,11-1,22-1	24,580	35.1	17
27G-2	2-80	P/Cr	1	700	30-1	13,860	19.8	26
34G-1	2-80	P/Cr	1	1,000	33-1	2,510	2.51	7
4G-1	9-84	K	3	6,972	3-1,8-1	18,680	2.68	16
27G-3	9-84	K	1	8,400	30-1	3,470	0.41	7
27G-2,3,4	5-85	K	2,2,1	25,200	30-1,30-2,33-1	25,470	1.01	30
15G-1	6-85	K	1	5,040		0	0	--
5G-1	8-85	K	1	6,720	6-1,19-1,27-1	14,250	2.12	27
13G-1	8-85	K	1	6,720	21-1,23-1	4,050	0.60	10
22G-2	8-85	K	2	6,720		0	0	--
4G-1	1-86	K	4	6,720	3-1,8-1	2,110	0.31	6
				76,840		114,660		

Treatment Types
P/Cr - polymer/chromium gel
K - Halliburton K-TROL VC gel

TABLE 5

GELLED POLYMER TREATMENT PERFORMANCE BY TYPE OF TREATMENT

Type of Treatment	Number of Treatments	Total Pounds of Polymer	Incremental Oil, STB	STB / lb. polymer	Average Lifetime, Months	Average Per Treatment: Pounds of Polymer	Average Per Treatment: Incremental Oil, STB
P/Cr	5	4,348	46,630	10.7	16.0	870	9,330
K	10	72,492	68,030	0.94	19.5	7,249	6,803
Totals	15	76,840	114,660	1.49	18.3	5,123	7,644

Treatment Types
P/Cr - polymer/chromium gel
K - Halliburton K-TROL VC gel

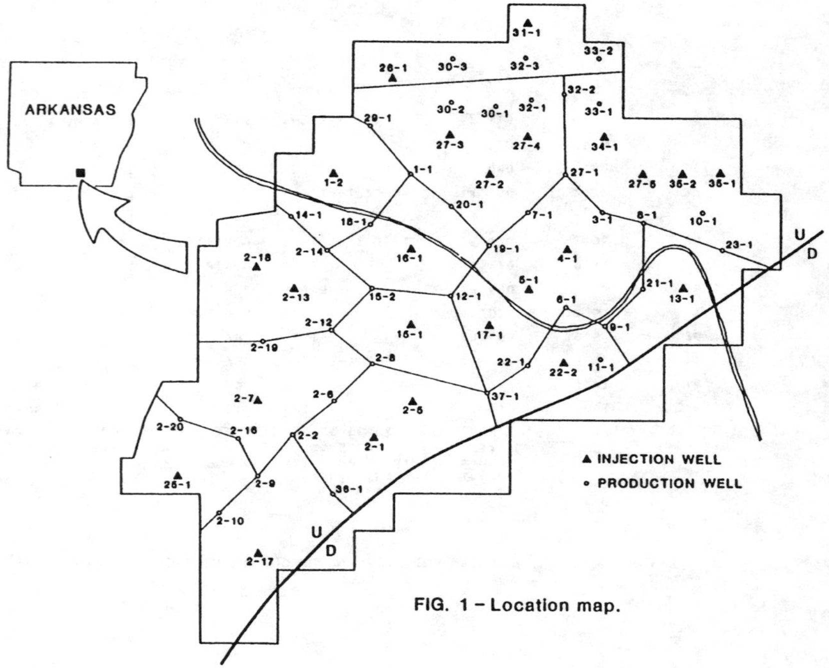

FIG. 1 – Location map.

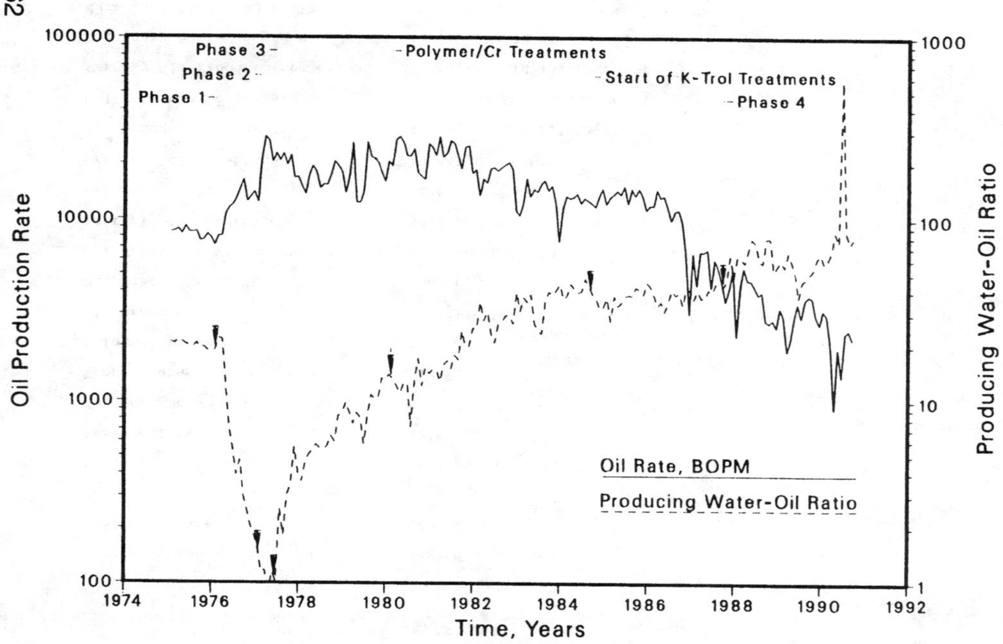

Fig. 2-Lick Creek Meakin Sand Unit, Total Unit Production Performance.

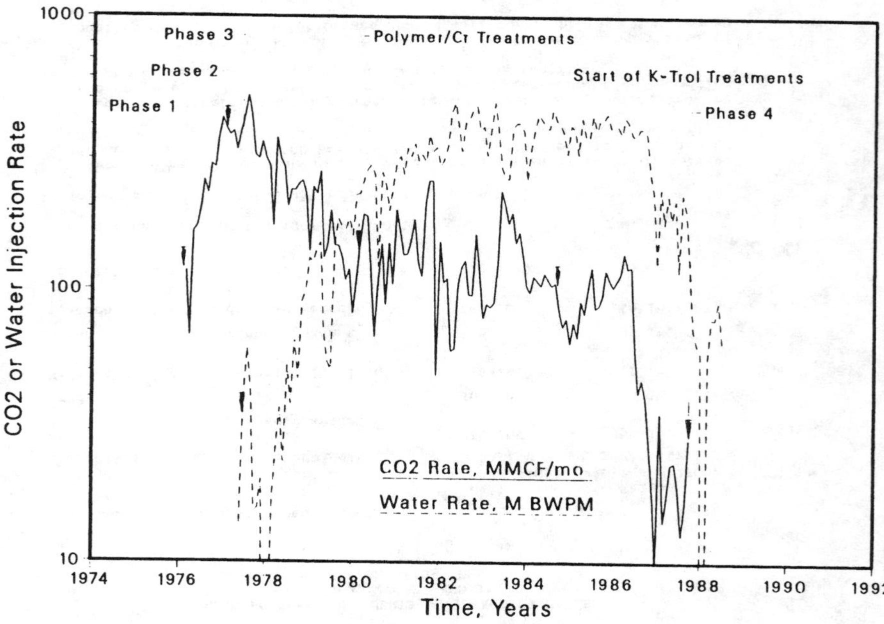

Fig 3-Lick Creek Meakin Sand Unit, Total Unit Injection Performance.

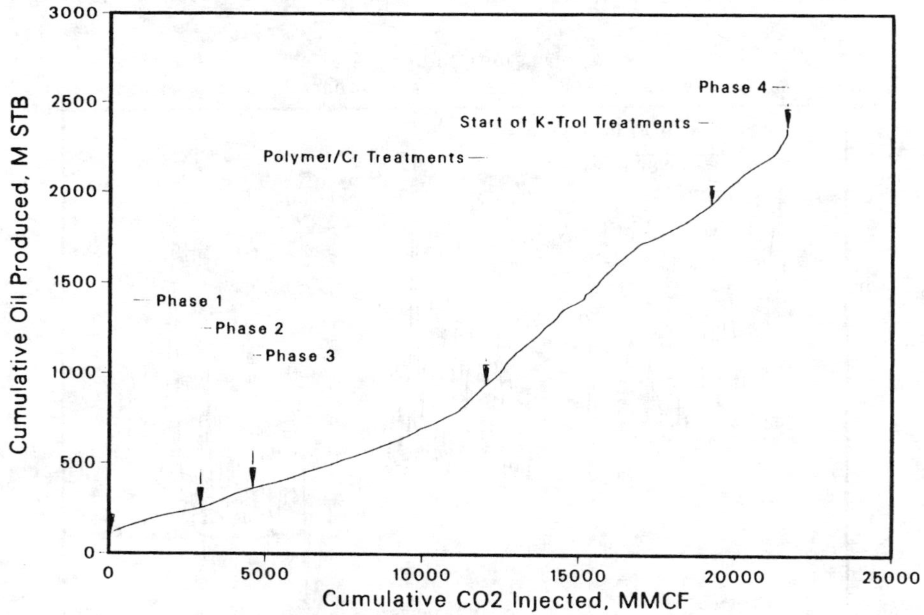

Fig. 4-Lick Creek Meakin Sand Unit, Total Unit Production and Injection Performance.

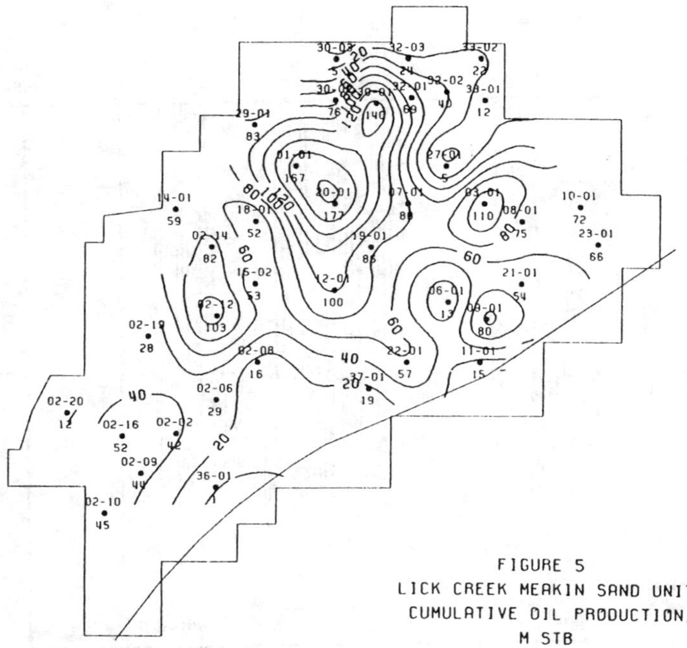

FIGURE 5
LICK CREEK MEAKIN SAND UNIT
CUMULATIVE OIL PRODUCTION,
M STB

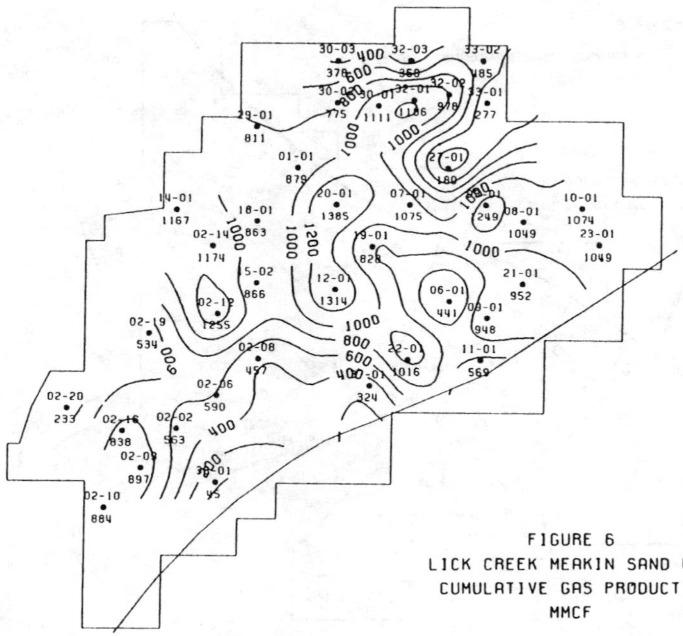

FIGURE 6
LICK CREEK MEAKIN SAND UNIT
CUMULATIVE GAS PRODUCTION,
MMCF

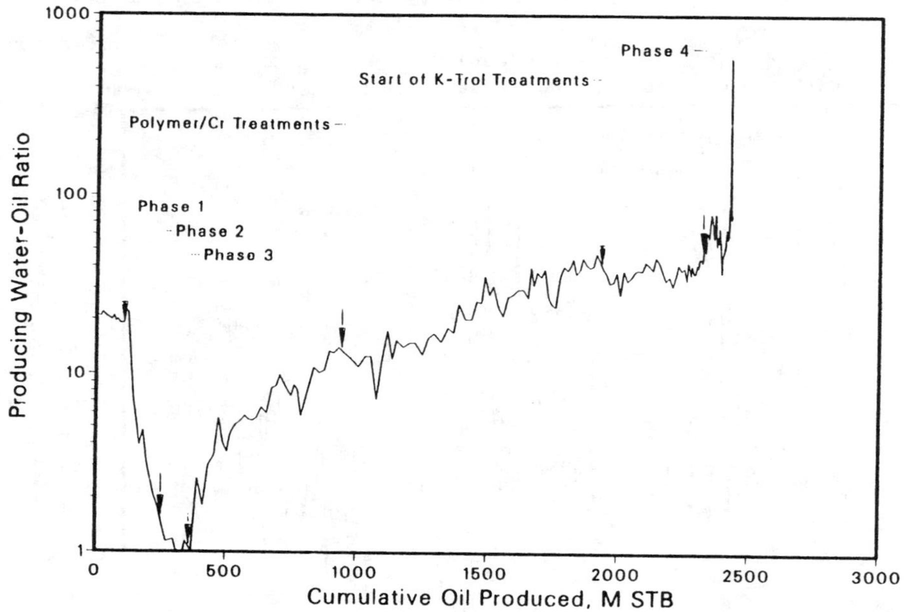

Fig. 7-Lick Creek Meakin Sand Unit, Producing Water-Oil Ratio Performance.

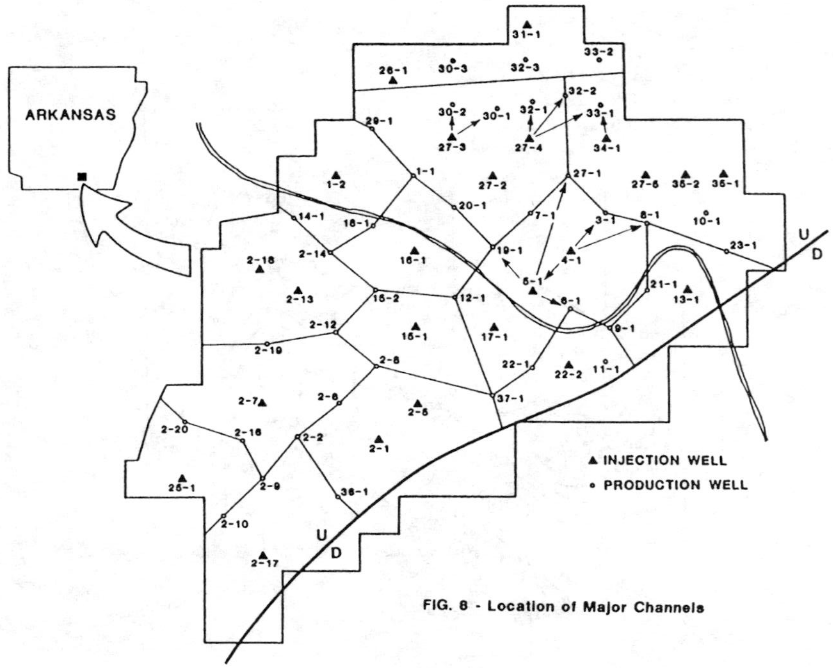

FIG. 8 - Location of Major Channels

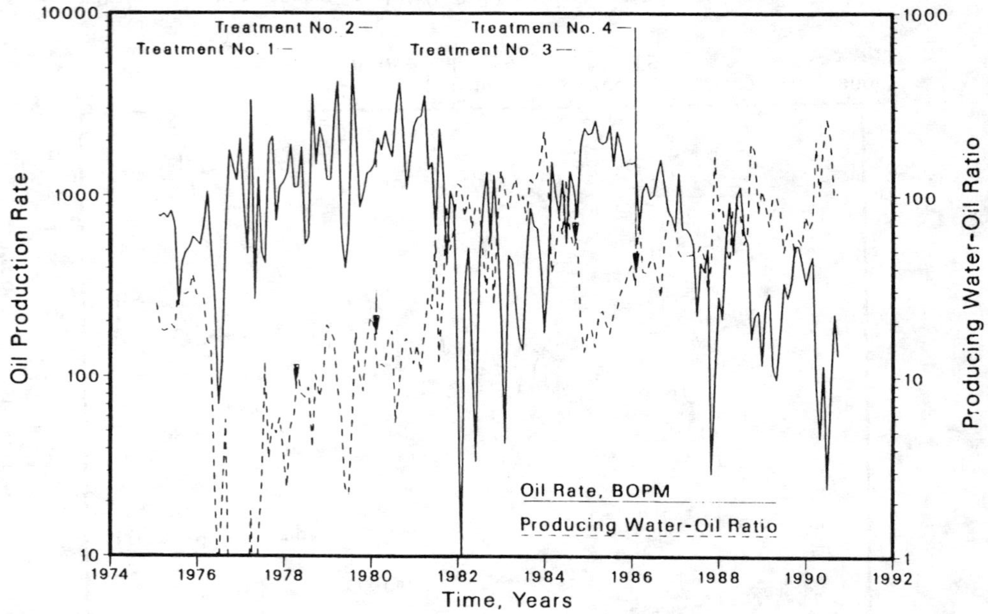

Fig. 9-Polymer Treatments on Injection Well 4G-1; Producing Wells 3-1,6-1, and 8-1 Response.

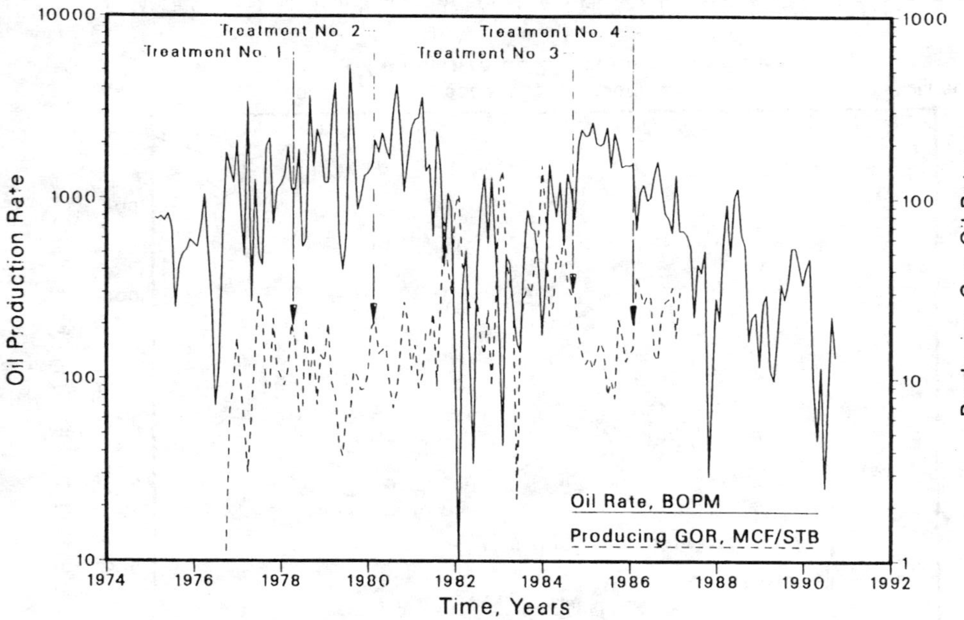

Fig. 10-Polymer Treatments on Injection Well 4G-1; Producing Wells 3-1,6-1, and 8-1 Response

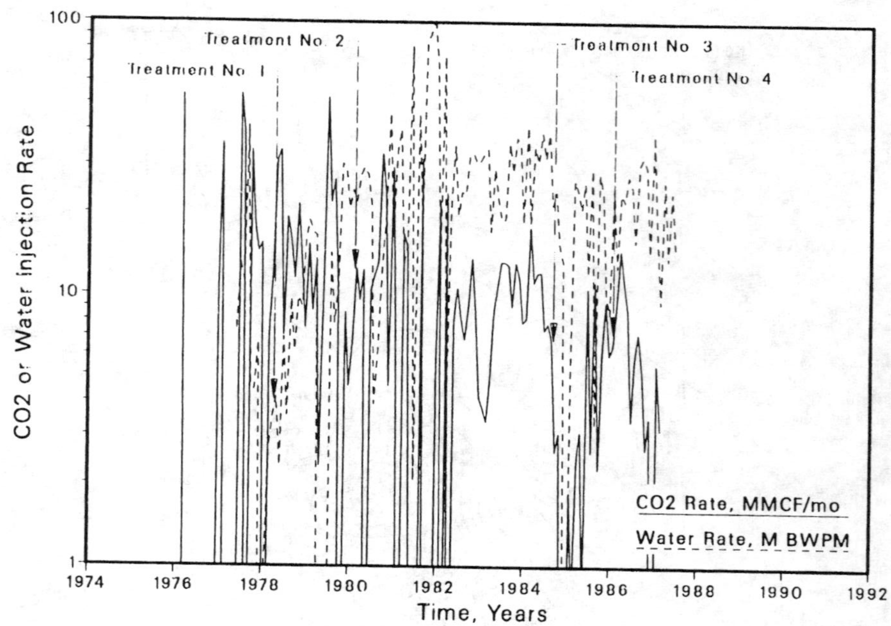

Fig. 11-Well 4G-1 Injection Performance.

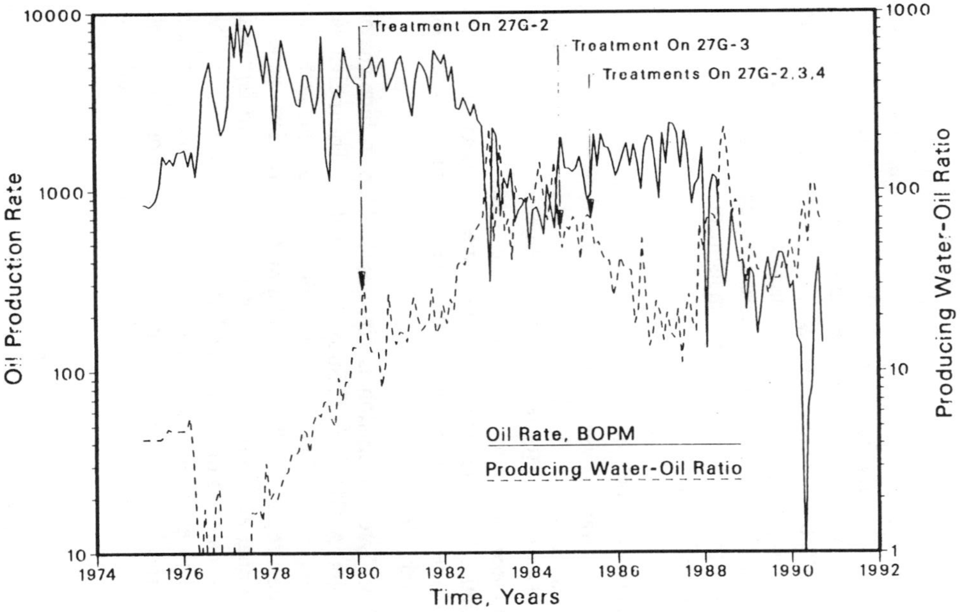

Fig 12-Polymer Treatments on Injection Wells 27G-2,3,4
Response of Surrounding Producing Wells

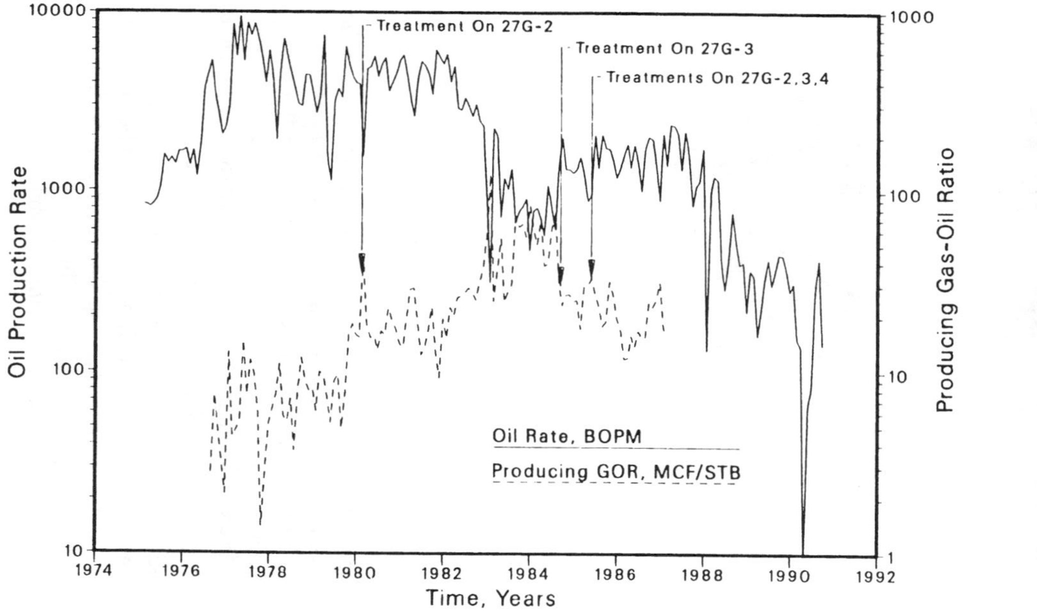

Fig 13-Polymer Treatments on Injection Wells 27G-2,3,4
Response of Surrounding Producing Wells

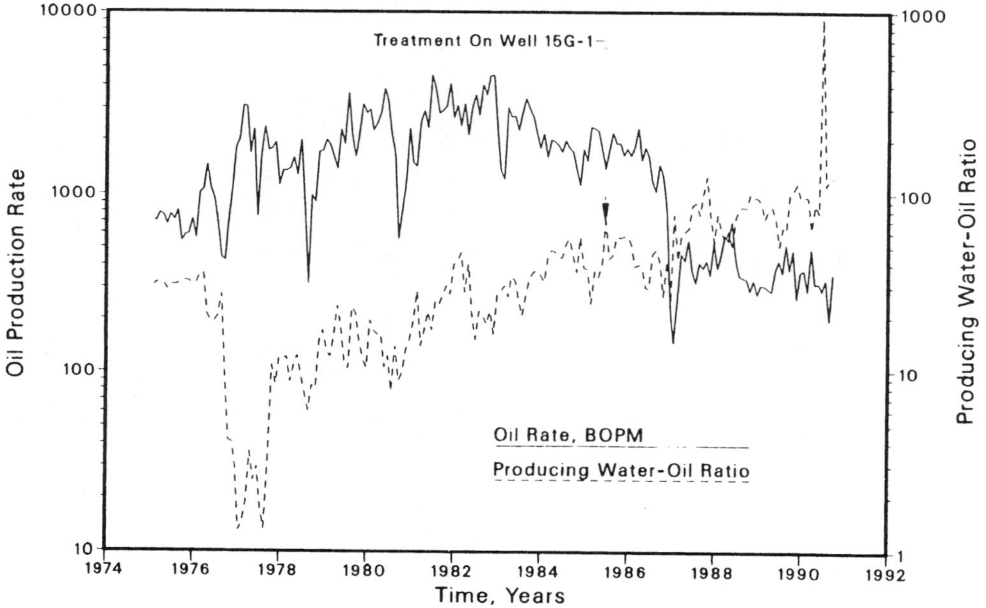

Fig. 14-Polymer Treatment on Injection Well 15G-1
Response of Surrounding Producing Wells

SPE 30729

An Overview of the North Ward Estes CO_2 Flood

J. N. Ring, SPE, Chevron Petroleum Technology Co., and D. J. Smith, SPE, Chevron U.S.A. Production Co.

Abstract

A CO_2 pattern flood was initiated in a west Texas (Permian Basin), sandstone reservoir as a tertiary recovery process in 1989. This paper is an overview of the development and management of the North Ward Estes CO_2 Flood. The primary objective of managing the CO_2 flood has been to maximize oil recovery for maximum profitability. This has been accomplished by maintaining a consistent water-alternating-gas (WAG) process, optimizing CO_2 utilization, and maintaining a proper balance between patterns. Reservoir modeling, which is emphasized in this paper, has also been critical in the development and management of this flood.

Introduction

The North Ward Estes Field, located in Ward and Winkler Counties, TX (**Fig. 1**), was discovered in 1929. The field is situated on the western flank of the Central Basin Platform, approximately 60 miles southwest of Midland, TX. The net productive area of the field is approximately 39,000 acres, and the original oil in place (OOIP) was approximately 1.1 billion bbl. Cumulative oil production is approximately 338 million bbl (30.7% OOIP).

Production is from the Yates and Queen Formations. The Yates, which is the dominant producing formation, is composed of up to nine major reservoirs. The individual reservoirs are very fine-grained sandstones to siltstones separated by very dense dolomite beds. The average depth of the Yates is 2,600 ft.

The field was initially developed on 20-acre spacing except for the most productive areas. These areas were developed on 10-acre spacing. In 1954 a pilot waterflood was initiated in the southern area of the field. Due to the favorable response of the pilot, approximately 26,000 acres were being waterflooded by 1965. The initial pattern configurations consisted of 40-acre five-spots in areas developed on 20-acre spacing and incomplete nine-spots in areas developed on 10-acre spacing.

Additional improved recovery projects have been conducted since 1975. These included in-situ combustion and caustic flood pilots, polymer-injection profile modification, pattern realignments, and infill drilling. Of these projects, drilling infill wells, pattern realignments, and increased water injection were the most economically successful processes.

In early 1989 a CO_2 pattern flood was initiated in six sections of the Hutchings Stock Association Lease, located in the southern area of the field (**Fig. 1**). The project area consists of approximately 3,840 acres and is limited to the Yates Formation. Rock and fluid properties of the Yates Formation are shown in **Table 1**. After six years of CO_2 injection, many of the patterns have been switched to chase water, and recycled CO_2 is being injected into most of the remaining patterns. This paper describes the development and management of the North Ward Estes CO_2 Flood.

Geological Description

The North Ward Estes Field is associated with an anticlinorium, approximately 18 miles long by 4 miles wide, on the western flank of the Central Basin Platform. Production is from the Yates and Queen Formations. However, the Yates is the primary productive formation within the field. Average reservoir properties of the Yates Formation within the CO_2 project area are described in **Table 2**.

The Yates Formation is composed of up to nine major reservoirs (A, BC, D, E, F, Stray, J_1, J_2, and J_3). The individual reservoirs of the Yates are very fine-grained sandstones to siltstones separated by dense dolomite beds. The general depositional environment was that of a tidal flat to lagoonal setting situated to the east of and behind the shelf margin. The reservoirs were deposited as sand and silt in the subtidal to beach environment and silt to clay in the supratidal environment. Depositional strike is inferred to be northwest-southeast, roughly parallel to the shelf margin. Individual reservoirs are laterally continuous in the strike direction.

The Queen Formation, which lies below the Yates, is similar in depositional environment to the Yates Formation. It is composed of 9 zones of fine-grained sandstones to siltstones separated by dense dolomite beds. These zones are composed of numerous thin (< 15 ft), lenticular sands with poor lateral continuity. Average reservoir properties are lower than the Yates.

The sands with the greatest reservoir capacity in the CO_2 project area are the BC, D, E, F, Stray, J_1, and J_2. The BC, J_1, and J_2 are the most productive sands. Production from the A sand is minimal, and the J_3 sand is poorly developed in this area of the field. The Queen Formation is difficult to waterflood and was excluded from the CO_2 flood.

The BC sand has the largest reservoir capacity of any sand in the project area due to its areal coverage, thickness (up to 60 ft), and reservoir quality. The BC is best described as a siltstone to fine-grained sandstone with detrital clay. The depositional environment was that of a shallow water tidal flat with an abundant amount of windblown sediments.

The D and E sands are similar to the BC. However, porosity and permeability are highly variable. The Stray section is composed of thin-bedded, lenticular, intertidal to subtidal siltstones and fine-grained sandstones with the highest clay content of any Yates interval. Because of this, permeability and reservoir continuity suffer while porosity remains high.

The J_1 and J_2 sands are composed of coarser sands with much less clay content and, therefore, higher effective porosities and permeabilities. The depositional environment was that of a beach to nearshore marine where turbulence winnowed finer silts and clays out of the strike-oriented sand deposits. These sands are best developed in the western sections of the project area.

CO_2 Project Area History and Development

The CO_2 project area consists of six sections (3, 6, 7, 8, 9, and 10) in the southern area of the North Ward Estes Field. The first producing well in this area was drilled in Section 7 in 1929. Primary production of the six-section area peaked in 1944 (**Fig. 2**) and was approaching the economic limit in the mid-1950's. A 960-acre pilot waterflood was initiated within Sections 6, 7, and 8 in 1954. Two years later the waterflood was expanded to include all of Sections 6, 7, 8, 9, and 10. Section 3 was not included until 1964. The prevailing flood patterns were 40-acre five-spots. Oil production peaked in 1960 and then declined at 11% per year until 1979. At that time, oil production began to stabilize as a result of drilling infill and replacement wells (**Fig. 2**). As a result, patterns within Sections 3, 6, 7, and 8 were converted from 40-acre five-spot to 20-acre five-spot patterns, and patterns within Sections 9 and 10 were converted from incomplete nine-spot to 20-acre line-drive patterns (**Fig. 3**).

After waterflooding this area for more than 30 years, approximately 23% of the wells had reached their economic limit. Cumulative oil production accounted for approximately 41.5% of the OOIP in this area by 1989. The performance of the waterflood resulted in a 2.3 ratio of ultimate secondary to ultimate primary for wells existing at the start of the waterflood.[1] The success of the waterflood was attributed to a good areal sweep efficiency, which was achieved because of a favorable mobility ratio. However, the vertical sweep efficiency was not as favorable, due to the permeability contrast among the major sands (i.e., a Dykstra-Parsons[2] coefficient of 0.85).[1]

CO_2 Flood Design

In 1985 an economic evaluation of the six-section area was made to determine the feasibility of a CO_2 flood. This evaluation was based on extensive laboratory work, an injectivity test, and reservoir simulation studies.

The laboratory work[1] included black-oil PVT and oil/CO_2 phase behavior studies of recombined separator oil and gas samples, slim tube experiments, and CO_2 core floods. The phase behavior studies were used to evaluate oil swelling, viscosity reduction, and phase transition pressures all as a function of CO_2 concentration. Slim tube experiments were utilized to determine the minimum miscibility pressure (MMP). Based on these experiments, the MMP ranged from 1,010 to 1,350 psia for CO_2/hydrocarbon gas mixtures and 937 psia for pure CO_2 at a reservoir temperature of 83°F. The core floods were conducted to evaluate the recovery of residual oil to water. The results indicated that a WAG ratio of 1:1 was the most efficient.

A CO_2 injectivity test[1] was conducted to identify whether a reduction in injectivity would occur either during or following CO_2 injection. As a result of this test, a total of 31 MMscf of CO_2 (1.3% HCPV) was injected into a well within a 50-day test period. No apparent reduction in injectivity or changes in the injection profile were observed during or after CO_2 injection. However, the CO_2 injection rate was about 20% higher than the water injection rate at the same bottomhole injection pressure.

3D pattern-element models[1] were developed for four of the six sections (Sections 3, 6, 7, and 8). The pattern-element models were $^1/_8$ five-spots, indicative of average patterns within the respective sections. These models were used to evaluate WAG ratios, CO_2 slug sizes, incremental oil recovery, and CO_2 utilization. The forecasts of these models were scaled-up to yield results more indicative of the project area. A four-component miscible simulator was used in the development of these models.

The results of these studies[1] indicated that the optimum CO_2 slug size was in the range of 38% to 60% HCPV, and the optimum WAG ratio was 1:1. The results also indicated that an incremental oil recovery of 8% of OOIP, in addition to waterflood, could be achieved with a CO_2 slug size of 38% HCPV over 10 years. Gross CO_2 utilization (ratio of purchased and recycled CO_2 to incremental oil recovered) was forecasted to be 12 Mscf/bbl, and net CO_2 utilization (ratio of purchased CO_2 to incremental oil recovered) was forecasted to be 4 Mscf/bbl. CO_2 utilization was expected to be high because of poor vertical conformance.

The CO_2 flood design was based on injection of a 32% HCPV CO_2 slug over 7 years, followed by reinjection of produced CO_2 for three years. A total CO_2 slug of 38% HCPV would be injected over 10 years. The WAG ratio for this design was 0.83:1, injecting 2.5% HCPV per WAG cycle. The pattern configurations were to exist as developed in the latter stages of the waterflood[3] (**Fig. 3**). Within the CO_2 project area, there were 26 lease-line injectors (water injectors), 112 20-acre five-spot patterns, and 56 20-acre line-drive patterns for a total of 168 CO_2 injection patterns.

CO_2 Flood Performance

The CO_2 flood was implemented in the six-section area in April 1989. After six years of CO_2 injection (April 1989 - March 1995), approximately 94.4 Bscf of CO_2 (21.0% HCPV) have been injected into the project area. The performance of the CO_2 flood as of March 1995 is shown in **Fig. 4**. Many of the patterns have been processed and switched to chase water. Recycled CO_2 is being injected into most of the remaining patterns.

Initially, the CO_2 flood was implemented as designed. After approximately 7 months, the CO_2 flood was fully developed, and CO_2 was being injected at an average rate of 67.3 MMscf/D into 168 patterns. The CO_2 concentration in the produced gas stream increased from 2.4% to 40.3% within the first three months and continued to increase thereafter. Oil response to CO_2 was evident within the first five months of injection. The average WAG ratio was 1:1, and the CO_2 half-cycle slug size was approximately 1.5% HCPV, which was slightly lower than the original design. WAG cycles were completed every two months, since they were based on time rather than volume.

The performance (oil production) of the project area peaked in September 1990 (**Fig. 4**) and was maintained through April 1992. The CO_2 flood was continuously monitored and managed to maximize economic oil recovery. As a result of this management process, two major adjustments were made to the CO_2 flood, one in July 1991 and the other in January 1992. In July 1991, 55 patterns were taken off of CO_2 injection and put on chase water injection, and 9 patterns were switched to continuous CO_2 injection. This reduced total CO_2 injection to an average rate of 49.6 MMscf/D. In January 1992 an additional 45 patterns were switched to chase water, and 20 patterns were switched to continuous CO_2 injection. This adjustment also reduced total CO_2 injection, to a rate of 43.8 MMscf/D.

Most of the patterns that were switched to chase water were located in Sections 3, 6, and 10. The Yates Formation within these sections contained less recoverable oil than Sections 7, 8, and 9 and did not respond as well to CO_2 injection. The patterns that were switched to continuous CO_2 injection, in general, had difficulties in achieving target water injection rates and responded better to continuous CO_2 than to WAG injection.

Concurrent with the CO_2 adjustment in January 1992, the CO_2 flood was switched from a time-based WAG to a volume-based WAG. The problem with the time-based WAG was that during a complete WAG cycle of 2 months, the CO_2 slug (1.5% HCPV) could be injected over a 1-month time frame, but in some instances a complete slug of water (1.5% HCPV) could not be injected within the following month. This was consistent with the injectivity test which indicated that the CO_2 injection rate was about 20% higher than the water injection rate at the same bottomhole pressure. The solution to this problem was a volume-based WAG whereby a 1.5% HCPV CO_2 slug is injected followed by 1.5% HCPV of water, regardless of the time frame. The physical requirements of a volume-based WAG were operationally possible with the reduction in patterns. After switching to a volume-based WAG, most cycles are completed approximately every 2 months.

After the two major adjustments, 39 patterns remained on WAG injection, and 29 patterns were on continuous CO_2 injection. Since January 1992 only 9 patterns have been switched to chase water. Currently, 59 patterns are being flooded with an average CO_2 injection rate of 22 MMscf/D, of which only 2 MMscf/D are being purchased. Two of the 59 patterns are on continuous CO_2 injection, and the remaining 57 patterns are on a volume-based WAG. The WAG ratios are currently ranging from 0.8:1 to 1.2:1, and the CO_2 half-cycle slug sizes are approximately 1.5% HCPV.

Gross CO_2 utilization, net CO_2 utilization, and incremental oil recovery are approximately 15.0 Mscf/bbl, 7.1 Mscf/bbl, and 4.3% of OOIP, respectively. These performance dimensions are based on a CO_2 slug size of 21.0% HCPV and are expected to improve as the CO_2 flood matures.

Reservoir Management

The objective of managing the North Ward Estes CO_2 Flood has been to maximize oil recovery for maximum profitability. This has been accomplished by maintaining a consistent WAG process, optimizing CO_2 utilization, and maintaining a proper balance between patterns. Being able to adequately model and forecast performance of the CO_2 flood has also been critical in the management of this flood.

The most challenging aspect of managing the CO_2 flood has been to utilize CO_2 in an optimal manner. This has been accomplished by maintaining a consistent WAG process and utilizing various injection schemes (such as WAG and continuous CO_2 injection) to optimize CO_2 utilization.

The WAG ratios and half-cycle slug sizes have remained essentially constant at 1:1 and 1.5% HCPV, respectively, throughout the life of the flood. However, localized variations in WAG cycles have occurred due to operational reasons. For example a WAG injector may be switched from a 1:1 ratio to a 2:1 ratio to mitigate gas breakthrough or to a 1:2 ratio to facilitate an associated flowing producer (i.e., to maintain a producer flowing rather than putting it on pump). The only permanent modification to the original WAG cycle has been in going from a time-based WAG to a volume-based WAG, which was insignificant in terms of CO_2 utilization.

The optimization of CO_2 utilization has been achieved primarily by using various injection schemes, since the WAG process is essentially fixed. The goal in optimizing CO_2 utilization is to maximize oil recovery while minimizing CO_2 purchases. The opportunity to optimize CO_2 utilization generally occurs when patterns are switched to chase water. The additional CO_2 that becomes available is either used for improving recovery in other patterns or for reducing total CO_2

purchases. In some patterns the additional CO_2 has been used to improve recovery by switching patterns from WAG to continuous CO_2 injection. Occasionally patterns that have been switched to chase water will be put back on WAG or continuous CO_2 injection; however this has not been a common practice. Most of these decisions are based on the objective of improving recovery at the expense of injecting CO_2.

An additional aspect in managing the CO_2 flood has been to maintain a proper balance between patterns. Balancing patterns maximizes areal sweep efficiency by ensuring that patterns are evenly swept. This has been accomplished primarily by adjusting injection rates. Establishing a balance between patterns was critical in the initial stages of the flood; however, once the proper injection rates were implemented, a proper balance has been easily maintained.

Reservoir modeling has been both successful and effective in making reservoir management decisions. Initially, these models were used to identify the advent and duration of peak performance and long-term expectations of the CO_2 flood. Currently, they are being used to forecast the outcome of switching patterns from CO_2 injection to chase water, and to answer questions concerning WAG management and CO_2 utilization on an individual pattern basis. Questions such as these need reasonable answers to plan for the future.

Reservoir Modeling

Several reservoir models have been developed throughout the course of the flood to facilitate the management process. Streamtube models were developed to visually observe the balance between patterns. These models were also coupled with the characteristic solutions of finite-difference cross-sectional models to forecast performance. This method of modeling reservoir performance is known as the hybrid technique.[4] A pattern element model was also developed to answer questions concerning WAG management and CO_2 utilization on an individual pattern basis.

Streamtube Models. Streamtube models were developed for each section of the project area, and have been used to visually identify imbalances in injection and production between patterns and to forecast performance. Streamtube models are 2D areal models which solve for fluid flow areally. Streamtubes are flow paths defined between two adjacent streamlines connected to an injector and producer pair.

2D areal streamtube models for Sections 8 and 9 are illustrated in **Figs. 5 and 6**, respectively. **Fig. 5** illustrates streamtubes for a five-spot pattern configuration (Section 8), and **Fig. 6** is a similar illustration of a line-drive pattern configuration (Section 9). The magnitude of injection and production rates is differentiated by the size of the circular regions around wells; the larger the region, the higher the rate. The lighter shaded regions (gray) are injectors, and the darker regions (black) are producers.

These models were constructed with approximately 6000 areal cells and a single layer using a vertically averaged thickness, porosity, and permeability for each areal cell. In general, no-flow boundaries were utilized where patterns do not exist adjacent to section lines, such as lease lines. The fluid rate necessary to generate the streamtubes is averaged over a specified time interval for each well. Injection rates are determined by summing averaged CO_2 and water rates, and production rates are determined by summing averaged gas, oil, and water rates, over a given period of time.

Although there are assumptions associated with these models (such as steady-state, incompressible flow), they are not as limiting for this particular application. The pressure and temperature under which CO_2 is being injected suggest a miscible process, and fluid-in is essentially equal to fluid-out (i.e., a voidage balance) within the flooded area.

Hybrid Models. Hybrid models were developed for sections of the CO_2 project area to forecast reservoir performance. These models have been used to identify the advent and duration of peak performance and to evaluate various management strategies such as switching patterns from WAG or continuous CO_2 injection to chase water.

The hybrid technique couples a 2D areal streamtube model with 2D finite-difference cross-sectional models to model reservoir performance. The vertical heterogeneity of the reservoir is represented by a limited number of high-resolution cross sections. Rock properties within these cross sections are usually determined by geostatistical techniques. A displacement program maps the characteristic solutions of the cross-sectional model to the areal streamtube geometries to give a 3D calculation. The characteristic solutions relate phase fractional flow at the producer to pore volume of fluid injected. The hybrid method is a low-cost and accurate method for developing displacement models subject to the restrictions of the method.

Construction of several hybrid models (Sections 3, 8, and 9) began after the CO_2 flood was initiated, such that enough observed data were available to history match actual performance. Initially, cross-sectional models were developed followed by streamtube models and displacement calculations. The history matches of the models were obtained from the displacement calculations.

Several cross-sectional models of the Yates Formation were constructed for each section of the CO_2 project area being modeled. The cross sections consisted of 5000 cells extending from an injector to a producer within the same pattern. A geostatistical technique was used to generate rock properties (porosity and permeability) between wells.

The cross-sectional models were initialized at initial reservoir conditions, and water was injected until a reasonable history match was obtained on field pressure and WOR. This was to ensure that a proper saturation distribution of oil and water existed in the model prior to CO_2 injection. The fractional flow of oil as a function of total pore volume of water injected (characteristic solution) was compared for each cross section of each section. From this evaluation, it was concluded that the characteristic solution of a single cross section from each section was sufficient for the displacement calculations.

CO_2 was then continuously injected concurrent with water into the cross-sectional models at a WAG ratio of 1:1. Fluid flow through the cross sections was modeled using a four-component miscible simulator. A total of six simulation runs were made for each cross section which consisted of injecting CO_2 slug sizes of 0.1, 0.2, 0.3, 0.4, and 0.5 total pore volumes and a significantly large slug size. In making the displacement calculations, several slug sizes are necessary to adequately model the CO_2 flood, because flooded patterns (streamtubes) generally undergo various stages of processing for a given total CO_2 slug size.

The fractional flow curves obtained from the cross-sectional models were then combined with the sectional streamtube models using an algorithm to make the necessary 3D displacement calculations. Initially, these calculations were made to match historical data. The history match consisted of matching gas, oil, and water production rates from April 1989 (start of the CO_2 flood) through December 1992. Also, approximately 2 years of the waterflood were matched prior to the CO_2 flood. The oil rate history match is shown in **Figs. 7, 8, and 9** for Sections 3, 8, and 9, respectively.

Results from the 3D displacement calculations were then utilized in making short-term and long-term forecasts. Forecasts are currently being used to determine the outcome of switching patterns from WAG or continuous CO_2 injection to chase water. An algorithm, known as a Field Project Scheduler, was developed for a personal computer to automate the generation of these forecasts. The forecasts illustrated in **Figs 7, 8, and 9** depict a simple scenario in which all remaining patterns on CO_2 injection within these sections are switched to chase water in 1996.

The original hybrid model has never been adjusted, since subsequent historical performance has tracked the forecasts. The changes that have taken place after the initial history match, such as patterns being switched to chase water and patterns that have been switched to chase water back to CO_2 injection, are taken into account in the current forecasts.

Pattern-Element Model. A pattern-element model was developed to answer questions concerning WAG management and CO_2 utilization on an individual pattern basis. Although hybrid modeling can be used to evaluate individual patterns, the hybrid models that had been developed were not well suited for evaluating individual patterns.

A 3D pattern-element model representative of a $^1/_2$ line-drive pattern was developed using a four-component miscible simulator. The grid consisted of 9 cells in the x-direction and 9 cells in the y-direction and encompassed 10 acres. The grid was aligned parallel to the pattern with 7 cells between the injector and producer. Initially the model was constructed with 155 layers representing the Yates Formation. The model was then scaled-up to 34 layers using an algorithm[5] developed in-house.

Cell properties (porosity and permeability) were determined using a geostatistical technique. Observed production and injection data for the history match were determined by averaging data from similar performing patterns. This resulted in a better representation of an average pattern.

The pressure and saturation distribution of the pattern-element model, prior to CO_2 injection, were established by initially waterflooding the model. WORs and pressures were used as history matching parameters, since a rigorous match was not attempted during the waterflood period. The objective of the history match during CO_2 injection was to match gas, oil, and water production rates.

Forecasts were made to evaluate total CO_2 slug sizes, WAG tapering, and continuous CO_2 injection. The results of these forecasts indicated that incremental recovery is more of a function of how much CO_2 is injected, rather than how it is injected at this stage of the CO_2 flood. Economic assessments of these forecasts are currently being made to guide future reservoir management decisions.

Conclusions

1. Gross CO_2 utilization, net CO_2 utilization, and incremental oil recovery are approximately 15.0 Mscf/bbl, 7.1 Mscf/bbl, and 4.3% of OOIP, respectively. These performance dimensions are based on a CO_2 slug size of 21.0% HCPV and are expected to improve as the CO_2 flood matures.

2. Reservoir management has consisted of maintaining a consistent WAG process, optimizing CO_2 utilization, and maintaining a proper balance between patterns. Most management decisions are based on the objective of improving recovery at the expense of injecting CO_2.

3. Reservoir models, such as streamtube, hybrid, and pattern-element models, have been both successful and effective in developing and managing the CO_2 flood.

References

1. Winzinger, R. *et al.*: "Design of a Major CO_2 Flood, North Ward Estes Field, Ward County, Texas," *SPE Reservoir Engineering* (Feb. 1991) 11.

2. Dykstra, H. and Parsons, R. L.: "The Prediction of Waterflood Performance with Variation in Permeability Profile," *Prod. Monthly* (1950) 15, 9.

3. Power, M. T., Leicht, M .A., and Barnett, K. L.: "Converting Wells in a Mature West Texas Field for CO_2 Injection," paper SPE 20702 presented at the 1990 SPE Annual Technical Conference and Exhibition, New Orleans, LA, Sept. 23–26.

4. Tang, R. W., Behrens, R. A., and Emanuel, A. S.: "Reservoir Studies with Geostatistics to Forecast Performance," *SPE Reservoir Engineering* (May 1991) 253.

5. Durlofsky, L. J. *et al.*: "Scale Up of Heterogeneous Three Dimensional Reservoir Descriptions," paper SPE 30709 presented at the1995 SPE Annual Technical Conference and Exhibition, Dallas, TX, Oct. 22-25.

SI Metric Conversion Factors

acre	x 4.046 873	E + 03	= m^2
°API	141.5/(131.5 + °API)		= g/cm^2
bbl	x 1.589 873	E − 0	= m^3
cp	x 1.0*	E − 03	= Pa · s
ft	x 3.048*	E − 01	= m
ft^3	x 2.831 685	E − 02	= m^3
°F	(°F − 32)/1.8		= °C
md	x 9.869 233	E − 04	= μm^2
psi	x 6.894 757	E + 00	= kPa

Table 1–Rock and Fluid Properties of the Yates Formation

Lithology	Sandstone
Porosity, %	16
Permeability, md	37
Initial water saturation, %	50
Oil gravity, °API gravity	37
Initial oil viscosity, cp	1.4
Initial solution GOR, SCF/STB	500
Initial oil FVF, rb/STB	1.2
Minimum miscibility pressure, psia	937

Table 2–Reservoir Properties of the Yates Formation

Depth, feet	2600
Original reservoir pressure, psia	1400
Current reservoir pressure, psia	1300–1700
Reservoir temperature, °F	83
Productive surface area, acres	3840
Net thickness, feet	108

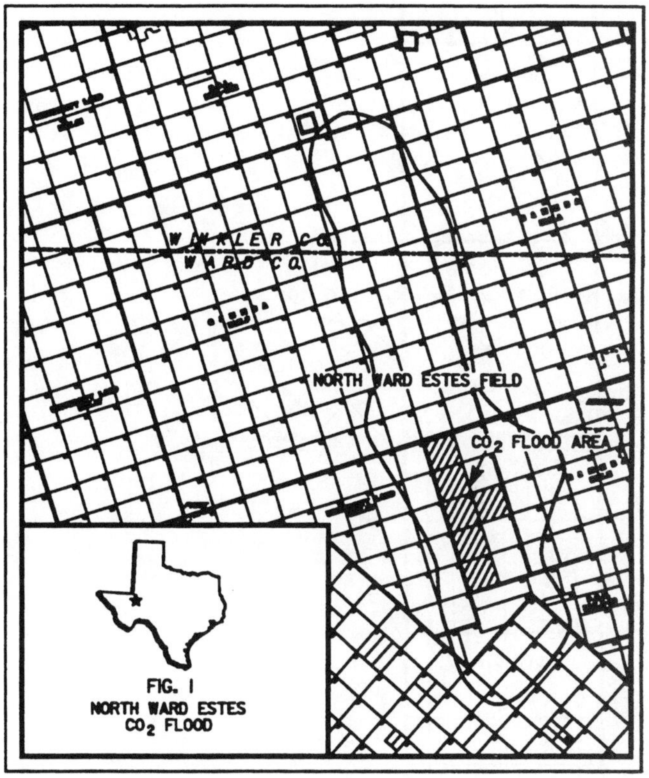

Fig. 1–North Ward Estes Field

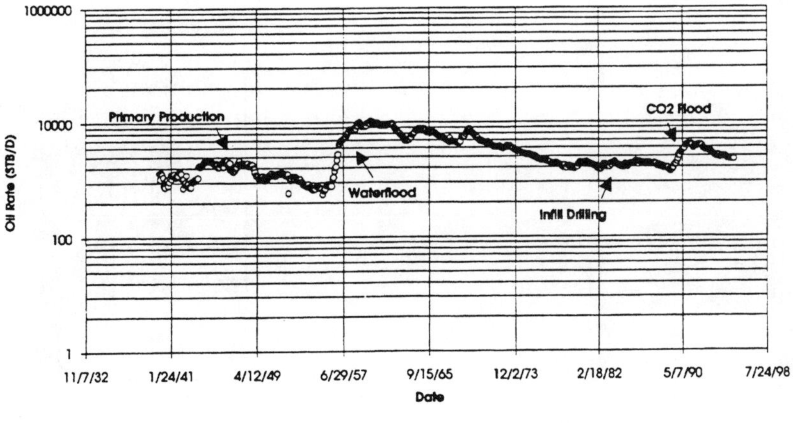

Fig. 2–CO_2 Project Area Historical Performance

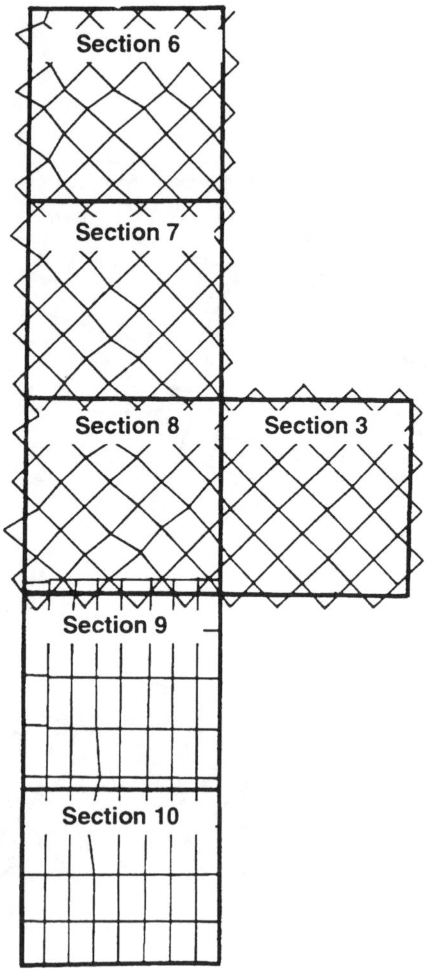

Fig. 3–CO_2 Project Area Pattern Configuration

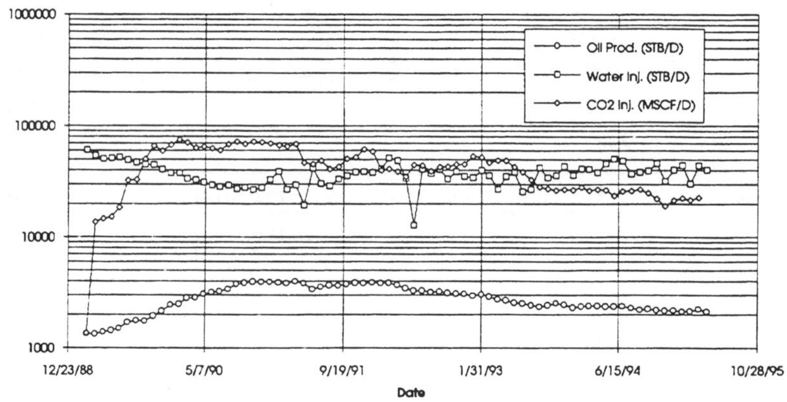

Fig. 4–CO_2 Flood Performance

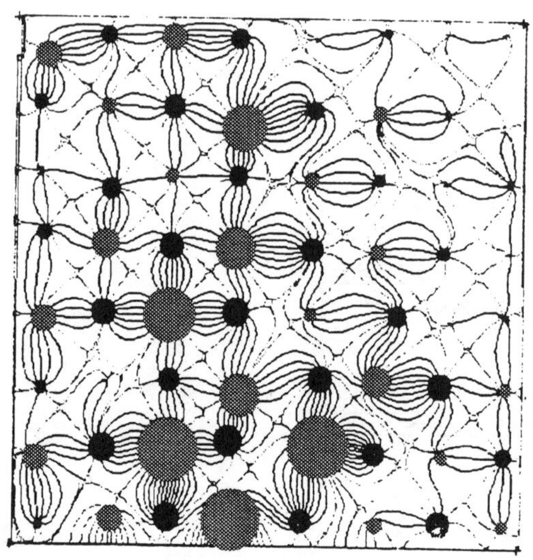

Fig. 5–Streamtube Configuration for Section 8

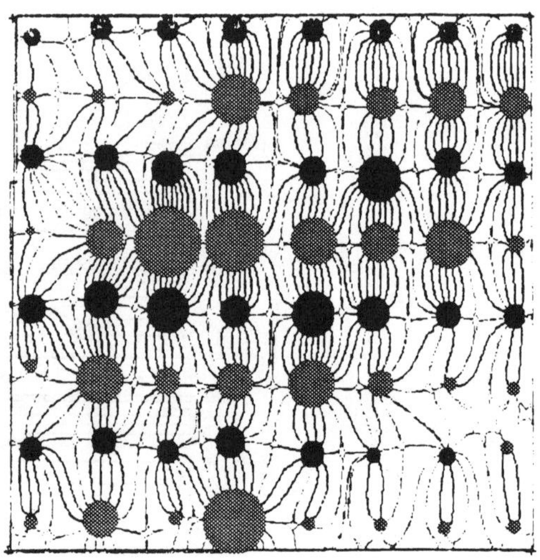

Fig. 6–Streamtube Configuration for Section 9

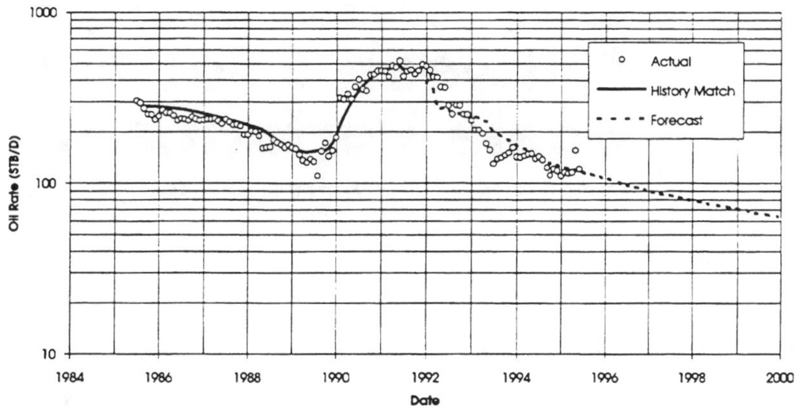

Fig. 7–History Match and Forecast for Section 3

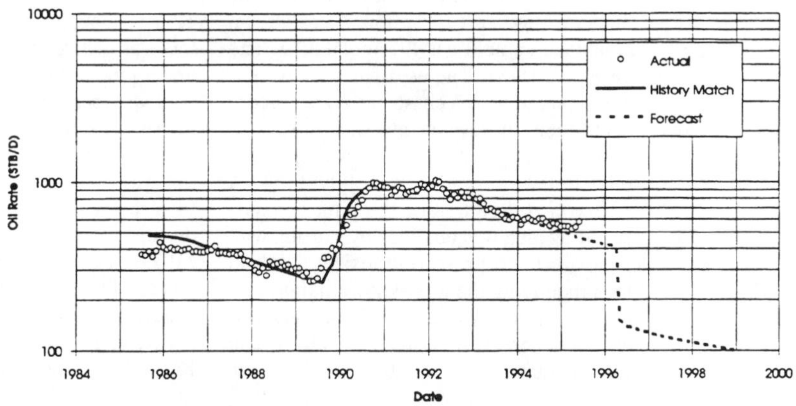

Fig. 8–History Match and Forecast for Section 8

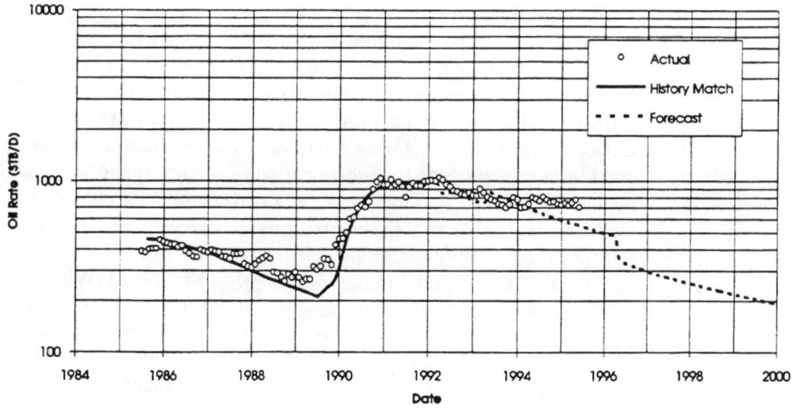

Fig. 9–History Match and Forecast for Section 9

Slaughter Estate Unit CO_2 Flood: Comparison Between Pilot and Field-Scale Performance

M.H. Stein, SPE, **D.D. Frey,** SPE, **R.D. Walker,** SPE, and **G.J. Pariani,** SPE, Amoco Production Co.

Summary. This paper describes the performance of a pilot and a unit-wide CO_2 flood in the Slaughter Estate Unit, Slaughter field, Hockley County, TX. The performance and design of both projects are compared to yield insight into the process, the impact of flood design variables, and the effects of project scale.

Introduction

Amoco Production Co. drilled an "oil-in-the-tank" pilot in the Slaughter Estate Unit to improve its understanding of the miscible-gas process. Oil-in-the-tank refers to a pattern in which the pilot producing wells are confined by EOR injection wells. The pilot producing wells are not influenced by any offset continuous-water-injection wells. The pilot was waterflooded during 1972–76. Miscible-gas injection was initiated in the pilot in Aug. 1976. The initial gas stream contained an acid gas (72 mol% CO_2 and 28 mol% H_2S). The acid gas was eventually replaced with a chase gas (various concentrations of residue gas and nitrogen). Water was injected alternately with gas. The pilot test was completed in July 1984. A reservoir description of the pilot and surrounding area was obtained by starting with geological and pressure-transient data and then modifying the least-known parameters to match primary and waterflood performance.[1] Reservoir description was important for quantitative interpretation of the tertiary performance. Pilot design,[2] performance[3] through 1981, and mechanical design considerations[4] are discussed in other technical papers.

The favorable tertiary oil response from the pilot provided comfort with the CO_2 flood process, and a unit-wide CO_2 flood was initiated in Dec. 1984. Previous feasibility studies were used to design a fieldwide CO_2 flood. Reservoir descriptions for various segments of the Slaughter Estate Unit had been developed by history matching primary and secondary (waterflood) performance. Once history matched, CO_2 predictions were made for a variety of conditions to determine the most profitable manner to inject CO_2. Ongoing CO_2 flood modeling has been part of the reservoir monitoring process. With changing oil prices, additional CO_2 flood predictions were made, resulting in changes in the slug size and grading of the gas/water injection ratio to improve vertical sweep.

This paper updates pilot performance through termination of the pilot, describes how the tertiary performance was matched, and compares pilot and unit performance. The procedure used for fieldwide design is also discussed.

Slaughter Estate Unit Miscible-Gas Pilot

The Slaughter Estate Unit tertiary pilot was one of several miscible-gas EOR pilots in the Permian Basin in west Texas. The pilot was located in the Slaughter field in Hockley County, TX (**Fig. 1**). The pilot, a 12-acre, double five-spot oil-in-the-tank pilot, was drilled in 1972 in an area that had undergone only volumetric depletion. The pilot was then waterflooded from Nov. 1972 to Aug. 1976. The volume of oil drained from the pilot predicted before waterflooding was 60.7 MSTB, or 9.4% original oil in place (OOIP).[1] The pilot oil recovery during waterflood operations was 191.9 MSTB, or 29.9% OOIP. The model predicted additional oil recovery for continued waterflood operations through pilot termination to be 72.0 MSTB (11.2% OOIP). From the predicted recoveries plus the actual waterflood oil recovery, the ultimate primary plus secondary oil recovery for the pilot is 324.6 MSTB, or 50.5% OOIP.

An alternate acid-gas (72% CO_2 and 28% H_2S)/water-injection project was initiated in Aug. 1976. Acid gas and water were injected over 10 cycles at a gas/water injection ratio (GWR) of 1 : 1 (1 res bbl/res bbl). A 25% HCPV slug of acid gas was injected. The pilot oil production rate increased from 37 to a peak of 152 STB/D in Feb. 1979 (**Fig. 2**). Chase-gas injection began in Oct. 1979. The chase gas was composed of residue gas or nitrogen, depending on the available supply. The chase gas was immiscible with the Slaughter Estate oil but was first-contact miscible with the acid gas. A 40% HCPV slug of chase gas was injected over 11 cycles. After five chase-gas cycles, the GWR was reduced to 0.7 : 1.0 to improve vertical sweep. Chase-gas injection was completed in June 1982. Chase-water injection continued until July 1985. The incremental tertiary oil recovery from the pilot was 125.9 MSTB (19.6% OOIP).

The reservoir description[1] of the Slaughter Estate pilot and the surrounding area was used with the contact-miscible solvent rela-

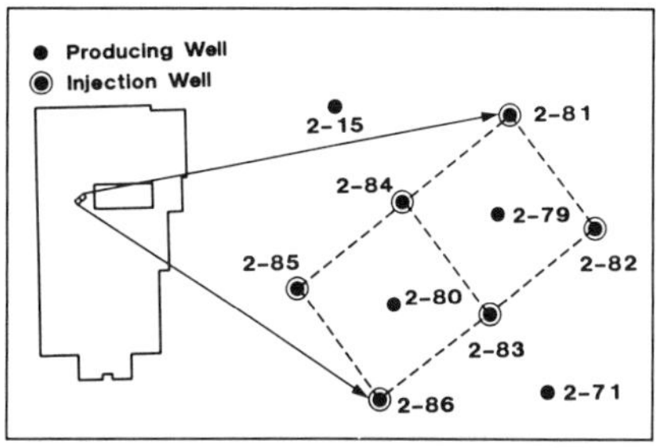

Fig. 1—Slaughter Estate Unit pilot.

"We studied the feasibility of implementing unit-wide CO_2 flooding in several west Texas properties. Preliminary pilot results were favorable, as were the potential economic incentives."

tive permeability model[5,6] to interpret tertiary pilot performance. The performance of the pilot was matched satisfactorily. **Figs. 3 through 6** show matches of tertiary performance typically obtained. In the model, fluid injection and production rates were specified on a reservoir-barrel basis to allow the pilot injection/withdrawal ratio to be met. Matching the injection/withdrawal ratio is required to ensure that the correct volumes of fluid enter and leave the pilot. The accuracy of the match is judged by the ability to match oil rate, GOR, WOR, and bottomhole pressures (BHP's). The reservoir description from the waterflood history match[1] was not changed. The predicted tertiary oil response and acid-gas response occurred simultaneously. This phenomenon, which is consistent with a high level of dispersion, occurred in the Slaughter Estate pilot and other Amoco CO_2 pilots.

The predicted peak tertiary oil response occurs early (Fig. 3). The error in matching the timing of the peak oil rate probably is caused by errors in the reservoir description in the vicinity of Well 280. The concept of solvent relative permeability[5] was used to match miscible-gas performance without any error in the timing of the peak oil rate. Furthermore, Chopra[7] found that predicted miscible-gas flood performance is more sensitive to reservoir heterogeneities than predicted waterflood performance is. Chopra's[7] result implies that the reservoir description could be modified slightly to fine tune the CO_2 flood match while not appreciably affecting the waterflood prediction. We did not perform this step during the pilot interpretation stage because our goal was to validate the physics used to predict miscible-gas performance from four different pilots. The simulator used for model validation did not account for viscous fingering,[5] which would tend to cause earlier gas breakthrough than predicted. Fig. 3 shows, however, that the timing of initial injected gas was predicted satisfactorily.

The concept of solvent relative permeability[5,6] was necessary to match the maximum 50% loss of water injectivity experienced (1979). The solvent relative permeability was also important in matching gas injectivity and gas production. A common assumption in first-contact-miscible simulators is that the solvent relative permeability is identical to the oil relative permeability. The concept of solvent relative permeability differs from this assumption when there is a final oil saturation after miscible-gas injection. The solvent-relative-permeability concept calls for the solvent relative permeability to be lower than the oil relative permeability curve if a final oil saturation remains after miscible-gas injection (see Fig. 4 of Ref. 5 for an illustration). The solvent relative permeability is reduced further if the final oil saturation is higher. A higher final oil saturation also leaves a higher residual solvent saturation, which in turn limits the maximum water saturation attainable and the maximum water relative permeability. The maximum water relative permeability attainable with the solvent-relative-permeability concept has been validated with pressure-falloff test data collected at the end of a water cycle.[5]

Because the average reservoir pressure of 2,000 psi was well above the minimum miscibility pressure of 1,050 psi, loss of miscibility had little effect on the predicted tertiary performance.

Water bottomhole injection pressures (BHIP's) ranged from 2,800 to 3,100 psi, depending on each well's fracture parting

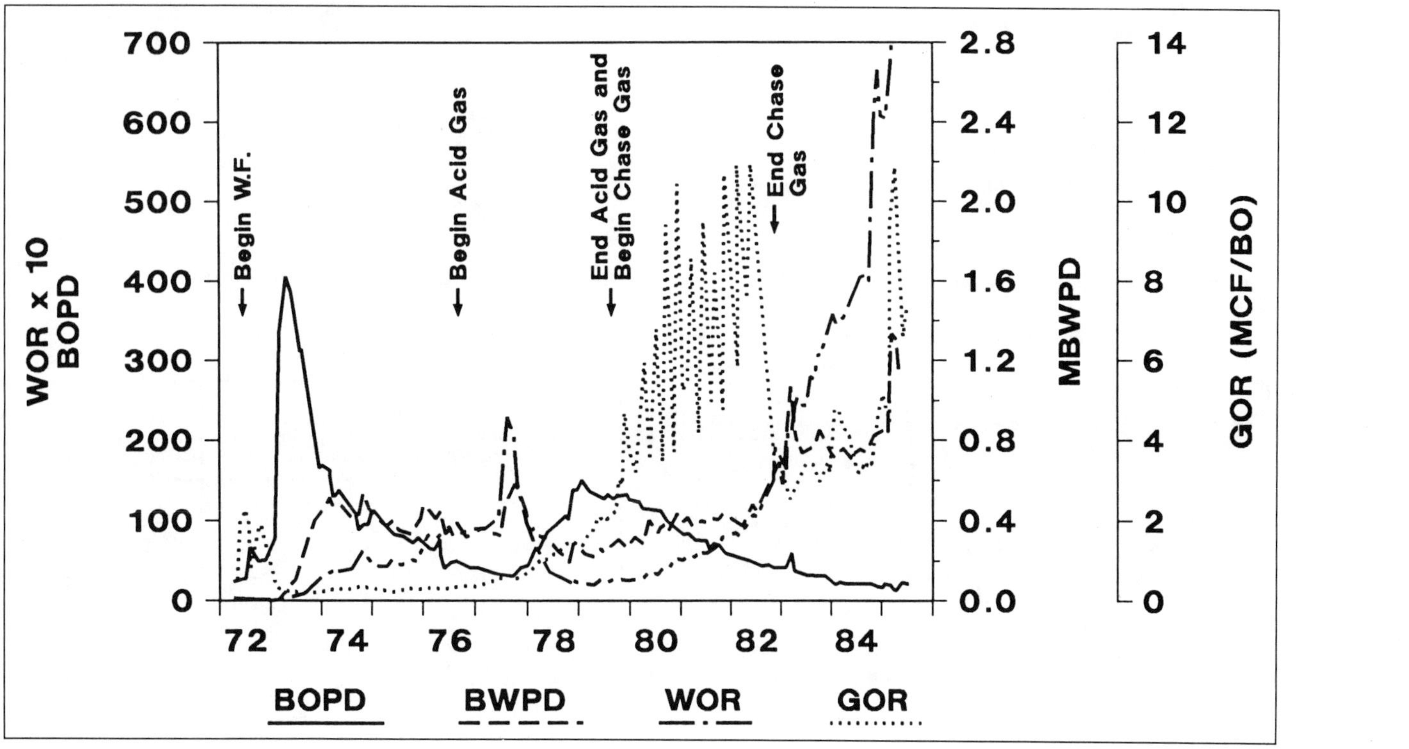

Fig. 2—Pilot performance.

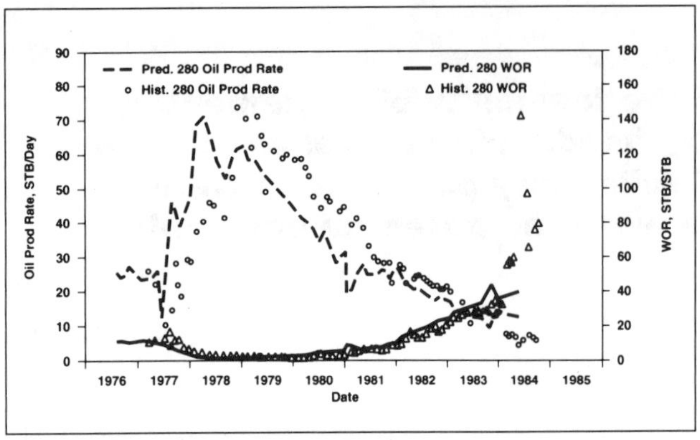

Fig. 3—Tertiary pilot oil and WOR matches, Well 280.

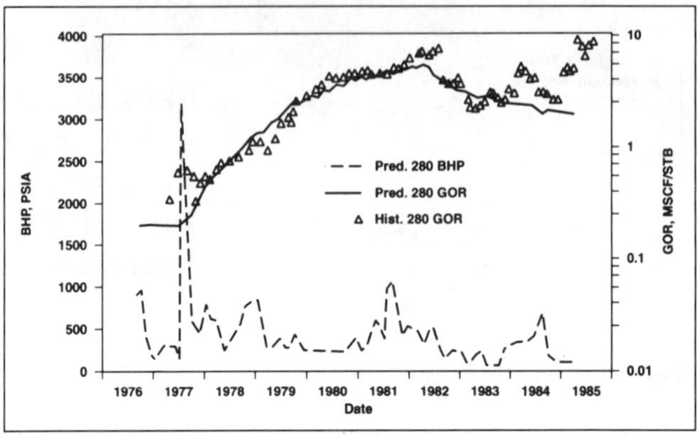

Fig. 4—Tertiary pilot GOR match, Well 280.

pressure. The pilot gas injection rate was kept equal to the pilot water injection rate to maintain a GWR of 1.1. The acid-gas BHIP's usually ran about 100 psi below the water BHIP. During chase-gas injection, the BHIP's were 2,500 psi. This lower pressure resulted in equal gas and water injection rates on a reservoir-barrel basis. Bottomhole producing pressures started out at about 200 psi but, near the end of acid-gas injection, climbed to 600 psi. High gas rates caused difficulties in pumping off the pilot producers. Producing BHP's eventually were lowered by removing vent strings and packers to allow the gas to flow up the annulus.

The chase gas was assumed to be miscible with the oil in the model.[5] This assumption, although not true, is allowable as long as the chase-gas bank does not overrun the acid-gas bank. This did not happen in the predictions. Based on numerical dispersion, the Peclet number along the diagonal between pilot injector/producer pairs was 16. With the excellent tertiary oil response, it is unlikely that the chase-gas bank overran the solvent bank.

The model in Ref. 5 assumes stabilized displacement (no viscous fingering). Fractional-flow calculations similar to that presented by Stalkup[8] were used to estimate the maximum GWR allowed for stable displacement. The maximum GWR was 19 for the Slaughter Estate pilot relative permeability (**Fig. 7**). This calculation uses the solvent relative permeability curve.[5]

The modeling of pilot tertiary performance indicated that the final oil saturation to the acid gas was 5% PV. This value was obtained by determining the specified final oil saturation that yielded the best match in cumulative oil recovered.

Because of the promising tertiary performance of the Slaughter Estate pilot a unitwide CO_2 flood was approved in 1982. H_2S was not considered for fieldwide injection with CO_2 because of safety reasons; it still is not considered.

Slaughter Estate Unit History

The Slaughter Estate Unit comprises 5,752 acres in the southeast portion of Slaughter field. Average net pay is 79 ft, with an average porosity of 12.0% and an average permeability of 4.9 md. **Table 1** gives other pertinent data.

The Slaughter Estate Unit was formed in Dec. 1963. Before unitization, limited waterflooding had occurred on various

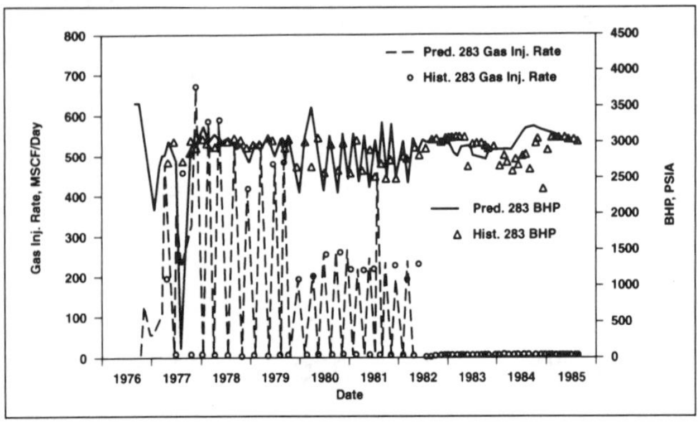

Fig. 5—Tertiary pilot gas injection match, Well 283.

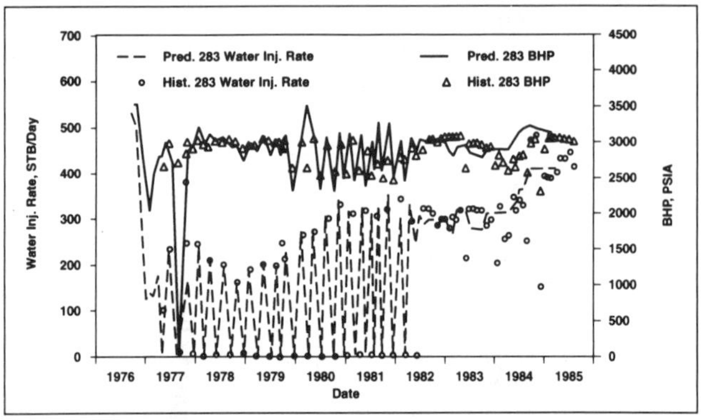

Fig. 6—Tertiary pilot water injection match, Well 283.

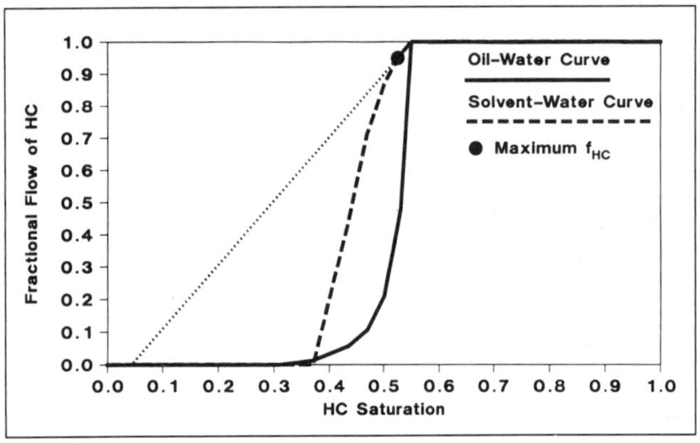

Fig. 7—Pilot maximum GWR (19).

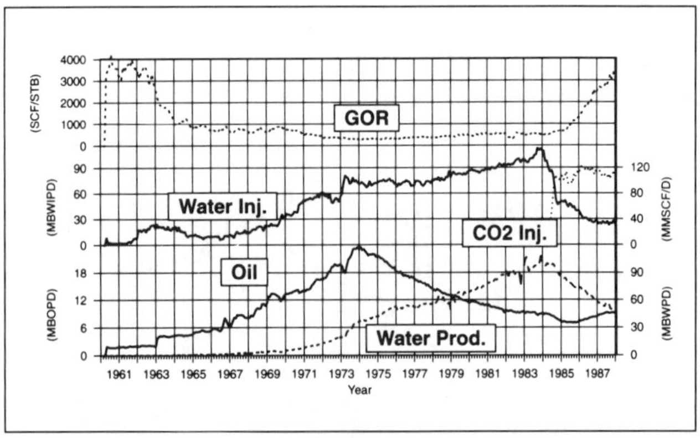

Fig. 8—Slaughter Estate Unit performance.

leases that were to comprise the unit. Leases were allowable-restricted, and oil production almost doubled as a result of the increased unit allowable. From unitization until 1968, some 16 injectors were added, with the allowable increasing accordingly. In 1968, a 25% injection bonus allowable was given to the unit, which put the unit capacity under the allowable for the first time.

About 125 wells were drilled during 1969–74, mainly in the southern two-thirds of the unit, and approximately 70 conversions to water injection were made. Unit production rose from around 8,000 to a peak of 23,500 BOPD in 1974 (**Fig. 8**).

Five-spot and chicken-wire patterns are dominant in the Slaughter Estate Unit. The southwest portion is drilled on 20-acre five-spot patterns. The southeast portion is drilled on 160-acre chicken-wire patterns (**Fig. 9**), while the northern portion contains 40-acre five-spots.

During waterflood, the unit was operated under the following philosophy: ensure that all the pay was open in both injectors and producers, keep the producing wells pumped off (<500-ft fluid level), and inject below formation parting pressure. The unit showed typical waterflood performance with high water injection as fill-up occurred followed by a decrease in water injection between fill-up and water breakthrough (Fig. 8). The GOR also collapsed during this time and water breakthrough occurred in 1967.

Oil response was observed beginning in 1965. From unitization in 1963 until 1968, it was difficult to notice response because the unit was allowable-limited. Several infill drilling packages were undertaken between 1969 and 1974 to enhance waterflood recovery. Oil production peaked in 1974 at 23,500 BOPD and began a 12% annual decline until 1983. During 1983 and early 1984, infill drilling was implemented in the northern portion of the unit to standardize the patterns and to prepare for CO_2 flooding. This infill drilling package halted the waterflood decline and held oil production flat.

CO_2 Model Study

We studied the feasibility of implementing unit-wide CO_2 flooding in several west Texas properties. Preliminary pilot results were favorable, as were the potential economic incentives. A method was formulated[9] to allow us to study the feasibility in a timely manner. Several rounds of studies were completed between 1978 and 1982. With the encouraging results from the pilot and feasibility studies, unit-wide implementation of CO_2 flooding was authorized in 1982. CO_2 injection began in Dec. 1984.

During CO_2 flooding, reservoir modeling was used as a tool to monitor performance. Several detailed geologic studies were completed and a new model developed for miscible flooding.[5] These new data and

TABLE 1—SLAUGHTER ESTATE UNIT PERTINENT DATA SHEET

Working interest, %	98.5
Producing area, acres	5,703
OOIP, million bbl	283
Formation	San Andres dolomite
Depth, ft	4,985
Number of wells	
Producers	202
Injectors	147
Producing mechanisms	
Primary	Solution-gas drive
Secondary	Waterflood
Tertiary	CO_2 miscible
Average pay, ft	
Gross	140
Net	79
Average porosity, %	12.0
Average permeability, md	4.9
Oil gravity, °API	32

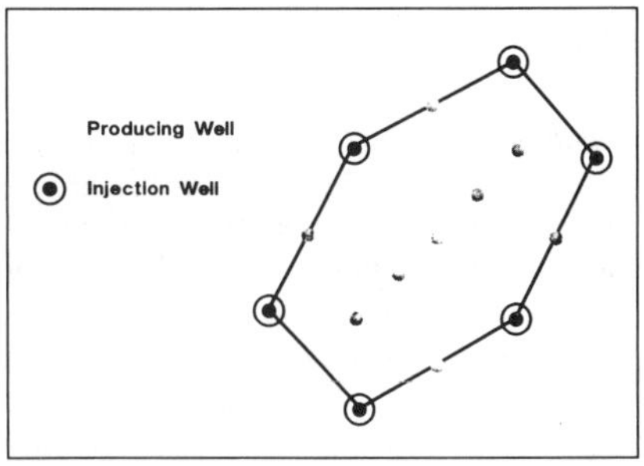

Fig. 9—Chicken-wire pattern (160 acres).

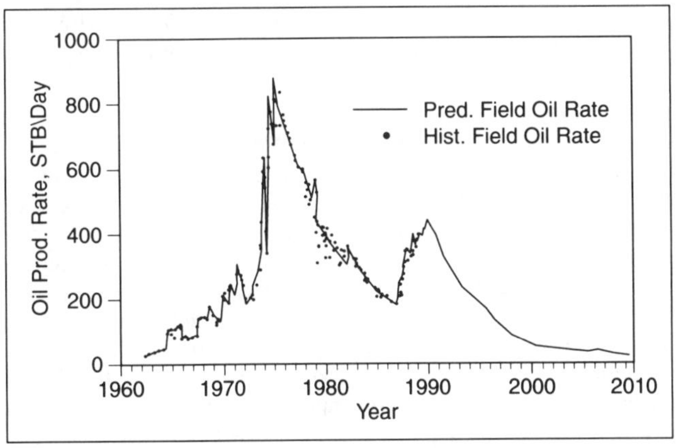

Fig. 10—Slaughter Estate Unit model Segment 1 oil rate.

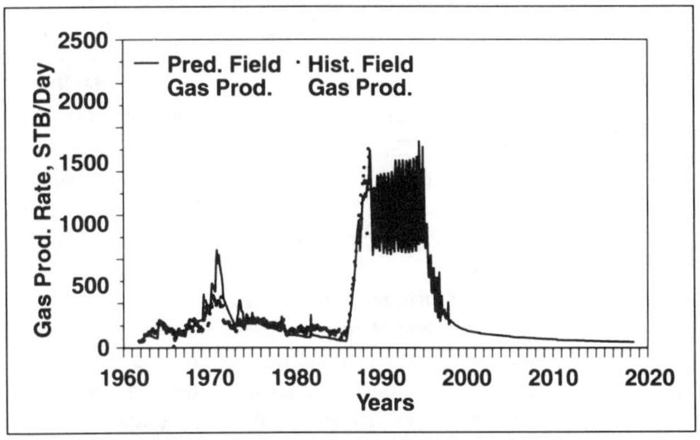

Fig. 11—Slaughter Estate Unit model Segment 1 gas production rate.

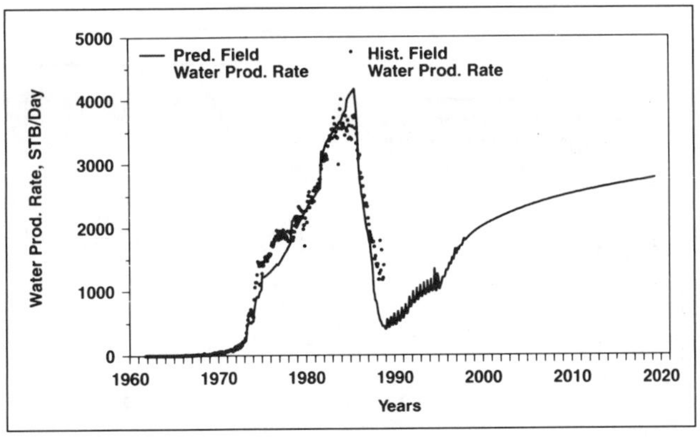

Fig. 12—Slaughter Estate Unit model Segment 1 water production rate.

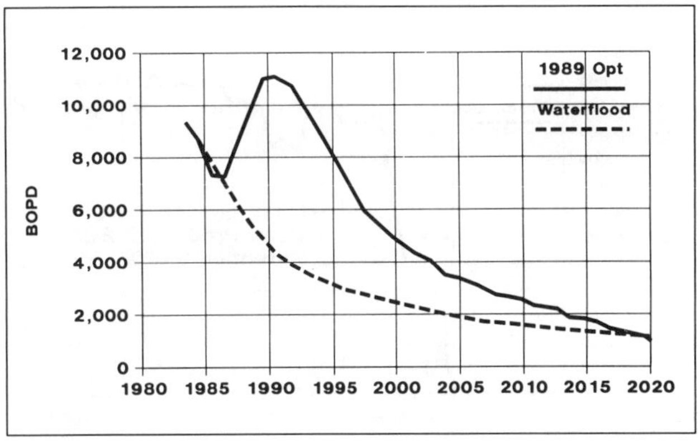

Fig. 13—Slaughter Estate Unit predicted waterflood and tertiary oil rates.

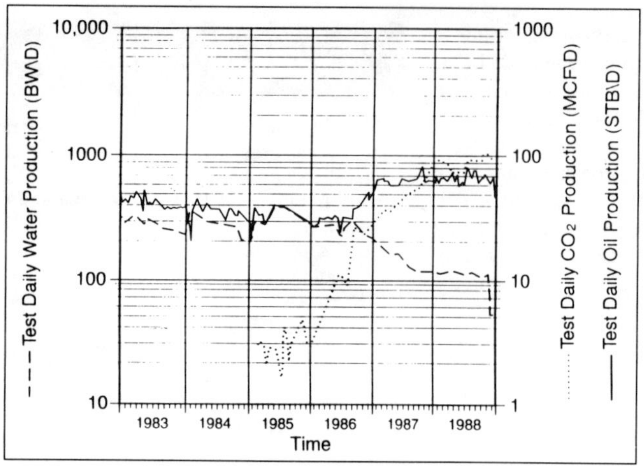

Fig. 14—CO_2 flood performance, Well A.

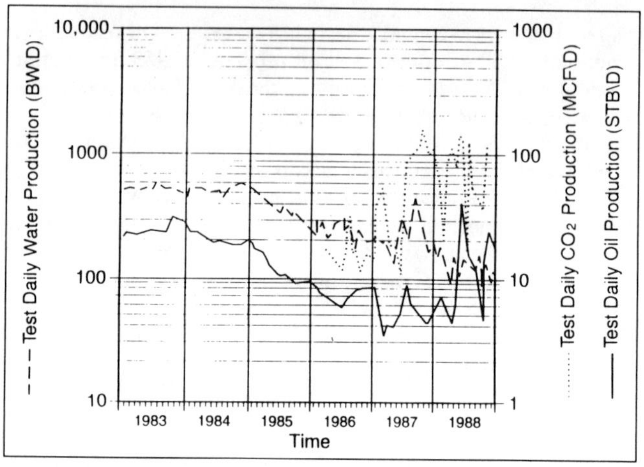

Fig. 15—CO_2 flood performance, Well B.

tools were used in the most recent model study.

History Matching

The unit was history matched for both waterflood and early tertiary performance with the latest geologic data and techniques for developing a reservoir description.[10] Relative permeability data for the unit were averaged and normalized to the unit average connate water saturation of 7.5% and average residual oil saturation of 20%. The hysteresis curve for average water relative permeability allowed for a trapped water saturation of 12%. The pilot relative permeability had higher residual oil and trapped water saturations (31% and 37%, respectively).

The field was modeled with an approach similar to the feasibility modeling.[9] Average patterns for the waterflood were based on similar development history, geology, and waterflood performance. The average pattern models were then scaled up for the full unit performance.

For the tertiary performance matching, the contact-miscible solvent-relative-permeability model[5] was used to match unit performance through 1988. Both waterflood and tertiary performance were used for the history matching, because matching only the waterflood performance did not yield an accurate enough reservoir description for tertiary performance. The contribution of some of the thin, high-k/ϕ layers can go unnoticed during waterflooding. Their effect is more pronounced during tertiary flooding because of the high mobility of the CO_2.[7] These high-k/ϕ layers account for the early gas responses. When only waterflood is history matched these critical high-k/ϕ layers may not be included.

The timing of converting wells to alternate gas/water injection (AGWI) was taken into account. Each well in a model area has various injection half-cycle times that were averaged for the model work. The initial GWR was 2 : 1 with a 1% HCPV half-cycle of CO_2 and a 0.5% HCPV half-cycle of water. The initial GWR was changed twice during the history-match period. In Aug. 1986, the GWR increased to 3 : 1 GWR with a 1.5% HCPV CO_2 half-cycle. In Jan. 1987, the GWR was again increased to 4 : 1 GWR with a 2.0% HCPV CO_2 slug. The water half-cycle remained the same at 0.5% HCPV during both increases. The GWR was increased to increase CO_2 injectivity and in turn accelerate tertiary oil response.

Several things became obvious during the history matching. Actual field performance was less than the model predictions using waterflood operating assumptions on the producing wells. That is, the wells have very low producing BHP's and are in an improved or negative skin condition. Because of problems inherent in pumping gassy wells, the actual measured producing BHP's varied from 140 to 300 psi. It was also found that wells tended to scale up faster than they had under waterflood. These two factors were taken into account in the model, and BHP's were increased on the producers and the effective wellbore radii reduced.

To match both the oil and gas performances of the tertiary period, the final oil saturation, S_{of}, had to be increased. The derived S_{of} for the unit is 12%. Using this value allowed both tertiary oil response and gas response to be matched. Matches of oil, gas, and water production for the 20-acre five-spot patterns (Model Segment 1) are shown in **Figs. 10 through 12**, respectively. These plots represent the average of four five-spot patterns. CO_2 flood performance starts in 1985. Prior performance represents waterflooding.

Tertiary Performance

The Slaughter Estate Unit is experiencing tertiary response resulting from CO_2 flooding (Fig. 8). The decrease in oil production seen during 1985–86 was expected because of the loss of injectivity experienced with AGWI floods.[3,5] The initial loss was not as great as expected, based on pilot experience, partly because of the staging of the CO_2 injection implementation. The current oil rate is about 4,000 BOPD above the predicted waterflood decline (**Fig. 13**).

The various areas of the field are in different stages of tertiary performance owing to

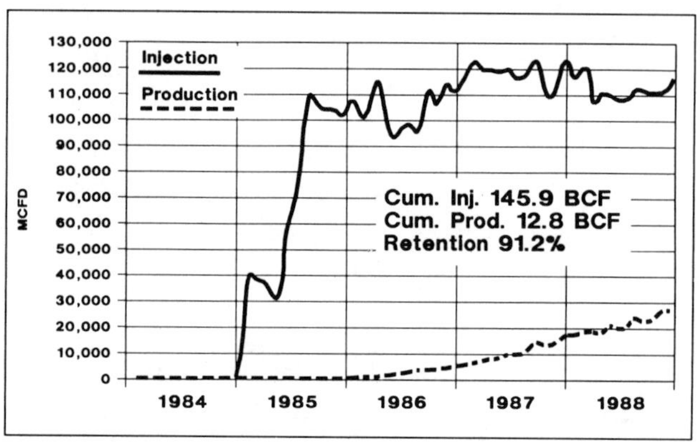

Fig. 16—Slaughter Estate Unit CO_2 production.

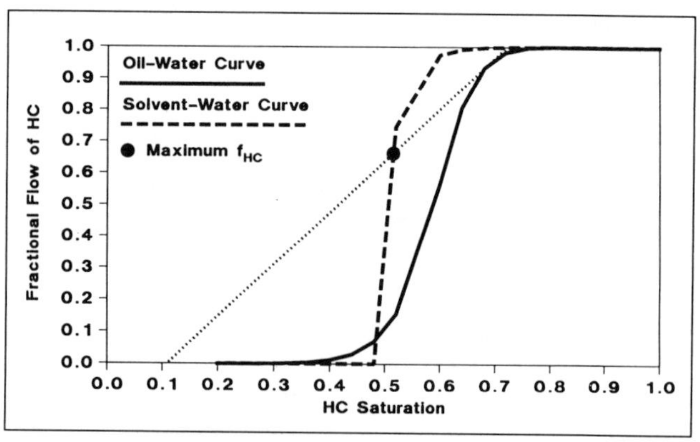

Fig. 17—Unit maximum GWR (19).

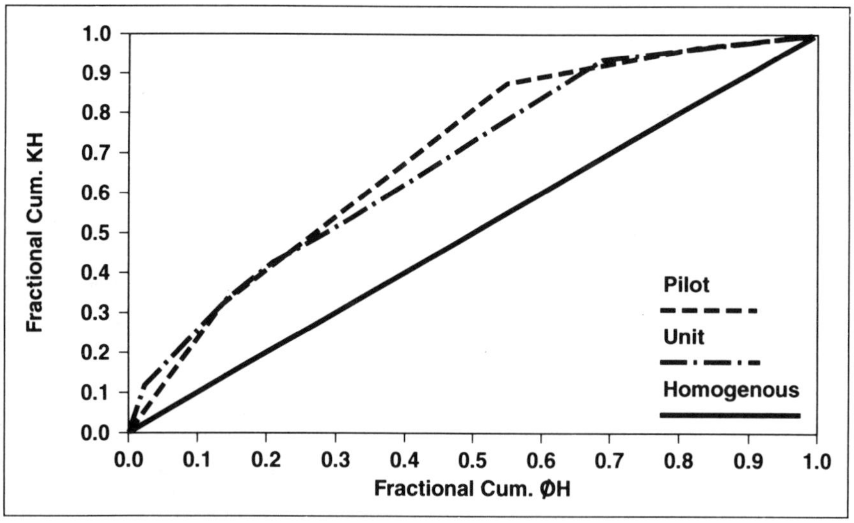

Fig. 18—Lorenz coefficient.

"The most mature patterns have experienced the greatest injectivity loss."

well spacing and startup of CO_2 injection. The 20-acre five-spot patterns in the southwest portion of the unit are the most mature, with an average of 26% HCPV CO_2 injected, although they were the last to be put on CO_2 injection. **Fig. 14** shows typical producing-well performance for this area. The CO_2 production seemed to lead the tertiary oil response, beginning in late 1986. The water production began declining when the CO_2 production rate began to increase.

The chicken-wire patterns in the southeast portion are the least mature area, with an average of 11% HCPV CO_2 injected. **Fig. 15** shows typical producing-well performance for this area. The characteristic shapes of the CO_2, oil, and water curves are similar to the performance of the five-spot pattern producer shown in Fig. 14. The tertiary oil response in Fig. 15 did not occur until 1988. Other areas of the field are between these two in maturity and response.

Table 2 shows the relative losses of injectivity for water and CO_2. These losses are relative to the pre-CO_2 water injectivity. The cycle numbers in Table 2 are based on a statistical sampling criterion of 30% of the wells achieving that cycle. The most mature patterns have experienced the greatest injectivity loss.

Injection profiles were not markedly different between waterflooding and CO_2 flooding. The profiles indicated that the same intervals were swept during waterflooding and CO_2 flooding.

CO_2 retention is very good, cumulative injection is 146 Bcf, and cumulative production is 13 Bcf for a retention of 91% (**Fig. 16**). The initial plans called for a total slug of 30% HCPV at a constant 2:1 GWR. Current plans call for reducing the GWR's from the 4:1 level to improve sweep and to inject a larger total slug. These changes were brought about by changes in oil prices, CO_2 prices, and recycle costs. The eventual total slug size and ultimate tertiary recovery will depend on future changes in the economics.

Comparison of Pilot and Field Tertiary Performance

The Slaughter Estate Unit and pilot both experienced tertiary response. The performances are similar. CO_2 breakthrough was observed at the first sign of tertiary oil response. The ratio of the peak tertiary oil rate to the peak waterflood oil rate was 0.5 for both the pilot and Model Segment 1. The unit chicken-wire patterns are expected to have a slightly lower ratio because the individual producers peak at different times. The decline in tertiary oil rate is expected to be slower in the chicken-wire patterns. The percent oil recovery will probably be less in the unit than in the pilot for several reasons. First, the pilot had a larger tertiary target. The minimum displaceable tertiary oil saturation (residual oil saturation to waterflooding minus model-estimated final oil saturation to CO_2) was estimated to be 26%. Unit-wide, the minimum displaceable tertiary oil saturation is estimated to be 8%.

The unit had higher GWR's, which helped increase the oil and gas production rates and may have led to some degree of viscous fingering. **Fig. 17** shows fraction-flow calculations for the average Slaughter Estate field relative permeability curves. The maximum GWR above which viscous fingering is estimated to occur is 1.9:1. The initial GWR in the Slaughter Estate Unit was 2:1, which was then increased to 4:1 to accelerate rates. Therefore, there may be some viscous fingering,[11-13] which could lead to a higher effective final oil saturation. Some oil might be bypassed because of the nonequilibrium effects associated with viscous fingering.[12]

The maximum loss in water injectivity was similar in the pilot and unit. The water injectivity loss in the pilot is caused partly by a larger trapped water saturation in the pilot. The unit loss in water injectivity results primarily from the higher final oil saturation to CO_2, which reduces the solvent relative permeability and the maximum water relative permeability.[5]

Gas injectivity is difficult to compare between the pilot and the unit. The acid-gas injection rates in the pilot were lower than the unit-wide CO_2 injection rates. Lower BHIP's for acid gas than water make this comparison difficult.

The degree of stratification differs only slightly between the Slaughter Estate pilot and the unit. The average layering of the Slaughter Estate Unit pilot (Table 2 of Ref. 1) is displayed in a Lorenz coefficient plot (**Fig. 18**), along with the layering used to describe Model Segment 1 of the Slaughter Estate Unit. A higher Lorenz coefficient (maximum value is 1.0 and a value of zero represents a homogeneous formation) implies that the formation is more heterogeneous. The Lorenz coefficients of 0.41 and 0.37 for the pilot and Model Segment 1, respectively, are not considered to be significantly different.

To match unit performance, the fastest layer was broken up into two layers to provide a small-PV faster layer (high k/ϕ). This may be caused by a lack of sensitivity of reservoir layering to waterflooding[6] or by

TABLE 2—LOSS OF INJECTIVITY

		Injection Reduction (%)	Cycle Number
Total unit	CO_2	31	15
	Water	49	15
20-acre five-spot patterns	CO_2	40	16
	Water	57	16
160-acre chicken-wire patterns	CO_2	16	9
	Water	49	9

Authors

Stein

Frey

Walker

Pariani

Michael H. Stein worked at Amoco Production Co.'s research center in Tulsa during 1978–87, where his major efforts were in miscible-gas pilot interpretation and reservoir characterization. Stein transferred to Amoco's Houston Region as a consulting engineer. He also has been a production engineer involved with daily operations in the Slaughter Estate Unit CO_2 flood. An Annual Meeting Technical Committee member, Stein also served on the Editorial Review Committee. He holds a BS degree from the U. of Missouri and MS and PhD degrees from Purdue U., all in chemical engineering. **Douglas D. Frey** started with Amoco Production Co. in Houston after graduating from Texas A&M U. with a BS degree in mechanical engineering in 1977. He has held assignments in primary and secondary operations, drilling, and reservoir analysis. Most recently, he has led a multidisciplinary team to exploit the Prentice field area in the Permian Basin. Frey has held several section offices, served on and chaired the Enhanced Oil Recovery Field Reports Committee, and was a member of the 1990 SPE Forum Series Committee. **Rodney D. Walker** is a senior staff petroleum engineering supervisor with Amoco Production Co. in Houston. Since joining Amoco in 1978, he has held several positions working with CO_2 and waterflooding operations in west Texas properties. He holds a BS degree in petroleum engineering from Texas Tech U. **Gary Pariani** is a staff petroleum engineer with Amoco Production Co. in Houston. He has more than 13 years' reservoir engineering experience, including 6 years of reservoir modeling CO_2 floods, waterfloods, and primary gas reservoirs. He holds a BS degree in petroleum engineering from the U. of Missouri.

viscous fingering. This faster layer was not always apparent from core or log data. Because the predictive model did not account for viscous fingering,[5] the high-k/ϕ, small-ϕh layer may represent an artifact to account for viscous fingering in the fastest layer. Larger cycle sizes in the faster layers may not fully stabilize the solvent/water bank. The effect of the small-PV faster layer on the Lorenz coefficient was small.

Conclusions

1. The Slaughter Estate Unit pilot tertiary oil recovery was significant. Incremental tertiary oil recovery was 19.6% OOIP. This is attributed to the success of acid-gas mobilizing tertiary oil.
2. The "oil-in-the-tank" pilot concept was valuable for advancing our understanding of miscible-gas performance. Data from the pilot were used to validate the contact-miscible solvent-relative-permeability model and was successfully used to evaluate unit-wide performance. The Slaughter Estate Unit pilot performance provided management comfort to initiate a unit-wide CO_2 flood.
3. The Slaughter Estate Unit is responding to CO_2 flooding. The current incremental oil rate is 4,000 BOPD.
4. The unit and the pilot CO_2 flood responses have been different. These differences appear to be caused by variations in residual oil saturation to waterflooding, the final oil saturation to miscible-gas flooding, and higher initial GWR's.
5. The higher GWR's used in the unit CO_2 flood may have caused some viscous fingering, which may lead to lower oil recovery.

Nomenclature

h = thickness, ft
k = permeability, md
S_{of} = final oil saturation to miscible-gas displacement, %PV
ϕ = porosity, %PV

Acknowledgment

We thank Amoco Production Co. for permission to publish this paper.

References

1. Ader, J.C. and Stein, M.H.: "Slaughter Estate Unit Tertiary Miscible Gas Pilot Reservoir Description," *JPT* (May 1984) 837–45.
2. Faulkner, B.L.: "Reservoir Engineering Design of a Tertiary Miscible Gas Drive Pilot Project," paper SPE 5539 presented at the 1975 SPE Annual Technical Conference and Exhibition, Dallas, Sept. 28–Oct. 1.
3. Rowe, H.G., York, S.D., and Ader, J.C.: "Slaughter Estate Unit Tertiary Pilot Performance," *JPT* (March 1982) 613–20.
4. Adams, G.H. and Rowe, H.G.: "Slaughter Estate Unit CO_2 Pilot—Surface and Downhole Equipment Construction and Operation in the Presence of H_2S," *JPT* (June 1981) 1065–74.
5. Chopra, A.K., Stein, M.H., and Dismuke, C.T.: "Prediction of Performance of Miscible-Gas Pilots," *JPT* (Dec. 1990) 1564–72.
6. Schneider, F.N. and Owens, W.W.: "Relative Permeability Studies of Gas-Water Flow Following Solvent Injection in Carbonate Rocks," *SPEJ* (Feb. 1976) 23–30.
7. Chopra, A.K.: "Reservoir Descriptions Via Pulse Testing—A Technology Evaluation," paper SPE 17568 presented at the 1988 SPE Intl. Meeting on Petroleum Engineering, Tianjin, Nov. 2–4.
8. Stalkup, F.I. Jr.: *Miscible Displacement*, Monograph Series, SPE, Richardson, TX (1983) **8**.
9. Drennon, M.D., Kelm, C.H., and Whitington, H.M.: "A Method for Appraising the Feasibility of Field-Scale CO_2 Miscible Flooding," paper SPE 9323 presented at the 1980 SPE Annual Technical Conference and Exhibition, Dallas, Sept. 21–24.
10. Chopra, A.K., Stein, M.H., and Ader, J.C.: "Development of Reservoir Descriptions to Aid in Design of EOR Projects," *SPERE* (May 1989) 143–50; *Trans.*, AIME, **287**.
11. Christie, M.A. and Bond, D.J.: "Detailed Simulation of Unstable Processes in Miscible Flooding," *SPERE* (Nov. 1987) 514–22; *Trans.*, AIME, **283**.
12. Crump, J.G.: "Detailed Simulations of the Effects of Process Parameters on Adverse Mobility Ratio Displacements," paper SPE 17337 presented at the 1988 SPE/DOE Symposium on Enhanced Oil Recovery, Tulsa, April 17–20.
13. Gorell, S.B.: "Modeling the Effects of Trapping and Water Alternate Gas (WAG) Injection on Tertiary Miscible Displacements," paper SPE 17340 presented at the 1988 SPE/DOE Symposium on Enhanced Oil Recovery, Tulsa, April 17–20.

SI Metric Conversion Factors

acres	× 4.046 873	E−01	=	ha
°API	141.5/(131.5+°API)		=	g/cm³
bbl	× 1.589 873	E−01	=	m³
cp	× 1.0	E−03	=	Pa·s
ft	× 3.048*	E−01	=	m
ft³	× 2.831 685	E−02	=	m³
md	× 9.869 233	E−04	=	μm²
psi	× 6.894 757	E+00	=	kPa

*Conversion factor is exact.

Provenance

Original SPE manuscript, **Slaughter Estate Unit CO_2 Flood: A Comparison Between Pilot and Field-Scale Performance,** received for review Nov. 12, 1989. Revised manuscript received April 30, 1992. Paper accepted for publication May 13, 1992. Paper (SPE 19375) first presented at the 1989 SPE Production Technology Symposium held in Lubbock, Nov. 13–14.

JPT

Mobility Control Experience in the Joffre Viking Miscible CO_2 Flood

Derril J. Stephenson and **Andrew G. Graham,** SPE, Westcoast Petroleum Ltd., and
Richard W. Luhning, Alberta Oil Sands Technology & Research Authority

Summary. This paper discusses mobility control in the Joffre Viking field miscible CO_2 flood. Since 1984, three injection strategies have been tried: water-alternating-CO_2 ($WACO_2$), continuous CO_2, and simultaneous CO_2 and water. The studies showed that simultaneous injection results in the best CO_2 conformance. CO_2-foam injection has also been investigated.

Introduction

Vikor Resources Ltd. operates Canada's first miscible CO_2 flood in the Joffre Viking pool northeast of Red Deer, Alta., Canada **(Fig. 1)**. The reservoir is a Cretaceous, Lower Colorado group, Viking formation sandstone deposited as overlapping sand lenses.[1] **Table 1** shows the Joffre reservoir properties.

The field was discovered in July 1953 and developed with over 400 wells on 16-ha/well spacing. Primary recovery was by fluid expansion and solution-gas drive. During 1957, the field was unitized and a linedrive waterflood was begun. By the mid-1960's, 42% of the original oil in place (OOIP) had been recovered, and the field had reached its economic limit and was abandoned. In the early 1980's, laboratory tests and reservoir simulation studies determined that the Joffre Viking was suitable for miscible CO_2 flooding.

The Joffre Viking Tertiary Oil Unit (JVTOU) was formed in April 1983. In June 1983, the Alberta Oil Sands Technology & Research Authority (AOSTRA) agreed to fund 75% of the cost to implement a miscible CO_2 experimental pilot. Vikor is operator of the miscible CO_2 project for AOSTRA and the working-interest owners. The pilot and subsequent experiments were successful, and commercial field development started in Oct. 1991.

Conditions that promote viscous fingering and gravity override were known to exist in the Viking formation before CO_2 flooding. The CO_2/oil and water mobility ratios at reservoir conditions are 22 and 23, respectively; CO_2 density is ≈80% that of oil and 60% that of formation water; vertical permeability averages 0.05 μm^2. Therefore, we expected that some form of CO_2 mobility control would be required to enhance CO_2 conformance in the reservoir. Water injection with CO_2 is the most common method used to improve CO_2 conformance; however, water injection can reduce oil recovery by shielding oil from CO_2.[2,3] The key to a successful flood is an optimum injection strategy that maximizes oil recovery with efficient use of available CO_2.

During operation of the project, we experimented with different injection strategies to maximize oil recovery efficiency, including continuous CO_2 injection, $WACO_2$ injection, and simultaneous CO_2/water injection. Laboratory and field experiments were conducted with surfactant to generate foams to control mobility.

$WACO_2$ Pilot Operation

Fig. 1 shows the $WACO_2$ pilot, which consists of two adjacent inverted five-spot patterns (Patterns A and B) drilled on 32-ha/well spacing.[4,5] Development began in late 1982 with 1 year of production to establish baseline productivity before CO_2 injection. Source-water injection increased reservoir pressure to more than minimum miscibility pressure (MMP) in the pilot area during the latter part of 1983. CO_2 injection started in Pattern A in Jan. 1984; three 10% HCPV slugs of CO_2 and water were injected alternately into Patterns A and B; while Pattern A was on water injection, Pattern B was on CO_2 injection, and when Pattern A was switched to CO_2 injection, Pattern B was switched to water injection. Because of the excellent quality of the Viking sand, injection of >20% HCPV/yr into the pilot was possible, permitting rapid pilot evaluation with commercial well spacing. In Sept. 1986, $WACO_2$ was completed. Continuous water injection into Patterns A and B was started and continued until Aug. 1988, when a CO_2-foam mobility test was initiated in Pattern B. Currently, the pilot is on waterflood and is nearing its economic limit.

The decision to use $WACO_2$ injection in the pilot was based on results of a 1981 simulation study.[6,7] That study suggested that a three-cycle, 1 : 1 CO_2/water, 10% HCPV injection/cycle would be optimum for mobility control and for reduction of oil shielding in the presence of high water saturation. Advantages of the $WACO_2$ process are that it offers a high degree of operational flexibility in the field, reduces the potential for corrosion, and reduces possible injectivity problems caused by simultaneous CO_2/water injection.

Pilot performance during the $WACO_2$ operation was analyzed through an extensive field monitoring program and reservoir simulation studies.[8,9] The analysis gave insight into how injection fluids flow in the reservoir and how the $WACO_2$ process controls CO_2 mobility. Fundamental changes to our conceptual CO_2 flow model were required to explain production behavior.

The pilot performance showed that the wells could be divided into two general categories on the basis of production response. Well 14-17 in Pattern B is representative of a Type 1 well. All pilot wells except one exhibited Type 1 behavior. Production of an oil bank that exhibits an initial spike of oil and then tails off is typical of Type 1 wells. **Fig. 2** is a plot of the fractional-flow history of oil and CO_2 for Well 14-17 and shows that the oil bank starts in Dec. 1984 and peaks at a 50% oil fractional flow in March 1985. Peak oil production rapidly decreases to 30% fractional flow by June 1985, after which there is an almost linear decline in oil flow until the end of CO_2 injection. No subsequent oil banks were observed during the second and third CO_2 cycles. A moderate oil production increase occurred after the third CO_2 injection cycle because of the sharp CO_2 production decrease and subsequent increase in wellbore oil and water flow capacity. Fig. 2 also shows that some mobile oil was in the reservoir around Well 14-17 before the start of CO_2 injection.

Type 1 wells also have a characteristic produced-CO_2 profile; in Fig. 2, note that significant CO_2 response occurs simultaneously with the oil bank arrival. Another Type 1 well characteristic is the CO_2 response during the second and third CO_2 injection cycles. In these cycles, CO_2 production occurs almost immediately after the start of CO_2 injection even though the injection well is 560 m away from the producing well and a 10% HCPV water slug is injected between CO_2 injection cycles. CO_2 production also drops dramatically after injection is switched from CO_2 to water. Type 1 well CO_2 response suggests that a separate flow path is established in the formation for CO_2 and that water does not flush the CO_2/oil phase out of this flow path.

A multilayer, 2D, radial cross-sectional model was developed to study CO_2 and water flow near the injection well during $WACO_2$ injection to clarify Type 1 production response. The 2D model used a refined grid around the injection wellbore instead of the large gridblocks used in the initial 1981 simulation study.[6,7] **Fig. 3** shows the predicted CO_2 saturation near the injection well at the time of CO_2 response in the production well. The figure shows that, just 15 m from the injection well, CO_2 contacts only the top meter of the formation, indicating significant gravity over-

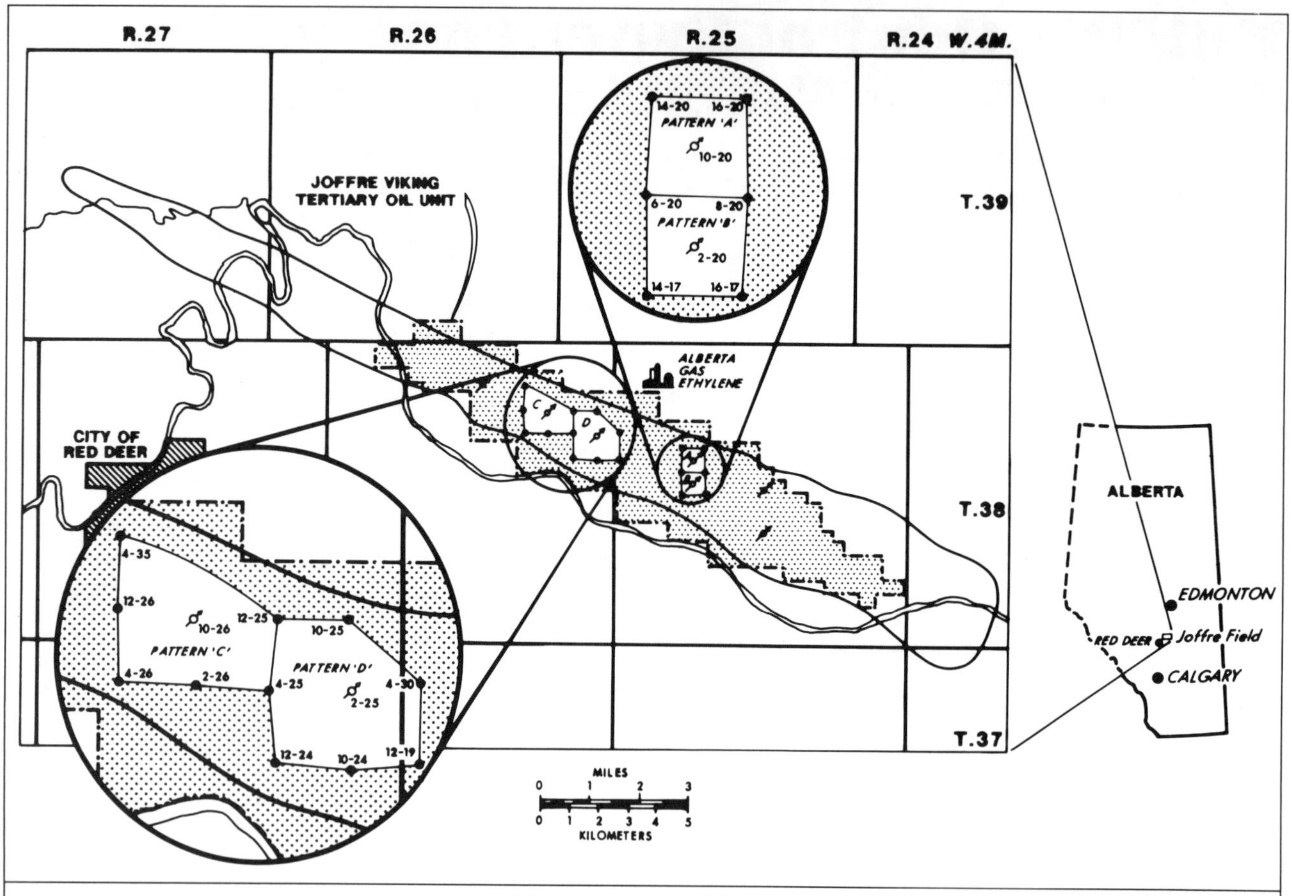

Fig. 1—Joffre field.

ride. A thin, high-CO_2-saturation flow path has been established at the top of the formation.

Using the 2D study findings and assuming a two-layer model with CO_2 injected into the top layer and water injected into the bottom layer, we were able to history match the pilot.

The pilot CO_2 response time gives another indication of the degree of gravity override and ineffectiveness of $WACO_2$ for mobility control. Pilot wells typically experienced CO_2 response in $\approx$6 months; the original $WACO_2$ process design study calculated that it would take 12 to 18 months. In contrast, Well 6-20 produced significant volumes of CO_2 within 3 months of CO_2 injection into Pattern A.

Fig. 4 shows that one pilot well, Well 16-17, had a dramatically different production profile than that of the Type 1 well. Well 16-17 is classified as a Type 2 well and displays a more classic oil/CO_2 fractional-flow profile with $WACO_2$ injection. Three banks of mobilized oil and three steadily increasing peaks of produced CO_2 are evident. The Well 16-17 production profile closely approximates the production profile predicted for a $WACO_2$ process in the initial Joffre modeling studies. In the reservoir segment between the Pattern B injection well and Well 16-17, water injection evidently improved CO_2 conformance.

The differences between Type 1 and 2 well production profiles can be explained by analysis of the pilot region geology; therefore, full understanding of the geology is necessary to determine $WACO_2$ effectiveness in a given field.

CO_2-Foam Pilot Test

The discovery that only the top one-third of the Viking formation was contacted by CO_2 led us to investigate use of CO_2-foam agents to improve CO_2 conformance. Laboratory studies[10,11] with CO_2 foams reported a significant decrease in CO_2 mobility. Sensitivity runs made during the 1987 history match of the pilot indicated that an additional 10% OOIP would be recovered if effective CO_2 viscosity could be increased to match the reservoir viscosity of the Joffre oil (i.e., an increase of 22 times the normal CO_2 reservoir viscosity).

During the first half of 1988, we undertook a laboratory study[12] to screen commercially available surfactants for possible use in Joffre. Two suitable candidates that were compatible with the reservoir fluids and that showed minimal adsorption characteristics were found.

Pilot Pattern B was chosen for a CO_2-foam test because it had the advantages of 5 years of injection/production operating experience; a simulation history match for comparison with the foam results; and continuous bottomhole pressure (BHP) monitoring equipment, which is important for evaluating a foam flood. A further reason for selection of Pattern B was that it was nearing the end of its economic life; therefore, exposure to lost production was limited if an irreversible injectivity loss occurred in the injection well.

We decided that a short-duration injectivity test in Pattern B should be conducted before a full-scale field test. The purpose of the injectivity test was to determine whether a full-scale field test was justified; to identify the magnitude of BHP increase under field operating conditions; to test simultaneous CO_2/surfactant solution injection and batch surfactant injection followed by simultaneous CO_2/water injection[13]; to establish full-scale field test design parameters; to establish minimum surfactant concentration requirements; and to determine which of the two surfactants screened would be better for the extended field test.

During the injectivity test, bottomhole injection pressure increased >2000 kPa with simultaneous CO_2/surfactant solution injection, indicating a substantial increase in CO_2 viscosity. In the injectivity test, simultaneous CO_2/surfactant solution injection maintained a higher BHP than batch surfactant-slug injection. Batch injection caused an immediate peak increase of 2000-kPa BHP when CO_2 and water were injected simultaneously after the surfactant slug. Within 6 hours, however, the pressure decreased rapidly, and there was negligible pressure effect after 24 hours.

A pressure response analysis of simultaneous CO_2/surfactant solution injection was made with a radial-flow model that assumed

TABLE 1—JOFFRE VIKING OIL RESERVOIR PROPERTIES

Discovery OOIP, m^3	14.8×10^6
Discovery pressure, kPa	7751
1981 pressure, kPa	9100
Miscible-flood operating pressure, kPa	14000
Reservoir temperature, °C	56
Depth, m	1500
Average porosity, %	13
Average thickness, m	3.0
Average permeability, μm^2	0.50
Connate-water saturation, %	30
Residual oil saturation to waterflood, %	35
Discovery oil FVF, reservoir volume/stock-tank volume	1.24
Current oil FVF, reservoir volume/stock-tank volume	1.17
Stock-tank oil gravity, kg/m^3	0.815
Discovery saturation pressure, kPa	7716
Oil viscosity, mPa·s	1.0
Discovery solution-gas ratio at saturation pressure, m^3/m^3	73
Current solution-gas ratio at saturation pressure, m^3/m^3	47
Dip, m/km	16

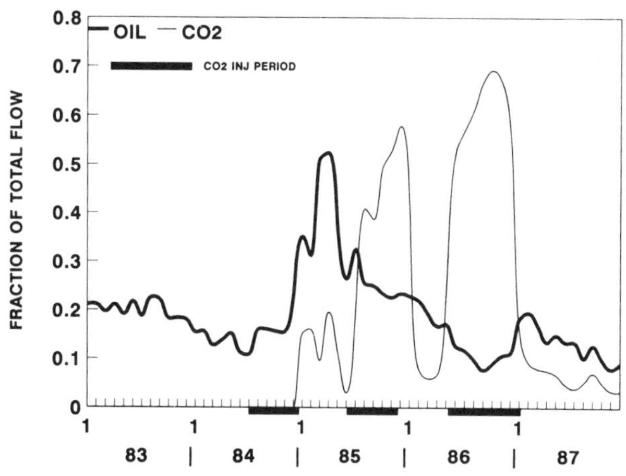

Fig. 2—Well 14-17 fractional flow (1 = January).

a cylinder of foam expanding out from the wellbore.[14] In the model, CO_2 and surfactant solution were considered one fluid and the mobility was adjusted to match experimental pressure buildup. After surfactant injection stopped, the model assumed an annular foam volume pushed by an expanding cylinder of CO_2/water. The flow model predicted a faster pressure drop and a lower BHP after surfactant injection stopped than that observed in the field test **(Fig. 5).** The reasonable match between the model and test results led us to assume that surfactant was propagating radially in the reservoir with simultaneous CO_2/surfactant solution injection.

Encouraged by the injectivity test results, we decided to conduct a more extensive field test.[15] For the extended field test, approximately 10 000 kg (100% active) of a 0.2-wt% surfactant solution was injected simultaneously with CO_2. A 900-kg flush of 0.5-wt% surfactant solution was injected first to condition the reservoir.

Fig. 6 shows the BHP response at the injection well during the first 5 months of the extended foam test. The BHP response compared with that of the injectivity test initially shows an increase of >2000 kPa. Further monitoring of the test indicated the following.

1. Within 3 weeks of injection, the BHP increase stopped and a pressure plateau was reached. The radial-flow model predicted that the BHP should continue to increase if a cylinder of foam was continuously expanding away from the injection wellbore.
2. The GOR in the production wells did not decrease during the test.
3. Sulfur hexafluoride tracer injected with the CO_2 foam was detected at Well 14-17 within 3 weeks and at the other Pattern B producers within 2 to 3 months.

We concluded from the extended-field-test results that foam did not propagate a significant distance into the reservoir. The modest rise in oil production observed in the pilot was attributed to the additional CO_2 injection. We still believe that foam has the potential to improve reservoir conformance; however, new chemicals or injection techniques are required to propagate CO_2 foam.

Pilot Performance. Oil recovery in the pilot has been excellent despite the less-than-expected CO_2 mobility control with $WACO_2$ injection and foam. Pilot oil recovery through 1991 was 21.4% OOIP. This represents an estimated 35.5% recovery of OOIP remaining at the start of the pilot. Cumulative CO_2 injected is 44.4% of pilot HCPV, with 40% CO_2 recovery. The 1981 simulation study indicated that water would push out a substantial amount of the previously injected CO_2 slug.[6,7] The cumulative water/CO_2 HCPV injection through 1991 is 165%. CO_2 utilization in the pilot has been an excellent (1.5 metric tons CO_2 injected/m^3 oil recovery).

Continuous CO_2 Injection

The project was expanded in 1985 because of the early pilot oil recovery response. The original intention of the expansion was to use the $WACO_2$ process in two adjacent, inverted nine-spot patterns developed on a 64-ha/well spacing (twice the size of the pilot) in an area of the unit where oil saturation was close to waterflood residual saturation. Pattern C (Fig. 1) was brought on production in mid-1985, but the 1986 oil-price collapse resulted in a postponement of the adjacent pattern development. It also became apparent

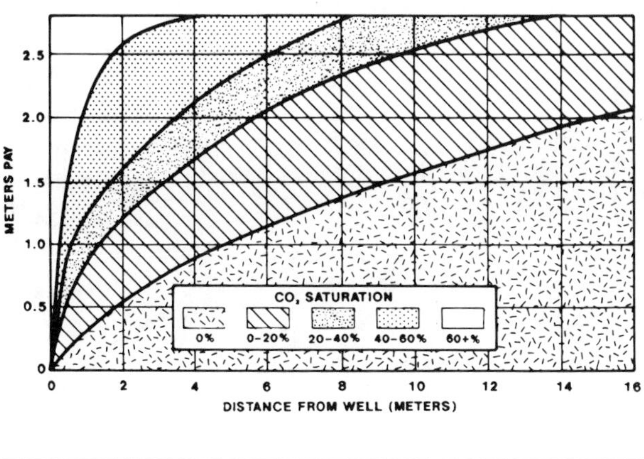

Fig. 3—CO_2 saturation profile at breakthrough.

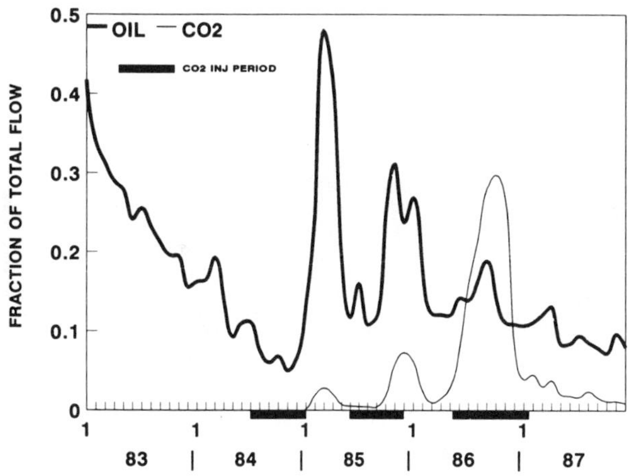

Fig. 4—Well 16-17 fractional flow (1 = January).

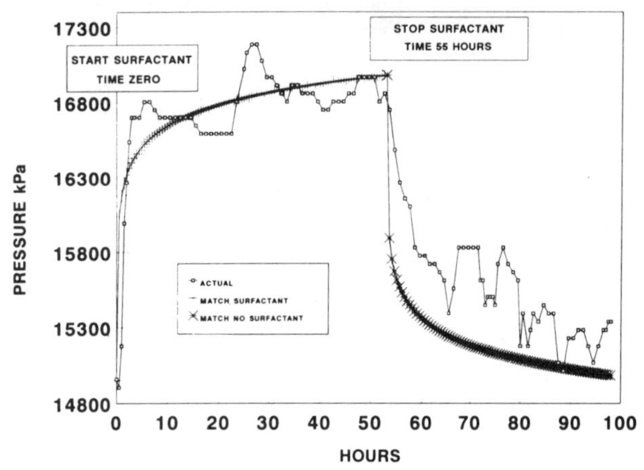

Fig. 5—Injectivity test BHP match.

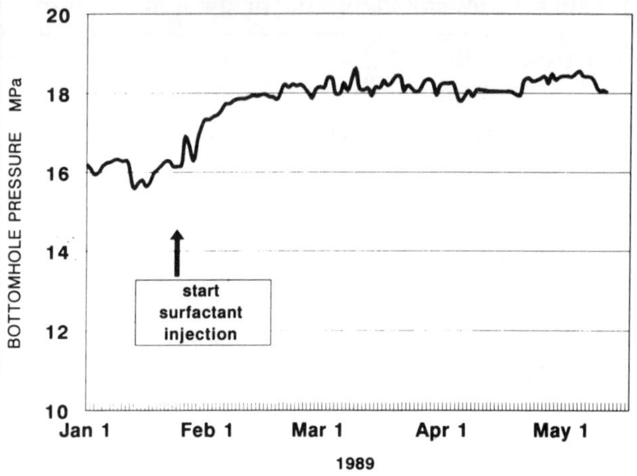

Fig. 6—Foam test BHP history.

that $WACO_2$ injection into the pilot was improving CO_2 conformance very little. We decided to inject a continuous CO_2 slug into Pattern C until further pattern expansion could be completed. The 1981 simulation study indicated that only a 0.5% OOIP recovery loss would result from the change from $WACO_2$ to continuous CO_2-slug injection followed by water.[6,7] A planned cumulative total of ≈30% HCPV CO_2 was injected into Pattern C until June 1988, when the pattern was switched to waterflood. The decline in oil production with the switch to water injection indicated that further CO_2 injection volumes would be required to reach the forecast recovery. In Aug. 1990, after 35% HCPV water injection, simultaneous CO_2/water injection was started in Pattern C, and oil production has responded.

Fig. 7 shows the oil and CO_2 fractional flows for Pattern C Well 2-26. A significant oil bank was mobilized by CO_2 injection like that seen in the pilot wells. CO_2 was detected in the produced-gas stream concurrently with the oil-bank arrival; however, CO_2 production did not rise rapidly until ≈4 months after the oil bank appeared. The oil bank in Well 2-26 peaked at 60% fractional flow, then declined rapidly. CO_2 fractional flow increased dramatically after initial response and reached 60% in <6 months. The increase in CO_2 fractional flow began to slow during the next 12 months and reached a maximum of 70% in June 1988.

A comparison of the Well 2-26 response in Fig. 7 with that of Well 14-17 in Fig. 2 shows many similarities. First, the shapes of the produced-oil banks are similar; both wells peak at ≈60% fractional oil flow, then tail off. The rapid fall in the CO_2 production rate after CO_2 injection stopped freed up flow space in the wellbore, causing the secondary oil peak. The CO_2 fractional-flow curves for both wells are also similar. Both curves exhibit very rapid CO_2 production increases immediately after response, and both show dramatic CO_2 rate decreases when CO_2 injection stops. Although the Well 14-17 fractional-flow curve in Fig. 2 shows a cyclical response to $WACO_2$ injection, Wells 14-17 and 2-26 reach an almost identical CO_2 fractional flow (i.e., 70% after a cumulative 30% HCPV CO_2 injection).

The response similarities between Pattern C wells and the Type 1 well suggest that CO_2 gravity override and establishment of a high-CO_2-saturation flow path are prevalent in Pattern C. The well response time in Pattern C is additional evidence that supports low CO_2 conformance. Although typical response times were 6 months longer than those experienced in the pilot, Pattern C spacing is twice that of the pilot and response occurred with <10% HCPV CO_2 injected in the pilot and Pattern C. Wells 12-25 and 12-26 in Pattern C showed significant CO_2 production within ≈4 months of CO_2 injection.

Approximately 13% OOIP was recovered and 41% HCPV of CO_2 was injected into Pattern C through Dec. 1991. Injected CO_2 recovery to date is 40%. Ultimate oil recovery for Pattern C is forecast to be 18% OOIP, with 6 more years of projected pattern life. Currently, CO_2 utilization is 2.3 metric tons gross CO_2 injection/m^3 oil recovered, and final CO_2 utilization is predicted to be 1.8 metric tons gross CO_2 injected/m^3 oil recovered.

Simultaneous Injection

During the second quarter of 1988, Pattern D was developed next to Pattern C (Fig. 1). Pattern D is a truncated, inverted nine-spot pattern developed on a 64-ha/well spacing. Production response from Pattern D was the litmus test for CO_2 miscible flooding in

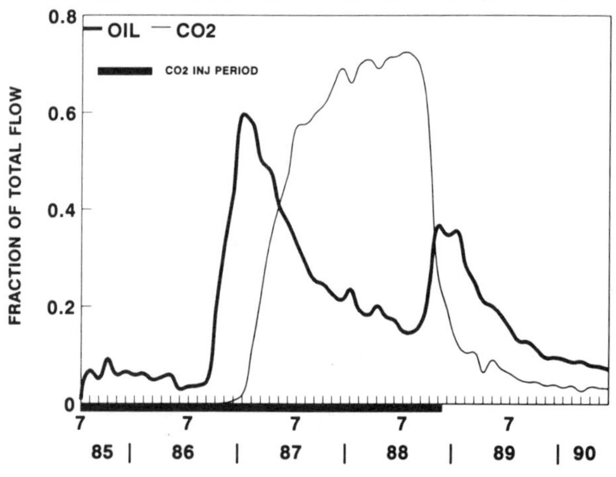

Fig. 7—Well 2-26 fractional flow (7 = July).

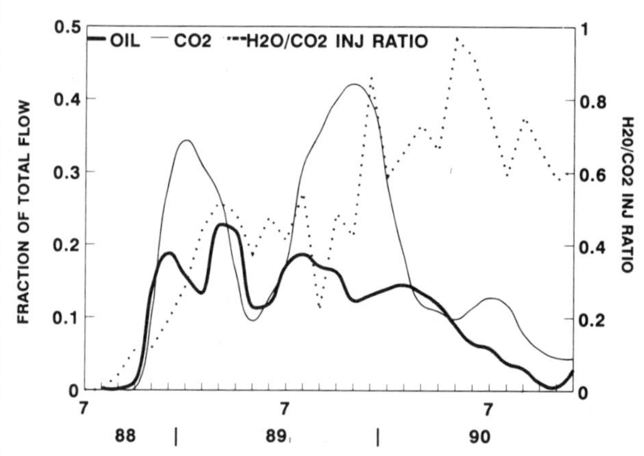

Fig. 8—Well 4-30 fractional flow (7 = July).

the Joffre Unit. Wells 4-30 and 12-19 of Pattern D were converted from water-injection to production wells. These two wells had been used as water injectors during the 1960's waterflood of the field and as pressure-maintenance wells for the CO_2 pilot. Cumulatively, $>85\times10^4$ m^3 water had been injected into the wells, ensuring that oil saturation in this reservoir region was at waterflood residual saturation.

The pilot and Pattern C production responses showed that CO_2 gravity override was significant in the Viking sand and that large 10% HCPV water slugs alternating with CO_2 had little effect on improving CO_2 sweep. By June 1988, we decided that the best method to improve CO_2 conformance was simultaneous CO_2/water injection.

Starting at the end of June 1988, a CO_2 slug was injected for ≈ 2 months into Pattern D. The purpose of this initial CO_2 slug was to enhance injectivity in the wellbore before simultaneous injection. Simultaneous injection began in Sept. 1988 at a water/CO_2 ratio of ≈ 0.1 on a reservoir volume basis. A small hydraulic fracture in the injection well in Nov. 1988 increased fluid injectivity. In Dec. 1988, the water rate was increased to give a water/CO_2 ratio of ≈ 0.5 on a reservoir volume basis. In Dec. 1989, dual tubing strings were installed in the well and the water injection rate was increased again. The average water/CO_2 ratio during 1990 was 0.77 on a reservoir volume basis.

Fig. 8 plots the Well 4-30 CO_2 and oil fractional flows and the Pattern D water/CO_2 injection ratio. In contrast to the oil banks seen in Figs. 2 and 7, Fig. 8 shows a broad, mobilized oil bank. The maximum oil fractional flow in Well 4-30 is only 20% compared with 60% for Wells 14-17 and 2-26. As would be expected with an initial oil rate of zero, there was no residual mobile oil around the well. The CO_2 fractional-flow curve in Fig. 8 shows two CO_2 peaks. The first peak is the result of the initial 2 months of continuous CO_2-slug injection into Pattern D. A rapid fractional-flow rise in CO_2 occurs after CO_2 breakthrough, and CO_2 fractional flow reaches 30% by the end of 1988, after which CO_2 production declines. This CO_2 production decline corresponds to the date that the water injection rate was increased after the hydraulic fracture. CO_2 production decline with the same CO_2 volume injected is the result of simultaneously injecting water, which reduces CO_2 mobility and thereby improves reservoir conformance.

A second peak in the Well 4-30 CO_2 fractional-flow curve occurs from about Aug. 1989 to Jan. 1990. This peak is the result of intentional overvoidage to contain CO_2 within Pattern D. In Dec. 1989, larger volumes of water were injected, and the pattern voidage replacement ratio increased to >1. A subsequent CO_2 production decrease occurred, and oil production declined. The impact of increased water injection with CO_2 on Well 4-30 production during 1990 is attributed to improved CO_2 conformance in Pattern D. Oil production is expected to increase when oil in newly contacted areas reaches Well 4-30.

Oil recovery in Pattern D through Dec. 1991 is 7.5% OOIP. Cumulative CO_2 injection is 39.5% HCPV. The Pattern D forecast recovery is 14% OOIP at 45% HCPV cumulative CO_2 injection. The lower Pattern D forecast recovery compared with pilot and Pattern C recoveries reflects the lower oil saturation in the pattern from previous waterflooding operations in the field. CO_2 utilization through Dec. 1991 is 3.8 metric tons gross CO_2 injected/m^3 oil recovered and forecast CO_2 utilization is 2.3 metric tons gross CO_2/m^3 oil recovered.

Injection Strategy Comparison. Joffre production experience and simulation studies show that CO_2 in the Viking sandstone contacts $<50\%$ of the formation because of gravity override and an adverse mobility ratio. Injecting large 10% HCPV water slugs in a $WACO_2$ process offers little improvement in CO_2 conformance over that achieved with CO_2 injection alone. Although it is still early in the production life of Pattern D, it appears that simultaneous CO_2/water injection at water/CO_2 ratios approaching one can improve CO_2 sweep in the reservoir.

The degree of CO_2 conformance in the reservoir affects the amount of injected CO_2 produced. **Fig. 9** shows the cumulative

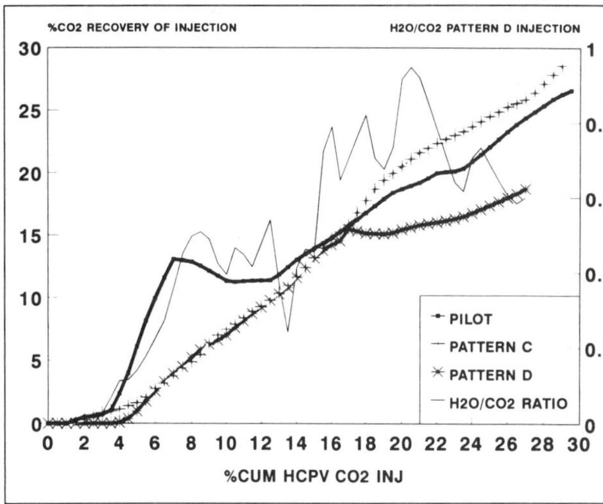

Fig. 9—Percent CO_2 recovered vs. percent CO_2 injected.

percentage of injected CO_2 recovered vs. the cumulative percentage of HCPV CO_2 injected for each pattern. Fig. 9 also shows that the curves for the pilot and Patterns C and D start to diverge at a cumulative HCPV CO_2 injected greater than $\approx 17\%$. Pattern C, with continuous CO_2 injection, shows the highest percentage of CO_2 recovery; the pilot with $WACO_2$ injection shows only a slightly lower percentage of CO_2 recovery. The anomalous peak in the pilot between 5% and 10% HCPV CO_2 injected can be explained by the rapid CO_2 response in Pilot Well 6-20. In Pattern D, the CO_2 recovery rate changes dramatically during simultaneous injection when the water/CO_2 ratio increases from an average of 0.4 to 0.77 on a reservoir volume basis. Before 17% injection, Patterns C and D track closely, indicating a critical water/CO_2 ratio is required to effect significant CO_2 mobility control during simultaneous injection.

Conclusions

1. Large 10% HCPV $WACO_2$ slugs in the pilot improved CO_2 reservoir conformance in Type 1 wells very little but provided good mobility control in Type 2 wells. Oil recovery and CO_2 utilization have been excellent despite an adverse mobility ratio and gravity override.
2. Simultaneous CO_2/water injection at water/CO_2 ratios approaching one resulted in improved CO_2 sweep compared with $WACO_2$ injection and continuous CO_2 injection.
3. CO_2-slug injection establishes a high-CO_2-saturation flow path at the top of the formation. The CO_2 remains virtually immobile during water injection, causing a sharp reduction in produced CO_2. Flow through the high-CO_2-saturation region is quickly re-established when CO_2 injection is restarted.
4. Gravity override effects should be analyzed carefully with a refined-grid cross-sectional model during performance simulations.
5. An extended foam test in the Joffre reservoir propagated a foam bank only within a few meters of the wellbore.
6. The significant volumes of injected CO_2 left in the reservoir reduce the amount of CO_2 emitted into the atmosphere.

Acknowledgments

We thank AOSTRA, JVTOU, and Westcoast Petroleum Ltd. for permission to publish this paper.

References

1. Love, A.M.: "Geology and Development—Joffre Oil Field, Alberta," Cdn. Inst. of Mining & Metallurgy, *Trans.* (July 1955) **48,** 433.
2. Shelton, J.L. and Schneider, F.N.: "The Effects of Water Injection on Miscible Flooding Methods Using Hydrocarbons and Carbon Dioxide," paper SPE 4580 presented at the 1973 SPE Annual Meeting, Las Vegas, Sept. 30–Oct. 3.
3. Walsh, M.P., Negahban, S., and Gupta, S.P.: "An Analysis of Water Shielding in Water-Wet Porous Media," paper SPE 19652 presented

Authors

Stephenson

Graham

Luhning

Derrll Stephenson is manager of planning and business development for Westcoast Petroleum Ltd. and president of Westcoast subsidiary, Vikor Resources Ltd. He previously worked as a consultant and for two major Canadian oil and gas companies in conventional oil and gas and heavy-oil production. Stephenson holds a BS degree in civil engineering from the U. of Alberta. **Andrew Graham** is a reservoir engineer for Westcoast Petroleum Ltd. He is responsible for monitoring and optimizing recovery in the Joffre CO_2 miscible flood, and he is involved in reservoir-simulation studies and horizontal-well development. Graham holds a BS degree in engineering chemistry from Queen's U. and an MS degree in chemical engineering from the U. of Alberta. **Richard W. Luhning** is vice chairman and executive director of AOSTRA in Calgary, Alta., where he directs field pilot plant operations for in-situ bitumen recovery from oil sands, heavy-oil recovery, and enhanced conventional oil production. He also directs operations of the Undergroup Test Facility at Fort McMurray; surface oil-sand mining, extraction, upgrading, and environmental activities; and technology transfer and commercialization efforts. Luhning holds a BE degree from the U. of Saskatchewan and a PhD degree from Imperial C. of Science Technology, both in chemical engineering.

at the 1989 SPE Annual Technical Conference and Exhibition, San Antonio, Oct. 8–11.

4. Isley, D.B., MacIntyre, K.J., and Stephenson, D.J.: "Joffre Viking Tertiary Carbon Dioxide Enhanced Oil Recovery Project," paper 9 presented at the 1983 AOSTRA Annual Advances in Petroleum Recovery Technology Conference, Calgary, May 30-31.
5. Stephenson, D.J.: "Carbon Dioxide Miscible Flooding Experience in the Joffre Viking Field," paper 20 presented at the 1990 AOSTRA Oil Sands 2000 Conference, Edmonton, March 26–28.
6. Ko, S.C.M.: "A Preliminary Feasibility Study of CO_2 Flooding a Watered Out Reservoir in Alberta, Canada," paper CIM-82-33-76 presented at the 1982 Annual Technical Meeting of the Pet. Soc. of CIM/Symposium on Engineering Applications of Mechanics, Calgary, June 6–9.
7. Ko, S.C.M., Stanton, P.M., and Stephenson, D.J.: "Tertiary Recovery Potential of CO_2 Flooding in Joffre Viking Pool, Alberta," *J. Cdn. Pet. Tech.* (Jan.-Feb. 1985) **24,** No. 1, 36.
8. Edmunds, N.R.: "Reservoir Engineering Studies for the Joffre Viking CO_2 Miscible Pilot Project, Jan. 1984–May, 1985," final report, Vikor Resources Ltd., Calgary (June 1985).
9. Graham, A.G. and Romauldez, L.A.: "1986/1987 Joffre CO_2 Pilot Simulation Study," final report, Vikor Resources Ltd., Calgary (July 1988).
10. Dellinger, S.E., Patton, J.T., and Halbrook, S.T.: "CO_2 Mobility Control," *SPEJ* (April 1984) 191.
11. Lee, H.O. and Heller, J.P.: "Laboratory Measurements of CO_2 Foam Mobility," *SPERE* (May 1990) 193.
12. Green, K. *et al.*: "Use of Surfactant Foams for CO_2 Mobility Control," final report, Alberta Research Council, Edmonton (July 1988).
13. Irani, C.A. and Solomon, C. Jr.: "Slim-Tube Investigation of CO_2 Foams," paper SPE 14962 presented at the 1986 SPE Oil and Gas Recovery Conference, Midland, March 13–14.
14. Graham, A.G. and McKibbon, J.H.: "CO_2 Mobility Control by Surfactants—Mini Field Tests," final report, Vikor Resources Ltd., Calgary (Jan. 1989).
15. McKibbon, J.H.: "CO_2 Mobility Control by Surfactants—Full Scale Field Test," final report, Vikor Resources Ltd., Calgary (Jan. 1990).

SI Metric Conversion Factors

acre	× 4.046 873	E−01	=	ha
°API	141.5/(131.5+°API)		=	g/cm³
bbl	× 1.589 873	E−01	=	m³
cp	× 1.0*	E−03	=	Pa·s
ft	× 3.048*	E−01	=	m
°F	(°F−32)/1.8		=	°C
lbm	× 4.535 924	E−01	=	kg
md	× 9.869 233	E−04	=	μm²
mile	× 1.609 344*	E+00	=	km
psi	× 6.894 757	E+00	=	kPa
ton (metric)	× 1.0*	E+00	=	Mg

*Conversion factor is exact.

SPERE

Original SPE manuscript received for review Aug. 22, 1991. Revised manuscript received Dec. 2, 1992. Paper (SPE 23598) accepted for publication May 20, 1992.

Refs. 8, 9, 12, 14, and 15 are available from the Alberta Oil Sands Technology & Research Authority, Calgary, Alta., Canada.

SPE/DOE
Society of Petroleum Engineers U.S. Department of Energy

SPE/DOE 24156

Production Performance of the Wasson Denver Unit CO_2 Flood

C.S. Tanner and P.T. Baxley, Shell Western E&P Inc.; J.G. Crump III, Shell Offshore Inc.; and W.C. Miller, Shell Western E&P Inc. (Retired)

SPE Members

This paper was prepared for presentation at the SPE/DOE Eighth Symposium on Enhanced Oil Recovery held in Tulsa, Oklahoma, April 22–24, 1992.

ABSTRACT

The Denver Unit, located in the Wasson Field of west Texas, is the world's largest Carbon Dioxide Enhanced Oil Recovery (EOR) project. Shell Western E&P Inc. initiated CO_2 injection in the Denver Unit in 1983, utilizing CO_2 from naturally occurring source fields in Colorado and New Mexico.

Production response to CO_2 flooding has been impressive, with over half of the current daily oil production attributable to CO_2 injection. In fact, oil production has substantially exceeded original project performance predictions.

The project was originally implemented with both a Continuous CO_2 injection area and a Water-Alternating-Gas (WAG) CO_2 injection area. A comparison of the performance of these two areas has clearly demonstrated the advantages of each injection method, and has led to the conversion of the the Unit to a "Denver Unit WAG" (DUWAG) injection process, which plays on the strengths of both injection schemes.

This paper will cover in more detail the observed production performance of the CO_2 flood to date, present a comparison of the Continuous and Water-Alternating-Gas injection processes, and review the design and implementation of the Denver Unit WAG project. The paper will be of interest to companies who are either operating or preparing to implement CO_2 Floods.

INTRODUCTION

Despite the softer oil prices of the 1990s, CO_2 flooding continues to maintain a leadership position in EOR technology. Nowhere has CO_2 Flooding been more popular and successful than in the Persian Basin of west Texas and eastern New Mexico. In the Permian Basin alone, there are at least 35 active CO_2 miscible floods ongoing, with an equal number of projects under consideration [1]. These projects, both active and planned, include a wide range of project size, depth, reservoir type, injection process, and other factors that result in a wide application of uses in the Permian Basin.

Several factors have contributed to the growth of CO_2 activities in the Permian Basin [2]. These include 1) the availability of competitively priced CO_2 throughout the basin, 2) the number of large reservoirs that have been successful waterfloods with high remaining oil saturations, and 3) the proven and improving CO_2 technology.

Although few CO_2 floods in the basin would be considered mature, a number of projects have progressed enough to be critically reviewed. Careful review of performance in these early projects can lead to improved technology. This improved technology can then be profitably applied to additional candidate reservoirs in the future. The purpose of this paper is to review the early performance of the Denver Unit CO_2 Flood in the Wasson (San Andres) Field.

This flood is an important benchmark for the industry, because: 1) the current oil production rate of 38,000 BOPD substantially exceeds original projections (Fig. 1), 2) different CO_2 injection processes have been applied, and 3) the recent expansion of the flood (both in areal extent and injection process) is nearing completion. Although the focus of this paper will be a review of the observed EOR performance, some potential areas for technology improvement will be identified.

References and figures at end of paper.

DENVER UNIT OVERVIEW

Although many fields within the Permian Basin have attractive CO_2 flooding potential, the Wasson Field leads all others with six active CO_2 floods. The huge Wasson field covers some 120 square miles in Gaines and Yoakum Counties, Texas, on the Northwest Shelf of the Permian Basin (Fig. 2). The Wasson San Andres reservoir is a particularly good CO_2 candidate because, despite successful primary development and waterflooding, significant oil targets remain. A summary of reservoir data for the Wasson San Andres is included in Table I.

The Wasson Field was discovered in 1936. Primary development peaked in the late 1930s, and supplemental recovery, by gas or water injection, was not seriously considered until the late 1950s. The Wasson Field Engineering Subcommittee, comprised of representatives of the majority operators in the Field, concluded that the Wasson (San Andres) was an excellent candidate for waterflooding [2], and recommended, in 1963, that the field be split into seven different supplemental recovery projects (Fig. 3). The largest of these units was the Denver (San Andres) Unit.

CO_2 DEVELOPMENT

Unitization efforts for the Denver Unit were successfully completed and water injection began in 1964. Initially only peripheral injectors were used, but by the early 1970s interior waterflood development was established through successful infill drilling and waterflood expansion programs [4]. Despite the lack of a uniform pattern, waterflooding of the Denver Unit San Andres reservoir, with its favorable mobility ratio and limited vertical permeability, was very successful, resulting in peak oil rates of 150,000 BOPD in 1975.

Due to the clear success of the Denver Unit waterflood and the high waterflood residual oil saturations (approximately 40%), EOR process studies were begun in the mid 1970s to determine the magnitude and economic feasibility of various EOR projects for the unit. Laboratory experiments concluded that miscible CO_2 injection could result in significant EOR potential in these reservoirs. Furthermore, CO_2 flooding had been successfully employed in other Permian basin fields (Kelly-Snyder [5] and Crossett [6]). With the (then) recent discovery of large naturally occurring CO_2 reserves in Colorado and New Mexico, a CO_2 pilot was designed and proposed for the Denver Unit.

Denver Unit CO_2 Pilot

A CO_2 pilot was initiated in 1978 and analysis of this pilot, 1) confirmed that adequate CO_2 injection and follow-up water injection rates could be attained, and 2) quantified the reduction in oil saturation resulting from CO_2 injection in a portion of the field at waterflood residual oil saturation. Following extensive coring and a brine preflood to establish baseline oil saturations and a uniform reservoir pressure, CO_2 was injected at miscible conditions. Throughout the CO_2 and brine postflood phases of the pilot, logging observation wells continuously monitored changes in oil saturation attributable to the CO_2 contacting, swelling, and displacement of the remaining oil in this watered out portion of the reservoir. Postflood cores confirmed the desaturation of oil interpreted from logging runs.

A successful history match of the CO_2 pilot was obtained using a four component (CO_2, water, and both light and heavy oil fraction), four phase, 3D, miscible, simulation model. The results of this history match were built into a pattern prototype element representing one quarter of a inverted 9-spot pattern. This pattern prototype was then scaled up to represent the potential for a field scale flood. The CO_2 flood was designed to be staged, with CO_2 produced from the initial (eastern) flood area used to flood the final (western) injection area, thus making most effective use of this valuable injectant.

EARLY CO_2 FLOOD PERFORMANCE

In preparation for the CO_2 Flood, the random waterflood pattern was "regularized" into inverted 9-spot patterns. In addition, reservoir pressure was reduced from 3200 psi to 2200 psi in order to improve the volumetric efficiency of the CO_2 (and yet maintain reservoir pressures above the MMP of 1300 psi).

The original flood design allowed for a side-by-side comparison of both Continuous and Water-Alternating-CO_2 (WACO_2) injection (Fig. 4) within the Initial Injection Area (IIA). In both cases, an ultimate slug of 40% HCPV (the hydrocarbon bearing pore volume of the reservoir at initial conditions) would be injected, followed by water injection until the economic limit was reached. The intent of this dual process test was to determine which injection strategy was best suited for Denver Unit operating conditions. CO_2 injection began in the WACO_2 Area (southern IIA) in April, 1983, and a year later, in 1984, in the Continuous Area (north IIA).

Continuous Area EOR Performance

The early production performance of the Continuous Area has been quite encouraging. Oil production response was observed soon after injection (and depressurization) began. Within four years of CO_2 injection, oil production had increased by 8000 BOPD (Fig. 5) and the oil cut had risen from a low of 14% to 31% . CO_2 response can clearly be seen on a plot of oil cut vs. cumulative oil production (Fig. 6). Within only months of CO_2 injection, the oil cut deviates sharply upward from what would have been expected under continued waterflood conditions, thus defining EOR response. Several other interesting phenomena accompanied EOR response in the Continuous Area.

First, there appeared to be an areal anisotropy in production response suggesting an east-west oriented permeability preference. This can be most clearly seen by comparing the oil response characteristics of Continuous Area producers relative to their location in the 9-spot pattern. Although each group of wells does show clear CO_2 response, wells located east-west of pattern injectors experience earlier EOR response (Fig. 7a). Wells located north-south of CO_2 injectors, or diagonally to the pattern injector, respond more slowly to CO_2 injection, yet appear to sustain their growth in EOR rate for a longer time.

Although each of the locations show response, the efficiency of their response, as measured by the producing CO_2/oil ratio (Fig. 7b), clearly illustrates that CO_2 travels faster to the east/west wells than other locations, suggesting a permeability anisotropy favoring east/west displacement. Recognition of this "non-radial" flood front was important in understanding performance in this and other parts of the Unit.

Injector-To-Producer Conversions

Another sign of EOR response, one unique to CO_2 flooding, is the oil response in former water injectors converted to production. The regularization to 9-spot patterns left as many as one hundred former water injectors in producer locations in the pattern. In order to maximize areal sweep, a number of these ex-injectors were converted to producers at the start of the CO_2 Flood. Although these wells had to produce a large water bank ahead of EOR response, the five original injector-to-producer (I-P) conversions have responded very well, and are currently producing at rates higher than originally predicted (Fig. 8).

Because of the success of these initial I-P wells, several others have recently been activated. One very recent I-P is producing well over 100 BOPD after only nine months of production. Contrasting this "response time" with the response time of the original I-P wells (almost 2 years) suggests that, after six years of CO_2 injection, response time is quite reduced. If there is any "silver lining" to the non-regular waterflood pattern, it is the opportunity that it has provided to make profitable adjustments to the CO_2 flood. In addition to the original I-P wells, 11 have been completed to-date with the potential for as many as 50 more.

Injection Performance

Injection performance in the Continuous Area has been quite encouraging. Desired injection rates (based on pattern OOIP) have been maintained and vertical distribution of CO_2, similar to the waterflood, is quite good. The computerized injection controllers have successfully maintained injection rates at desired levels, while guarding against injection pressures that will result in fracturing. Figure 9 shows the steady overall injection performance in the Continuous Area.

CO_2 injectivity, while starting at levels equivalent to pre-CO_2 water injectivity, has risen slowly throughout the CO_2 injection cycle. This can be observed in Figure 10, where even in a pressure limited injector, CO_2 injection rates have risen steadily as cumulative injection increases. This phenomenon is also observed in rate constrained injectors, where injection pressure decreases with increasing cumulative CO_2 injection. This rising injectivity is consistent with CO_2 pilot observations, and is attributed to the displacement of liquids (both oil and water) away from the injectors.

Gas-Oil Ratio Trend

Another indicator of EOR response in the early years of the CO_2 Flood is the rising hydrocarbon gas-oil-ratio (GOR) as the flood has progressed (Fig. 11). The producing GOR prior to CO_2 flooding was approximately 650 SCF/BO. As CO_2 contacts oil in the reservoir, it strips the lighter hydrocarbon components out of the remaining oil and displaces it towards the producing wells. In addition, CO_2 evaporates more of these components in the surface separators.

Therefore, not only has the hydrocarbon GOR increased, but due to the higher concentrations of the valuable liquid components (butane and propane), the gas stream is enriched as well, resulting in higher liquid recovery.

CO_2 Production

Upon closer inspection of Figure 5, it can be seen that along with increasing oil rates, CO_2 production rates rose steadily during the early years of the CO_2 flood. It should be noted that one of the operating constraints of the CO_2 flood has been the limited capacity to process produced CO_2 through the Denver Unit CO_2 Recovery Plant (DUCRP). The original plant began operation in 1986 with an inlet capacity of 140 MMCF/D. As gas rates directed to this field facility increased, modifications to the plant were initiated to provide additional gas processing capacity. This was necessary in order to prevent curtailed oil production while providing CO_2 for recycle injection. Although the CO_2 production rates were only slightly higher than originally predicted (Fig. 12), this produced CO_2, coupled with the increased hydrocarbon gas rates (attributable to higher than anticipated oil rates and rising GOR), became a consideration as the gas rates approached plant inlet capacity.

Flowing Wells

Another impact of the increasing CO_2 and gas rates was the growing number of flowing wells in the Continuous area [7]. There are currently over 125 flowing wells in the Unit. These flowing wells, often occupying infill or east/west pattern locations, presented unique operational challenges.

Due to the formation of hydrates, some flowing wells experienced freezing at the wellhead during the night and periods of cold weather, often remaining frozen for days at a time. Once thawing occurred, the build-up of near wellbore gas would result in short bursts of extremely high gas rates, shutting down surface facilities. Wells that produce adequate volumes of liquids were not problems, as the liquids tended to displace the hydrates down the flowline. Only a few wells were quite troublesome to handle. As gas/liquid ratios continued to rise, it was apparent that either a change in operating policies be made, or these wells would be shut in.

WACO_2 AREA EOR PERFORMANCE

In order to adequately compare the performance of the Continuous and WACO_2 Areas, injection rates (on a HCPV basis) were maintained at comparable levels despite the lower injectivity of the WAG process. The original injection plan in the WACO_2 Area involved injecting alternating six month slugs of CO_2 and water until a cumulative 40% HCPV slug of CO_2 had been injected.

Early oil production response in the WACO_2 Area was disappointing. Oil production continued to decline with only a marginal improvement over waterflooding for a number of years after CO_2 injection began (Fig. 13). It is important to note that not only was oil rate response not immediately observed in the WACO_2 Area, CO_2 production rates remained quite low.

In the early WAG injection cycles, considerable problems arose in attempting to maintain injection rates in parity with the Continuous Area. This was particularly challenging on the water injection cycle, where loss of injectivity further reduced injection volumes. In order to achieve desired water injection rates in the WACO_2 area, injection pressures were allowed to exceed fracture extension pressure. Although these high pressures were allowed *only* on the water cycle, material balance calculations suggest that losses of as much as 25% of injected fluid resulted from this practice.

In order to increase throughput rates to desired levels *without* exceeding fracture pressure, a pattern conversion from the existing 9-spot pattern to a line drive pattern was installed in 1988. This conversion (Fig. 14) increased the number of injectors in the WACO_2 Area from 33 to 60. By spreading the desired injection volume among more injectors, desired rates were then attainable without exceeding fracturing pressures. Although about 700 BOPD were lost due to well conversions, the oil production decline was arrested and oil rates slowly began to rise.

Several factors have contributed to the poorer EOR performance in the WACO_2 Area, including; 1) lower WAG injectivity, 2) out-of-zone injection losses, 3) structural continuity, and 4) waterflood induced fractures. While the geologic setting cannot be changed, by converting to the line-drive pattern, injection can be accomplished without fracturing. In addition, all injection profiles are closely reviewed and profile improvement workovers undertaken as required. Further, to improve injectivity, WAG cycle lengths have been extended from every six months to yearly.

DENVER UNIT WAG DEVELOPMENT

Although early EOR performance of the Continuous Area was very encouraging, gas production rates continued to rise steadily. Further, many wells had begun to flow and several east/west wells were "gasing out" and were forced to be shut-in.

A careful comparison of the performance of the Denver Unit Continuous and WACO_2 Areas with other Wasson Area CO_2 floods demonstrated the advantages of both Continuous and WAG injection. These reviews confirmed the encouraging early response of the Continuous process, while suggesting the long term flood manageability advantages of WAG injection.

Numerical models were refined based on actual observed CO_2 flood performance in each area of the Denver Unit, then these models were used to investigate various flood options. The history match of early CO_2 flood performance for the Continuous Area, WACO_2 Area, and Unit total is shown in Figure 15. Simulation studies suggest that the "Denver Unit WAG" (DUWAG) injection process, in which four to six years of continuous CO_2 injection is followed by 1:1 WAG, has advantages over both continuous CO_2 and conventional WAG processes. Figure 16 compares the EOR recovery for continuous CO_2, conventional 1:1 WAG, and DUWAG.

The DUWAG process combines the early EOR response of continuous injection and the higher ultimate recovery of WAG injection. The WAG portion of the process provides improved areal sweep efficiency and the prolonged economic life of the east/west and infill wells, which would have "gased out" under continuous flooding. Further simulations predict that a larger slug size (60% HCPV) could result in additional EOR recovery, using existing wells and facilities, without increasing peak gas production rates.

To test the DUWAG process in the field, four patterns in the heart of the Continuous Area were converted to this process. These patterns were characterized by high CO_2 injectivity, high CO_2 producing rates, and included several flowing wells. During this test, water injection rates sufficient to match production were successfully attained at pressures below fracture pressure. Despite the lower injectivity of water, these water injection volumes were achievable because of the large

CO_2 slug already in place. On the production side, within a month of initiating WAG injection in these patterns, CO_2 production rates began to drop (Fig. 17), while oil and water production rates remained constant.

Implementation of the Denver Unit WAG process has not only reduced anticipated peak gas rates, but has also allowed accelerated expansion of the CO_2 Flood to the Final Injection Area (FIA) by freeing up CO_2 from the IIA. This more rapid expansion provides increased oil production as these patterns respond to CO_2.

RECENT CO_2 FLOOD PERFORMANCE

The two major components of the DUWAG project were 1) the conversion of the Continuous Area to this injection process, and, 2) the accelerated expansion of the CO_2 flood to the FIA. Field installation of the project began in late 1989.

Continuous Area

As of October, 1991, fifty-six patterns (of ninety-four) patterns in the Continuous Area have been switched to WAG injection (Fig. 18). Since mid-1991 the DUCRP has been operating at maximum capacity. Therefore, as producing gas rates continue to rise, some Continuous Area wells have been shut-in pending completion of the plant expansion in early 1992. Figure 19 shows the production and injection performance of a 15 pattern area where DUWAG injection has been operational since late 1990. Notice that similar to the field test, CO_2 production rates declined soon after WAG injection began.

Because the prediction of follow-up water injectivity is very important to the success of the project being able to accurately match and predict fluid injectivities is of vital importance. Field evidence from the Denver Unit suggests a water injectivity reduction of approximately 30% following the CO_2 injection cycle (Fig. 20), apparently due to mobility and relative permeability effects, including CO_2 trapping. Over the period of the water injection cycle, this trapped gas is slowly dissolved, resulting in increased injectivity during the water portion of the WAG cycle.

$WACO_2$ Area

Due to the more encouraging response observed in the Continuous Area, a slug of continuous CO_2 was injected into the $WACO_2$ Area. This work began in mid 1990, and production response has been encouraging to date. Figure 21 shows the recent oil production response for the $WACO_2$ Area. Notice that after the continuous CO_2 injection slug was initiated, oil response began to increase dramatically. Although WAG injection has been resumed in a number of wells, EOR production response continues to grow.

Final Injection Area

Expansion of the CO_2 flood to the FIA has progressed rapidly. Since the DUWAG project was approved in late 1989, ninety-two new FIA patterns have begun CO_2 injection. By the end of 1991, patterns including 91% of the Unit oil column OOIP were under CO_2 injection (Fig. 18). The only patterns yet to begin CO_2 injection are a few patterns in the center of the unit (due to special considerations in equipping CO_2 injection wells within city limits), and patterns associated with leaseline areas, which are planned to be developed concurrent with offset operators.

Early indications of EOR production response have been encouraging in the FIA. Several of the more mature FIA CO_2 injection patterns have been switched to DUWAG injection in an effort to reduce gas production rates (due to DUCRP curtailment).

SUMMARY

Although it is still very early in the life of the Denver Unit CO_2 Flood, performance to-date has been quite encouraging. The very positive indications of the success of the CO_2 Flood include: 1) the early EOR response observed in the Continuous Area, 2) the excellent EOR response in I-P wells, 3) the good injection performance in the majority of wells and overall good vertical distribution of injected fluids, 4) the increasing hydrocarbon GOR, 5) the quick response (decline) of CO_2 production rates following the start-up of WAG injection, and 6) early EOR response in the FIA are all very positive indications of the success of the Denver Unit CO_2 Flood.

Injection losses in the $WACO_2$ Area, lower water injectivity, problems associated with flowing wells, and other less successful aspects of flood performance, should not be viewed as threats to the overall success of the project but rather as opportunities to improve performance. Similarly, as future concerns arise, they will be analyzed for potential impact and addressed.

Work to further refine the strategy in the Denver Unit is ongoing. Current areas of investigation include: 1) determining the best time to switch from continuous to WAG injection 2) the optimum WAG cycle length, 3) the best ultimate WAG ratio, and 4) optimum CO_2 slug size. It is likely that the answer to these varies across the field. For example, Figure 22 shows the oil response from a group of mature patterns in the south portion of the Continuous Area. Although this area has a cumulative CO_2 injection of 39% HCPV, thus making it quite mature, CO_2 production rates have been moderate. DUWAG injection began in mid 1991. These patterns' EOR performance indicate that, although they are part of what is referred to as the Continuous Area, the reservoir quality is quite different than that observed in the northern parts of the Continuous Area. By tailoring WAG strategy independently for different parts of the Unit, the utility of injected CO_2 can be maximized.

"Denver Unit WAG" installation and DUCRP expansion are proceeding as planned. The WAG process has lowered gas production rates and is expected to allow additional flexibility to improve long term flood performance. However, in addition to adjusting WAG ratios for different areas, additional opportunities to improve performance also exist including: 1) formalizing agreements with neighboring floods to cooperatively develop leaseline areas, thus adding significant reserves and preventing the formation of "dead zones" between units, 2) possible expansion of the CO_2 flood to a number of patterns on the eastern edge of the unit, that have only recently (through outstep drilling) been fully developed, and 3) increased EOR potential for a larger ultimate CO_2 slug size, all of which are examples of areas where existing technology can improve both ultimate recovery and economic performance.

Acknowledgments

We would like to thank the Management, Operations staff, and Engineering staff of SWEPI's Western Division Production group who lent their enthusiastic support to this project and to the preparation of this paper.

References:

[1] Moritis, G.,"CO_2 and HC Injection Lead EOR Production Increase," Oil & Gas Journal Special (April 23, 1990) pg. 49-82.

2] Taber, J.J.,"Current Status of the Use of CO_2 for Enhanced Oil Recovery," U.S.D.O.E. & Argonne Nat. Lab - Recovery & Use of Waste CO_2 in Enhanced Oil Recovery Workshop, Denver - June 1988, Proceedings pp 65-99.

[3] Holm, L.W., "Evolution of the Carbon Dioxide Flooding Processes," J Petroleum Tech V 39, No 11 pp 1337-1342, Nov 1987.

[4] Ghauri, W.K.; Magnuson, W.L.; Osborne, A.O., "Changing Concepts In Carbonate Waterflooding, West Texas Denver Unit Project, An Illustrative Example," 48th Annual SPE of AIME Fall Mtg, Reprint # SPE-4683.

[5] Langston, M.V.; Hoadley, S.F.; Young, D.N., "Definitive CO_2 Flooding Response in the SACROC (Scurry Area Canyon Reef Operators Committee) Unit," 6TH SPE/DOE EOR Symposium, Tulsa April 1988, Proceedings pp 27-34.

[6] Pontious, S.B.; Tham, M.J.: "North Cross (Devonian) Unit CO_2 Flood - Review of Flood Performance and Numerical Simulation Model", EOR Field Case Histories, SPE Reprint Series No 23, pp 107-115, SPE 1987.

[7] Fleming, E.A.; Cook, R.L.; Brown, L.M.; "Overview of the Production Engineering Aspects of the Denver Unit CO_2 Flood", prepared for - SPE/DOE 8th Symposium on Enhanced Oil Recovery, Tulsa, April 1992.

TABLE I

WASSON (SAN ANDRES) FIELD RESERVOIR PROPERTIES

Reservoir Structure:	Anticline
Original Drive Mechanism:	Depletion
Avg. Depth:	5200 feet
Avg. Porosity:	12 %
Avg. Permeability:	5 md.
Reservoir Temperature:	105°F
Avg. Net Pay thickness	137 feet
Oil gravity:	33 °API
Initial B.H. Pressure:	1805 psi
Bubble Point Pressure:	1805 psi
Original Formation Volume Factor:	1.312 rb/STB

FIG. 1

DENVER UNIT CO2 FLOOD

ACTUAL PERFORMANCE vs ORIGINAL PREDICTION

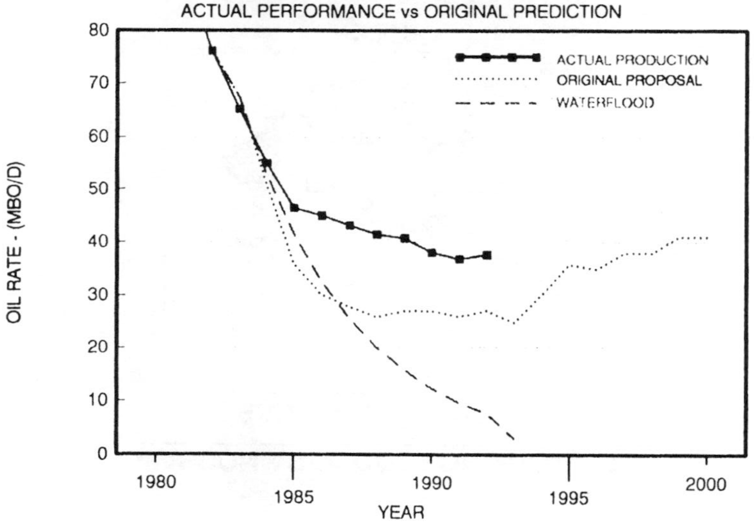

Fig. 2

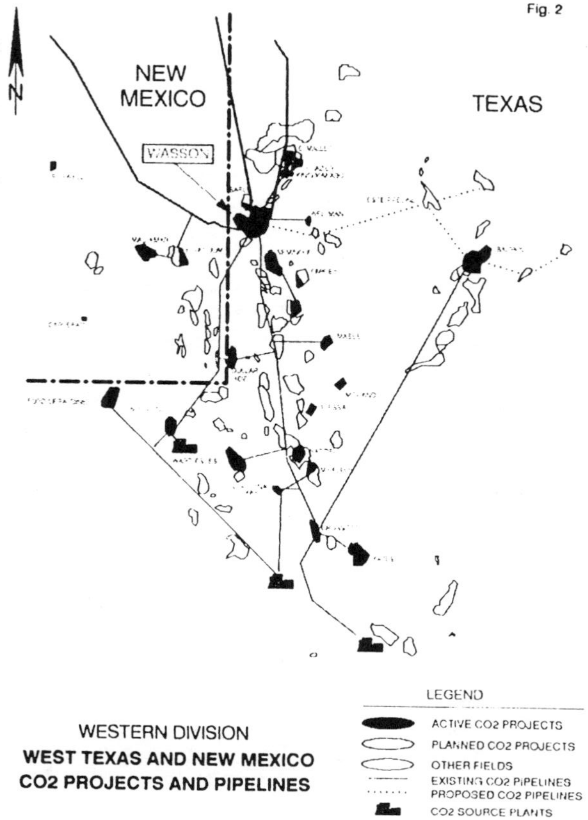

WESTERN DIVISION

WEST TEXAS AND NEW MEXICO CO2 PROJECTS AND PIPELINES

FIG. 3

WASSON AREA CO2 EOR PROJECTS

SAN ANDRES UNITS

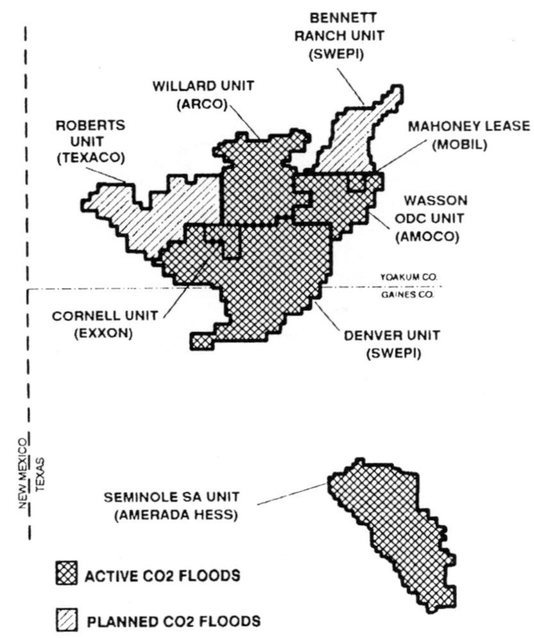

FIG. 4

WASSON FIELD - DENVER UNIT

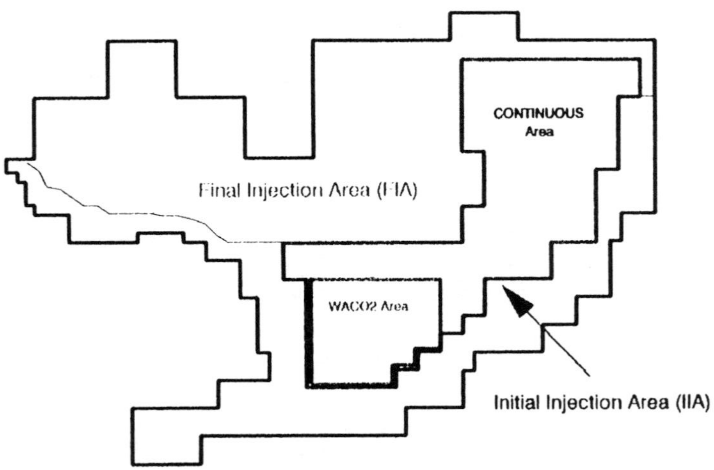

FIG 5

DENVER UNIT - CONTINUOUS AREA REVIEW

PRODUCTION PERFORMANCE HISTORY

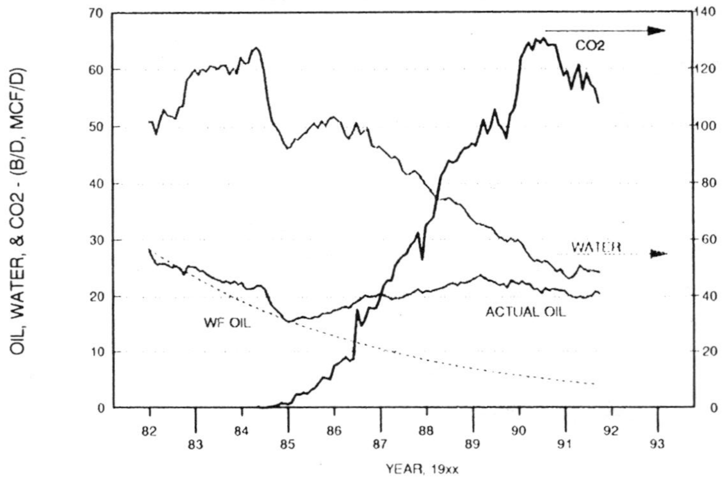

FIG 6

DENVER UNIT - CONTINUOUS AREA REVIEW

OIL CUT vs. CUM OIL (SINCE 1975)

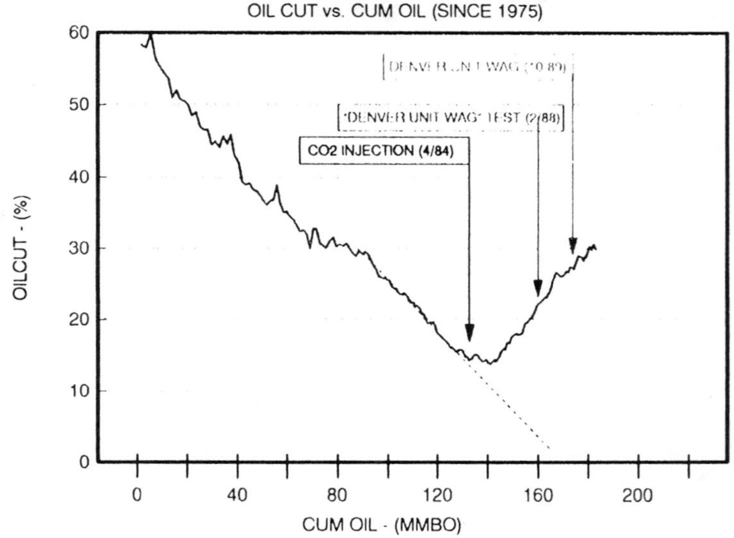

FIG 7

DENVER UNIT - CONTINUOUS AREA

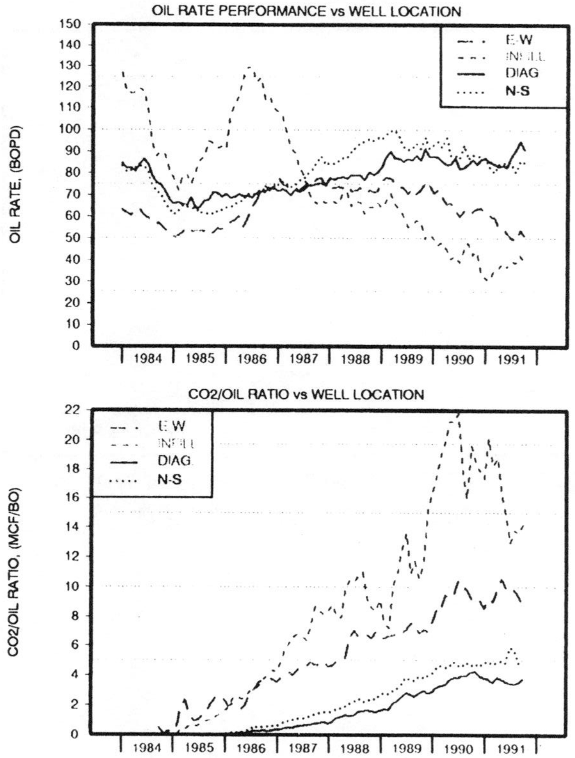

FIG. 8

FIVE ORIGINAL IIA I-P CONVERSIONS

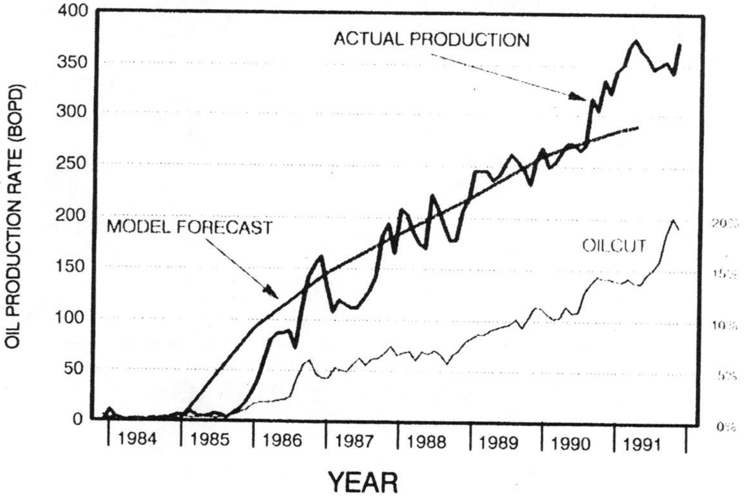

FIG. 9

DENVER UNIT - CONTINUOUS AREA

INJECTION HISTORY

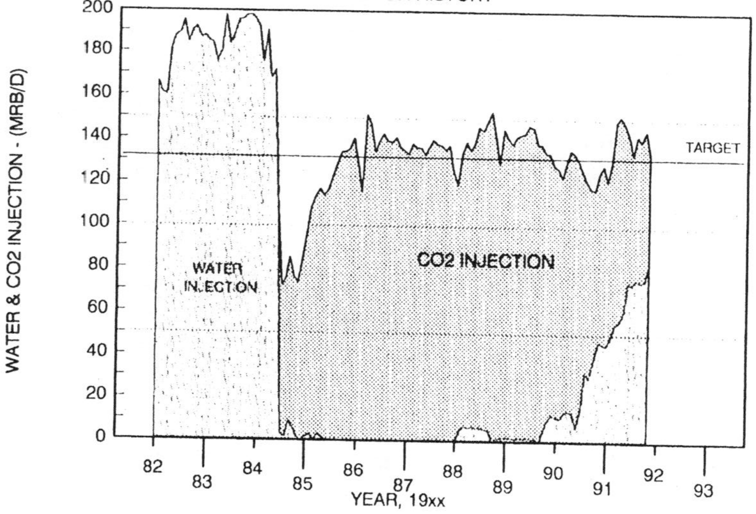

FIG. 10

DENVER UNIT - PATTERN 6740WC

INJECTION HISTORY

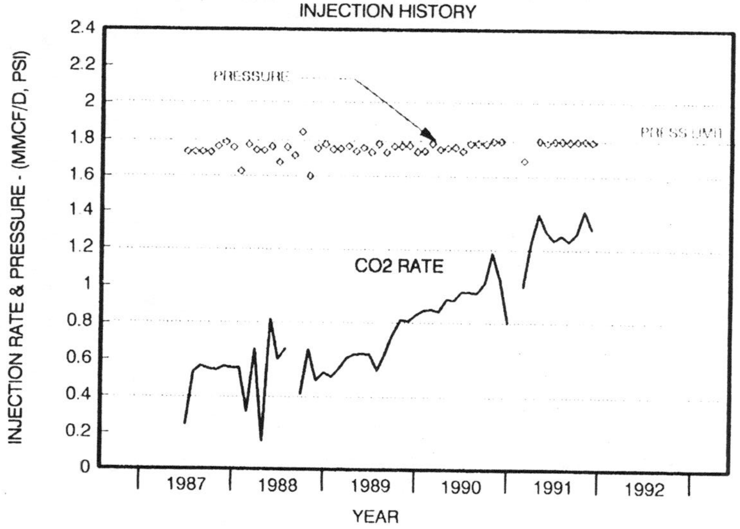

FIG. 11

DENVER UNIT CONTINUOUS AREA

HYDROCARBON GAS-OIL RATIO

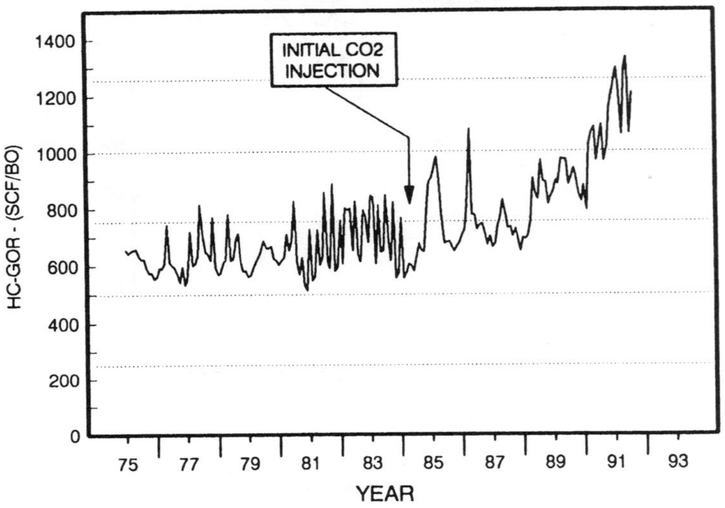

FIG. 13

DENVER UNIT - WACO2 AREA

OIL PRODUCTION HISTORY

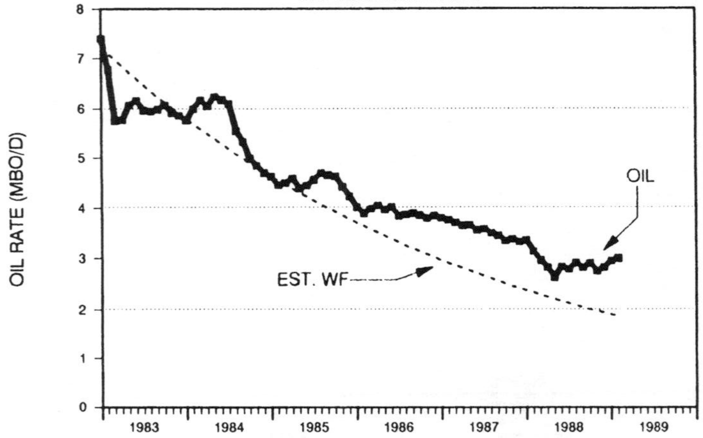

FIG. 12

DENVER UNIT CO2 FLOOD

ACTUAL PERFORMANCE vs. ORIGINAL PREDICTION

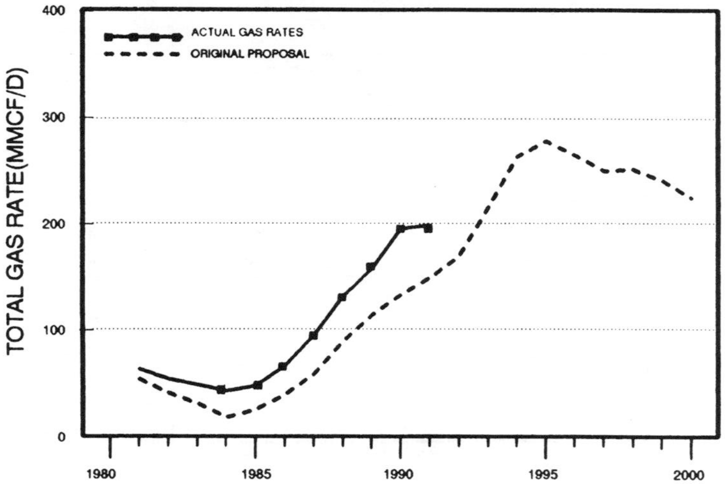

FIG. 14

DENVER UNIT - WACO2 AREA

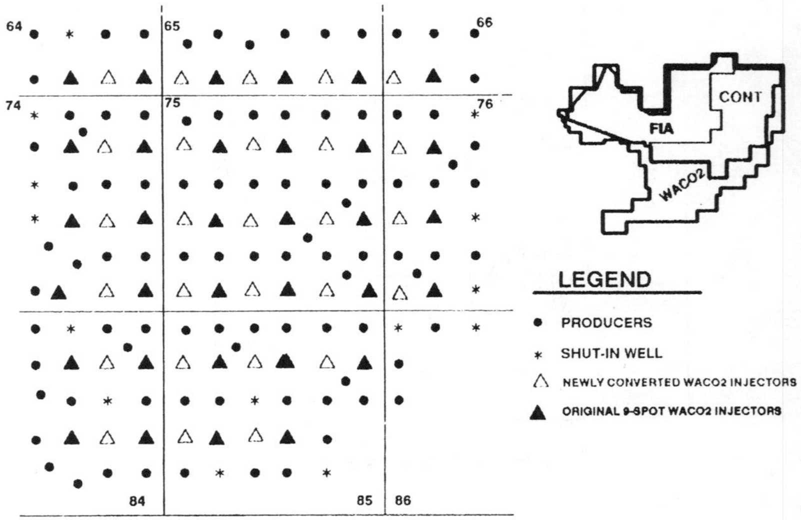

FIG. 15
DENVER UNIT OIL COLUMN CO2 FLOOD
CONTINUOUS AREA HISTORY MATCH

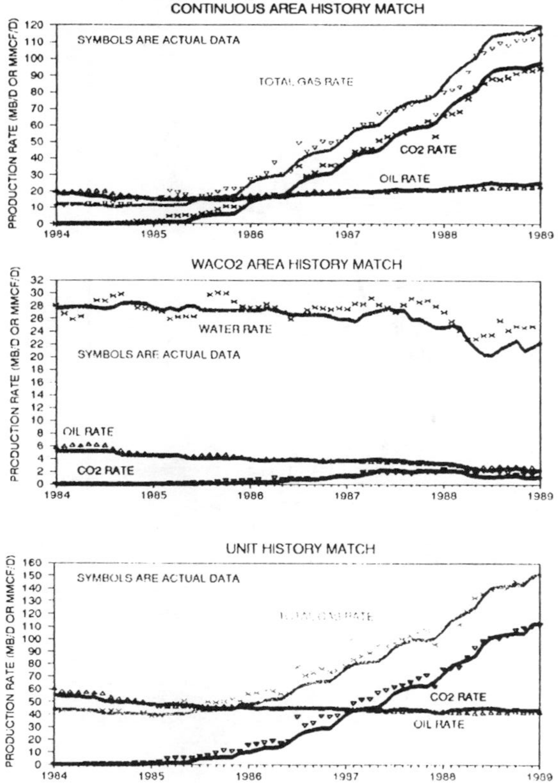

FIG. 17
CONTINUOUS AREA - "DENVER UNIT WAG" TEST
FOUR PATTERN PRODUCTION HISTORY

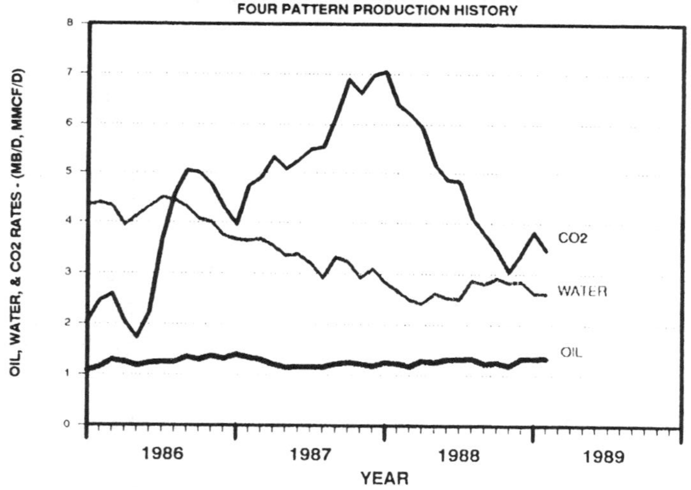

FIG. 16
CUMULATIVE INCREMENTAL EOR RECOVERY vs TIME

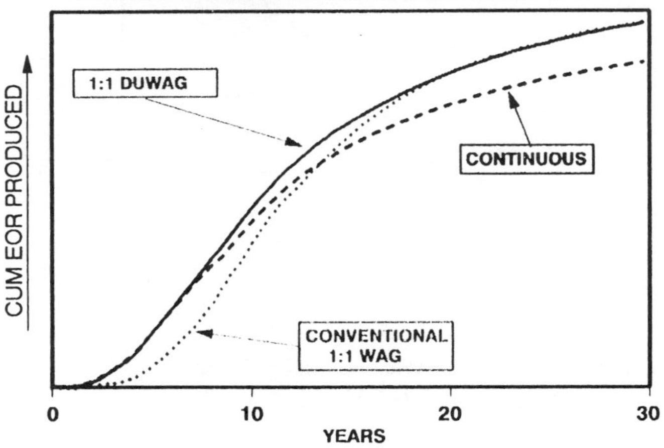

FIG. 18
DENVER UNIT - INJECTION STATUS
DECEMBER 1991

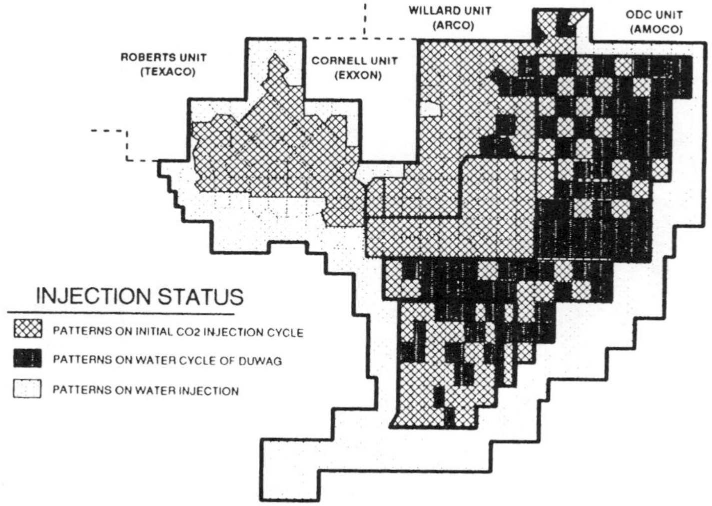

FIG. 19

DENVER UNIT WAG

PRODUCTION PERFORMANCE FOR 15 PATTERN AREA

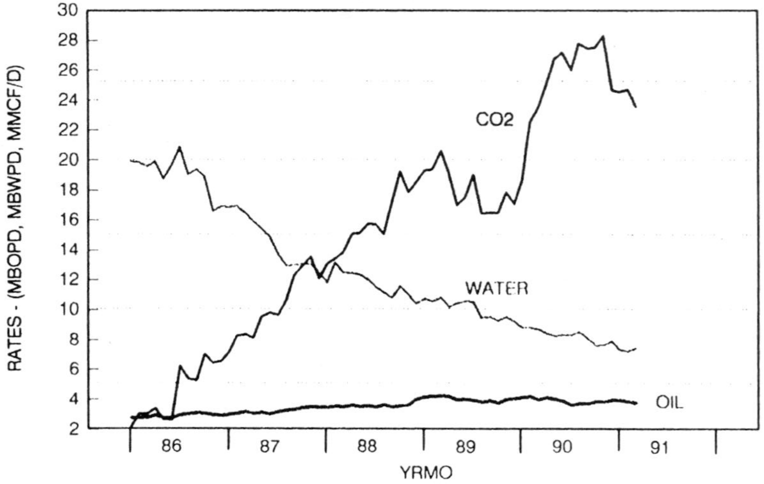

FIG. 21

DENVER UNIT - WACO2 AREA

OIL PRODUCTION HISTORY

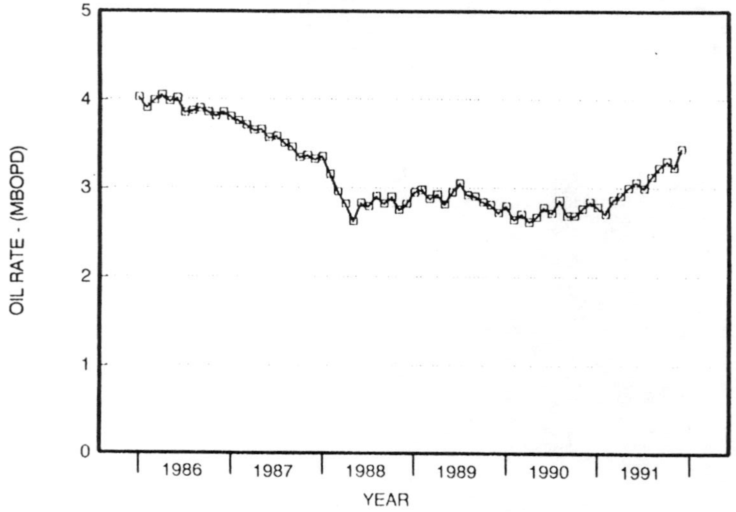

FIG. 20

WASSON DENVER UNIT

DENVER UNIT WAG TEST AREA

(THREE WELLS ON SECOND H2O CYCLE)

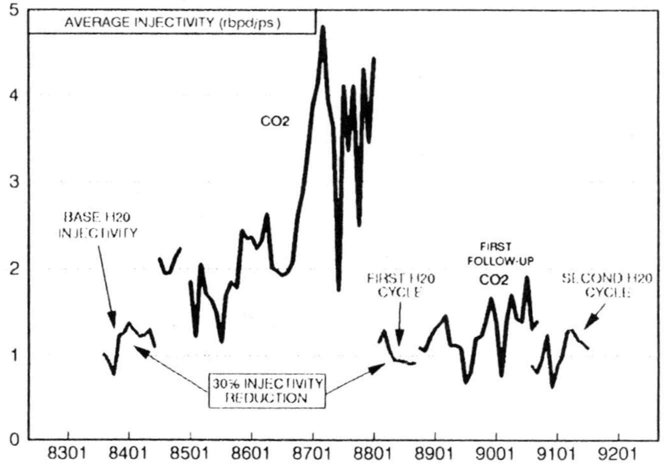

FIG. 22

DENVER UNIT - 6 PATTERN MATURE AREA

OIL RATE HISTORY

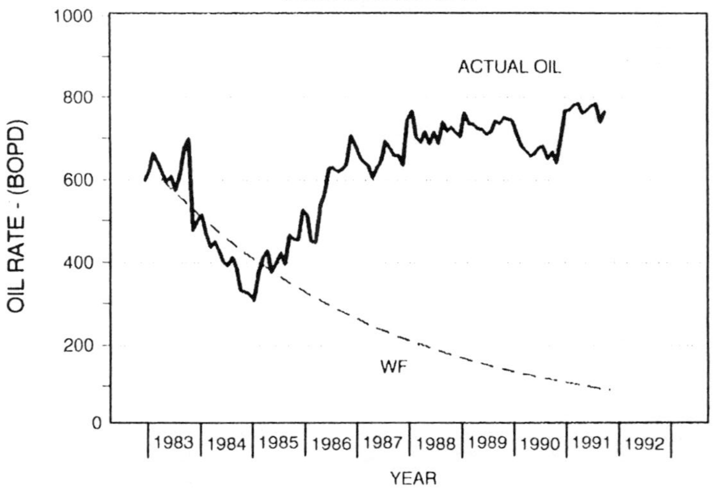

SPE/DOE
Society of Petroleum Engineers
U.S. Department of Energy

SPE/DOE 24157

Overview of Production Engineering Aspects of Operating the Denver Unit CO_2 Flood

E.A. Fleming, L.M. Brown, and R.L. Cook, Shell Western E&P Inc.

SPE Members

This paper was prepared for presentation at the SPE/DOE Eighth Symposium on Enhanced Oil Recovery held in Tulsa, Oklahoma, April 22-24, 1992.

ABSTRACT

Shell Western E&P Inc. has been operating the Denver Unit CO_2 flood, the world's largest CO_2 enhanced oil recovery project, since 1983. The unit is in the Wasson Field in Gaines and Yoakum Counties in West Texas. While operating the flood, SWEPI has developed successful technology and operating methods for carbon dioxide flooding.

The paper is structured in three parts: 1) physical properties of CO_2, 2) physical effects of CO_2, and 3) relative economic considerations associated with CO_2. Operational considerations associated with each topic are identified and SWEPI's resulting modifications in operating policy are discussed. Some of the topics included in the paper are equipment and materials, artificial lift, well control, wellbore impairment, injection surveillance, and operations in a populated area. Discussions also include issues where our operating experience has varied significantly from our original expectations. These include the lower than expected corrosion rate of materials, and the higher than expected number of wells that flow naturally.

This overview will be of interest to companies who are either operating or preparing to implement CO_2 floods. It will also interest service companies who may have or wish to develop products to service CO_2 operations.

INTRODUCTION

Shell Western E&P Inc. operates several CO_2 floods in the United States. This paper gives a brief background on CO_2 properties and behavior, the physical effects of CO_2, and relative economic considerations associated with CO_2 operations, focusing on the practical operating experience of our largest flood, the Denver Unit.

References and illustrations at end of paper.

BACKGROUND

The Denver Unit is located on the southern edge of the North Basin Platform of the Permian Basin in far West Texas (Figure 1). The San Andres reservoir at the Denver Unit is a dolomite formation existing 4700 to 5100 feet below the surface (Table 1). The oil accumulation is controlled by its anticlinal structure. The Denver Unit contains both gas-oil and oil-water contacts. The reservoir is subdivided vertically into "First Porosity" and "Main Pay" zones, with the Main Pay containing the better quality reservoir rock (Figure 2).

Primary depletion drive production began in 1936 with single well production rates in excess of 1500 barrels of oil per day being common. In 1964, the Denver Unit was formed and waterflooding began. Carbon dioxide (CO_2) injection began in 1983 when 9 inverted nine-spot patterns were placed on CO_2 injection. Today over 500 million cubic feet per day (200,000 reservoir barrels per day) of CO_2 are injected in 185 injectors (Figure 3). Daily production in 1991 averaged 37,000 barrels of oil and 36 MMCF of gas per day.

PHYSICAL PROPERTIES OF CO_2 AND ASSOCIATED IMPACT

At the Denver Unit, CO_2 is injected as a supercritical fluid at a surface pressure of about 1800 psi and a reservoir pressure of about 2200 psi (Figure 4). The critical pressure of CO_2 is 1071 psi and the critical temperature is 87.9 °F. Above these critical points, CO_2 behaves as a vapor whose density increases as pressure increases. Below these critical points, CO_2 may exist as a gas or a liquid.

CO_2 is an efficient injectant because it dissolves readily in both oil and water, with CO_2 solubility increasing with increasing pressure (Figure 5). The dissolution of CO_2 in oil increases or swells the crude volume and decreases the crude's viscosity which results in a more mobile oil bank and improved sweep efficiency.

When dissolved in water, CO_2 forms a weak acid called carbonic acid (H_2CO_3; pH of 3.3-3.7 in pure water) which readily reverts to CO_2 and H_2O with increasing temperature or decreasing pressure.

$$CO_2 + H_2O \rightleftharpoons H_2CO_3 \rightleftharpoons H^+ + HCO_3^-$$

The fluctuations in phase, temperature, pressure, and pH, which occur as a result of the addition of CO_2 to the production stream, result in modified operating conditions downhole and at the surface.

CO_2 Flooding Results In Reduced Wellbore Temperature

In CO_2 response wells, the bottom hole temperature drops from 105 °F due to the Joule-Thompson cooling effect as CO_2 enters the wellbore. Bottomhole temperatures as low as 40 °F have been reported in high rate CO_2 response wells. As the fluids move up the wellbore, pass through choke points at the wellhead, and enter into the flowline, additional temperature drops occur as the pressure decreases. The reduction in temperature results in increased scaling tendencies, paraffin/asphaltene formation, and wellhead freezing.

At the Denver Unit, the decreased bottom hole flowing temperature increases gypsum scaling tendencies and suppresses calcite scale formation at the wellbore. Calcite scale is more likely to be found at low pressure surface treating conditions. Calcite solubility increases as temperature and pH decrease (Figure 6). Fluid pH is reduced due to increased CO_2 solubility (Figure 7). Gypsum solubility is at its peak at 105 °F, the average reservoir temperature at the Denver Unit, and solubility decreases as temperature decreases (Figure 8). Thus, gypsum scaling tendencies increase in the CO_2 response wells due to the decrease in temperature. Additional scale inhibition squeeze and gypsum removal treatments are performed. Consideration should be given to the selection of a scale inhibitor which is effective at the reduced reservoir pH of 4.7 to 4.9. We have found phosphonate type inhibitors to be more effective than polymers at low pH.

The decrease in downhole temperature also causes an increase in paraffin and asphaltene formation. Paraffin forms when the temperature drops below the crude's cloud point, which ranges between 60 °F and 65 °F. More paraffin and asphaltene problems occur in areas that are being CO_2 flooded than in neighboring waterflooded areas. Backpressure valves are used to keep more pressure on the backside of the well, thereby keeping larger amounts of light end hydrocarbons and CO_2 in solution and reducing paraffin/asphaltene formation downhole. Higher pressure wellhead equipment such as the electrical feed through (EFT) submersible wellhead and high pressure stuffing boxes also hold more back pressure on the well, reducing downhole paraffin formation.

Wellhead freezing on flowing wells sometimes occurs as the CO_2 expands as it enters the wellbore and moves up the tubing. As the temperature drops, produced fluids may freeze or hydrates may form (Figure 9). This either restricts or completely stops the flow of produced fluids into the flowline. This operational problem is addressed by injecting warm produced water into the flowline upstream of the choke to warm the wellhead and by adjusting chokes to increase backpressure.

CO_2 Flooding Results In Increased Gas Production

Increased total gas production is experienced as a result of CO_2 flooding. The CO_2 is injected as a supercritical fluid at 1800 psi, but is produced as a gas at pressures well below the critical pressure of 1071 psi. As water production is displaced by CO_2 production, increasing numbers of wells will flow or operate in a condition between flowing and pumping (flump).

The installation of flowing trees is possible on some wells, resulting in a substantial savings in lift costs and maintenance time. However, a greater drawdown may be placed on a well that is pumped, as opposed to flowing. Therefore, production can be optimized by delaying the conversion to flow. Modifications can be made to the wellhead to accommodate the flumping stage. EFT connections can be installed on submersible pumped wells to contain gas at the wellhead and provide a higher pressure system (Figure 10). High pressure double pack stuffing boxes can be installed on beam pump wells to perform a similar function (Figure 11). A ram type rod BOP is installed below the pumping tee to shut-in the well for stuffing box changes.

Other than gas separators for ESP equipment, no significant changes in downhole equipment are made as a result of increased gas production. CO_2 resistant teflon based pipe dope is used for pipe connections. Packers -are installed in flowing wells, but no modification to the packer design is made. However, a CO_2 compatible packer rubber should be considered. Standard durometer hardness nitrile rubbers, which would normally be run at bottomhole temperatures of 105 °F, are susceptible to extrusion in CO_2 service. The harder 90 hardness nitrile rubbers better resist extrusion, but may not be soft enough to deform and seal at 100 °F. We have had successful operating experience with running a stacked multidurometer 90/70/90 seal. The 70 rubber assures a good seal and the 90 rubbers back up the 70 rubber, preventing excessive extrusion.

CO_2 Flooding Results In Increased Shut-In Wellhead Pressures

Increased shut-in wellhead pressures are experienced as a result of CO_2 flooding, regardless of whether the reservoir pressure increases, due to the phase behavior of CO_2. As the supercritical CO_2 warms and the pressure drops below the critical pressure in a shut-in well, the CO_2 becomes a gas and expands (Figure 4). This expansion begins as the CO_2 enters the wellbore and continues as it rises, resulting in an increase in wellhead pressure. Shut-in pressures as high as 1200 psi have been experienced in Denver Unit producing wells.

Increased wellhead pressures have prompted a review of producer wellhead designs. The Denver Unit produces sour gas containing 500-20,000 ppm H_2S, but NACE wellhead equipment has not been routinely installed during waterflood operations because shut-in pressures were very low.

However, the increased shut-in wellhead pressures which can be experienced due to CO_2 flooding combined with high concentrations of H_2S make it possible that wellheads will be subjected to conditions which would exceed the NACE 250 psig multiphase threshold, thus prompting the consideration of NACE equipment. No changes in downhole equipment are made.

The Combination Of Increased Wellhead Pressures And Increased Gas Production Prompts Consideration of Additional Well Control Precautions

Increased wellhead pressures and increased gas production result in wells capable of flow. Therefore, more emphasis is placed on well control precautions to ensure a safe operation. Well control has been addressed in three ways:

1. Detailed documentation of CO_2 workover procedures have been prepared.
2. A comprehensive training program has been implemented to prepare operations personnel for CO_2 workovers.
3. Pulling a well is avoided when possible and through-tubing workover techniques are being pursued.

Detailed documentation of CO_2 workover procedures have been compiled in the form of a manual. The manual provides a guide for gaining and maintaining primary and secondary control of various types of wells in a CO_2 flood. Each section includes an overview of the unique or important aspects of well control for a specific type of well, a list of equipment needed for well control, a step-by-step procedure for gaining and maintaining control of the well, and pertinent illustrations.

Supervisors of CO_2 well servicing operations complete a curriculum of formal training, a supervised apprenticeship training period of two to six months, and a written examination which covers the manual prior to supervising workover operations.

An increased emphasis has been placed on through-tubing workover operations to avoid pulling the well whenever possible. Stimulations using coiled tubing, standard diverting agents such as ball sealers, and foamed acid as a diverting agent are being applied when practical.

When killing a well is necessary, the appropriate weight fluid to provide 100-300 psi overbalance is calculated and utilized. Many times, sodium chloride brine is not adequate in weight and a calcium carbonate mud or calcium chloride brine is used. The use of heavier weight kill fluids can be eliminated by establishing a preliminary shut-in period to let the well stabilize, thereby allowing a lower and more accurate calculation of bottom hole pressure. Switching the well to water injection for a period prior to working over the well can also eliminate the need for the heavier kill fluids.

PHYSICAL EFFECTS OF CO_2 AND ASSOCIATED IMPACT

Carbon dioxide is not generally considered to be a toxic or poisonous gas. It is present in the atmosphere at a concentration of about 0.03 percent (300 ppm). However, exposure to high concentrations of CO_2 for extended periods of time can be hazardous. The primary effect of CO_2 is a change in respiration and pH of the blood. CO_2 blocks the absorption of oxygen and death can occur in time due to asphyxiation. The short-term physical effects of even very high concentrations of CO_2 can be reversed by providing exposure to fresh air. If breathing has stopped, the appropriate response is to move to fresh air and give artificial respiration.

The physical effects of CO_2 are gradual, based on concentration and time of exposure (Figure 12). At low concentrations of CO_2, up to 1.5 percent, there are no noticeable physical effects for about one hour. At concentrations of 3 to 5 percent, breathing quickens and deepens resulting in lightheadedness. At concentrations of 5 to 9 percent, nausea and dizziness are also experienced. Above 9 percent, unconsciousness can occur in 5 to 10 minutes. At concentrations above 20 percent, death can occur in 20 to 30 minutes.

The current OSHA standard for maximum allowable concentration of CO_2 in the air is 1.0 percent for eight continuous hours of exposure. OSHA defines 3.0 percent as the maximum concentration that operating personnel can be exposed to for a short period of time.

While it is important that these effects be thoroughly understood and accounted for to insure safe operations, the effects are not as severe as those commonly dealt with in handling the H_2S gas associated with waterfloods. Therefore, most associated modifications apply only to operations in populated areas.

Operation Of A CO_2 Flood In A Populated Area Prompts Additional Consideration

The Denver Unit operates 37 injection wells and 112 producers in populated areas of Denver City. The primary safety considerations as a result of operating a CO_2 flood in a populated area are injecting CO_2 at high pressure (1800 psi) and the presence of CO_2 in the produced gas. The presence of H_2S remains a consideration, but as the CO_2 content of produced gas increases, the H_2S content generally declines resulting in no significant net change in the H_2S radius of exposure. Shell dispersion modeling based on Quiet Pool and Jet Release models indicates that CO_2 leaks in the Denver Unit pose no public danger at maximum anticipated rates.

Additional safety precautions to address increased wellhead pressure have been taken for all CO_2 injectors and producers located within a city limits or within 200 feet of a public area.

Populated Area Wells

The injection and production wellhead/tree have been designed to provide optimum integrity in higher pressure gas service. The tubing head/spool is studded or flanged with studded or flanged outlets. Both active and temporarily abandoned wellheads are upgraded. To isolate the public from any potential CO_2 release, the entire well location of all populated area wells is fenced. In addition, all wells are monitored 24 hours a day via computer assisted alarms.

Injectors and flowing wells are equipped similarly (Figure 13). An extended neck mandrel tubing hanger with back pressure threads is installed. The back pressure valve enables isolation of the tree from formation pressure, allowing removal of the tree while maintaining secondary well control. Valve redundancy is provided with flanged dual master valves. A flanged wing valve is installed which is equipped with an actuator to function as an emergency shut down device (ESD). A crown valve is installed to provide easy access for wireline work and setting back pressure valves.

Additional control is also exercised on flowing and injection wells located in populated areas. Injector ESD (actuated wing) valves are programmed to close upon detection of low tubing pressure, low injection line pressure, or high casing pressure. Flowing well ESD valves are programmed to close upon detection of high or low flowline pressure. In addition, each CO_2 injection line is equipped with a second ESD valve upstream of the point where the line enters a populated area. The valve is programmed to close upon detection of increased flow rate, which may indicate a leak or rupture in the injection line.

Beam pumped wells have a wrap-around seal installed around the tubing below the slips to provide an improved annular seal. A second ram-type BOP is installed above the pumping tee to provide additional well control. The additional BOP isolates the stuffing box while allowing the well to flow or permitting the pumping of kill fluids as necessary. A double stuffing box with a leak detector and polished rod BOP is installed above the pumping tee. The leak detector switches off the pumping unit in the event of a leak (Figure 14). The polished rod BOP shuts in the well in the event of a polished rod failure. The use of polished rod liners is prohibited because they prevent operation of the polished rod BOP. A vibration switch is installed on the walking beam to detect excessive vibration resulting from a rod failure or a wrist pin failure. The pumping unit motor is switched off if such vibration is detected. No modification to downhole equipment is made as compared to remote CO_2 operations.

No modification to the submersible pump equipment is made for operation in populated areas other than the installation of a studded or flanged tubing head/spool. Submersible pumped wells are equipped with electric feedthrough type connectors as previously mentioned which allow higher wellhead operating pressures (Figure 10). No modification to downhole equipment as compared to remote CO_2 operations is made.

Populated Area Operating Practices and Procedures

All ESD valves and other shut down devices are routinely tested. The casing on flowing wells and injection wells is annually tested to the maximum anticipated shut-in pressure. Lease operators to visit each production and injection well daily.

Some additional precautions are also taken during workovers in populated areas. Any time that tubing is to be pulled, the well is equipped with an annular BOP, pipe rams, and blind rams. CO_2 qualified supervisors are present on location for all critical stages of the servicing operation if a well is capable of flow. Production flowlines are used to circulate producing wells during a workover, but injection well fluids are circulated through a manifold and scrubber which vents the gas 30 feet in the air to assure adequate dispersion of the CO_2 gas.

RELATIVE ECONOMIC CONSIDERATIONS OF CO_2 OPERATIONS

CO_2 is a cleaner injectant than either produced or fresh water. As a result, fewer injector clean outs and acid stimulations are performed when a well is injecting CO_2. Because of the decreased impairment, the injection profile rarely changes. The frequency of injection profile surveys can be decreased accordingly. (As an operational note, alcohol-based surveys are used for CO_2 injectors instead of water-based to more closely match the gradient of CO_2). However, as previously noted, CO_2 injection does result in increased gypsum and paraffin formation in producers. This results in an increased frequency for wellbore stimulations. Producers are treated, on average, once every two years as compared to once every three years for waterflood portions of the field. A majority of these jobs are either acid treatments or acid fractures and result in sustained average production increases of over 25 BBLS of oil per day. The increased oil production associated with these jobs pay for the stimulations, on average, in less than two months.

CO_2 is significantly more expensive than water on a reservoir BBL basis (see Figure 15 for MCF to reservoir bbl conversions). The economics of alternating water and gas cycles (WAG) and recycling produced gas should be considered when planning a flood. In addition, a greater emphasis on injection profile control is warranted when injecting CO_2. Profile corrections to prevent out of zone losses and improve sweep efficiency become much more economic when a more expensive injectant is used. Previously uneconomic profile control work may now be justifiable.

Cased Hole Completions Improve Profile Control

Because of the early difficulties in optimizing injectant placement for optimum sweep efficiency in openhole injectors during the original Denver Unit waterflood, a policy was adopted during the waterflood's implementation of running casing across pay intervals on injectors (Figure 16). By having the injectors cased across pay intervals, quantities of injectants going into individual zones can be better controlled through selective stimulations and injectant shut-offs, thereby optimizing the displacement efficiency of the injectants used. This philosophy is even more important for CO_2 floods, in light of the increased cost of the flood's injectant. We have found neat cement to be effective for cement squeezing in CO_2 intervals. Because of the increased penetration, micro-fine cement is used for squeezing off small leaks (ie. collar leaks).

Generally, wells which were originally open hole completions have had fiberglass liners run across the open hole and use a tandem packer assembly to minimize injectant contact with downhole steel components to provide corrosion protection. A cup type packer is run down into the fiberglass liner on tail pipe hung below a regular slip type packer. The slip type packer is set inside the steel casing (Figure 17). The internal components of the injection equipment are plastic coated to protect them against corrosive produced water injectants.

Newly drilled injectors use a continuous mixed string of steel and fiberglass casing. A normal configuration for a new injector would include 5.5" steel casing run from surface to 30'± of the injection interval. A 20' joint of corrosion resistant alloy (CRA) 5.5" casing is then run between the steel casing and the 5.5" fiberglass casing run across the pay interval. A standard slip type injection packer is then set in the CRA joint above the fiberglass.

The fiberglass is cemented across the pay zones with high strength Class C cement. The well is then perforated using hollow carrier, select-fire, mechanically decentralized, jet perforating guns (surface perforating tests have shown that the hollow carrier gun will absorb the energy of the shot and not damage the fiberglass, while expendable-type guns will damage the pipe). The perforated intervals are then selectively acidized with hydrochloric acid using closely spaced straddle packer assemblies. Unlike producers, injectors are not fracture treated in order that control of injectants exiting the wellbore can be maintained (ie. fractures, connecting various pay zones, would limit our ability to selectively stimulate or shut-off zones for redistribution of injectants). It is also important to inject below fracture pressure, while injecting CO_2 as well as water, to slow breakthrough and improve sweep efficiency.

Implementation Of Water Alternating Gas (WAG) Injection Strategy Prompts Additional Consideration

In order to maximize flood profitability through improved sweep efficiency and minimize the amount of the more expensive CO_2 required, Denver Unit injection programs follow WAG cycles. After predetermined pore volumes of CO_2 have been injected into a flood pattern, the pattern injector is switched to water where again a predetermined pore volume of water is injected before the injector is again returned to CO_2. Each of these cycles last months in the Denver Unit design.

WAG Increases Corrosion Concerns

Dry CO_2 is not corrosive. Tubing and packer failures are almost nonexistent in wells which have only been on CO_2 injection. However, special precautions have been taken to insure that corrosion problems do not occur as a result of mixing CO_2 and water (which forms carbonic acid) and to minimize corrosion due to injection of produced water. To date, produced water injection has proven to be by far the biggest corrosion problem experienced.

Injection wellhead designs predominantly use 316 stainless steel master valves and Inconel 625 tubing hangers to minimize corrosion in these critical surface components.

While bare steel tubing is acceptable for CO_2 injection service, an internally plastic coated (IPC) or lined tubing is used in WAG service due to produced water corrosion concerns. Many injectors have IPC injection strings with 8 round connections. Our record with these tubing strings has been disappointing. We have experienced an unacceptably high frequency of connection leaks in CO_2 service in these strings as a result of corrosion when these wells were previously on produced water injection. Connection corrosion problems on the produced water cycles often don't appear until after the well is switched to CO_2. At this point, the corroded connection seal, which would still hold pressure with water, will fail as the CO_2 more easily leaks through the corroded thread area. Even after improvements were made in handling and makeup procedures, problems persisted. This has resulted in the recent practice of using modified tubing connections. Fiberglass lined tubing is also being tested. While our success with fiberglass lined tubing has always been good for use in water injectors, the potential for swelling of elastomer seals as the result of the expansion of absorbed CO_2 when wells are blown down, has resulted in an ongoing testing program of this tubing.

As mentioned previously, downhole packer assemblies have successfully utilized plastic coating on internal components exposed to injectants. Again, the main corrosion concern is associated with components exposed to produced waters and the use of plastic coatings downhole has proven both effective and economical. Failure of downhole injection equipment accounts for only a small percentage of leaks to the tubing/casing annulus.

To date, CO_2 related corrosion has not been a problem in producing wells. This is attributed to the low production temperatures, the natural inhibition from the oil production and paraffin deposition, and periodic batch corrosion inhibitor treatments of many producers. Exposure to produced water remains the predominant corrosion concern. On producing wells, both pumping and flowing, bare 8 round tubing strings are used with good success. Rod strings and pumps have shown no additional corrosion problems as a result of CO_2 production. Tapered KD rod strings are typical and insert pumps typically have corrosion resistant features such as chrome-over-bronze barrels, plungers with nickel coated pins, stellite balls, and tungsten carbide seats.

WAG Reduces Water Injectivity

Field data has shown that after the first CO_2 cycle, water injectivity is approximately 70% of the original waterflood injectivity. In multiple cycles, the water injectivity closely resembles that of the first subsequent water cycle (Figures 18 & 19). Because of the reduced water injectivity, it becomes difficult to meet water injection set points in WAG areas. Injection near (but below) fracture pressure, an aggressive stimulation schedule, and a current knowledge of fracture pressures are increasingly important factors.

CONCLUSIONS

There are many operational considerations which should be taken into account when implementing a CO_2 flood; but once considered, operation of a CO_2 flood may be efficiently and effectively managed. The main points of consideration are:

1) Wellbore temperature will decrease, which may result in modified scaling tendencies, increased paraffin/asphaltenes and wellhead freezing;

2) Gas production will increase and water production will decrease, which may result in more flowing and flumping wells;

3) Wellhead pressure will likely increase, which may prompt modifications in design;

4) Increased training in well control for rig personnel may be considered as a result of increased gas production and increased wellhead pressures;

5) The physical effects of CO_2 can be hazardous, which may prompt additional operating considerations in populated areas;

6) CO_2 is a cleaner injectant, which may result in less frequent injector wellbore clean-outs. However, increased producer treatments may be justified due to increased impairment resulting from CO_2 production;

7) CO_2 is a more expensive injectant than water, which may modify the economics of profile control, and injectant cycling (with water) or recycling;

8) The mixture of water and CO_2 is very corrosive, which may result in additional corrosion concerns in producers and WAG injectors.

Managing the operations of a CO_2 flood is a challenging and dynamic task. We continually evaluate our operating policy, making modifications as appropriate, to strive for continuous improvement in operating conditions. We have found this process of continuing examination to be the key to anticipating the dynamics of a CO_2 flood.

ACKNOWLEDGEMENTS

We would like to thank L. N. Morgenthaler, G. N. Iken, J. L. Morris, J. B. Lawson, D. C. Berkshire, R. J. Faircloth and K. M. Cowan, for their significant contributions to this paper.

REFERENCES

1. "Application for Waterflood Response Allowable Wasson Denver Unit", Railroad Commission of Texas - Exhibit 1 - Application, February 1972.

2. "Application for Jurisdictional Certification as a Qualified Tertiary Oil Recovery Project Denver Unit Wasson Field", Railroad Commission of Texas Oil and Gas Division, Docket Number 8A-76,232, February 2, 1981.

3. GPSA Engineering Data Book, 9th ed., by Gas Processors Suppliers Association.

4. Goodrich, J., "Target Reservoirs for CO_2 Miscible Flooding, Task One, Draft Final Report, Volume I: Discussion of CO_2 Injection Process", work performed for the Department of Energy - Contract DE-AC21-79MC08341, Gruy Federal, Inc., Houston, Texas, 1980.

5. Carbon Dioxide for the Recovery of Crude Oil, The Department of Petroleum Engineering, University of Southern California, 1979.

6. Patton, Dr. C. C., Oilfield Water Systems, 2nd ed., Campbell Petroleum Series, Norman, Oklahoma, 1981: p. 54.

7. Newton, L. E., Jr. and Hausler, R. H., NACE Task Group T-13, CO_2 Corrosion in Oil and Gas Production, National Association of Corrosion Engineers, Houston, Texas, 1984: p. 7, 131, 178, 365, 660, 626, and 633.

8. "Carbon Dioxide in the Gas and Oil Industry," Amerigas, Carbon Dioxide Division.

9. "Standard Material Requirements - Sulfide Stress Cracking Resistant Metallic Materials for Oilfield Equipment," National Association of Corrosion Engineers, MR0175-90, 1990.

10. "Carbon Dioxide Effects on People," Technical Bulletin No. TBG-9, by Cardox Production, Division of Chemetron Corporation, Chicago, Illinois, 1972.

11. Saylors, S. E. and Basile, P. F., "CO_2: Properties and Handling with Safety Considerations," paper from Dowell Schlumberger.

12. Boone, S. A., III, CHCM, "CO_2 and H_2S Safety Regulations and Recommended Practices," paper from Professional Oilfield Safety, Inc.

13. OSHA, 29 CFR, 1910.1000, Table Z-1-A.

14. Tinker, S. J., "Equilibrium Acid Fracturing: A New Fracture Acidizing Technique for Carbonate Formations," paper SPE 18883, presented at the 1989 SPE Production Operations Symposium, Oklahoma City, Oklahoma (March 13-14, 1989).

15. P. T. Baxley, J. G. Crump, W. C. Miller and C. S. Tanner, "Production Performance of the Wasson Denver Unit CO_2 Flood," paper SPE 24156, to be presented at the 1992 SPE/DOE Eighth Symposium on Enhanced Oil Recovery, Tulsa, Oklahoma (April 21-24, 1992).

TABLE 1

DENVER UNIT
SUMMARY OF PROJECT DATA

Formation	Permian San Andres Dolomite
Structure	5200'
Average Depth	-1325'
Gas-oil contact	-1400' to -1650'
Oil-water contact	12%
Average permeability	5± md
Average net oil pay thickness	137'
Oil gravity	33° API
Reservoir temperature	105°F
Total acres in entire reservoir	62,500 acres
Number of acres in Denver Unit	27,848 acres
Number of productive acres in Denver Unit	25,505 acres
Date reservoir discovered	April 15, 1936
Date TRRC approved injection	October 14, 1964
Date of first injection	November 1, 1964
Date unitization effective	November 1, 1964
Primary producing mechanism	Solution gas (depletion)
Number of wells	
a) producers	763
b) injectors	427
c) plugged and abandoned	107
Original reservoir pressure	1805 psi
Bubble point pressure	1805 psi
Average pressure at start of secondary recovery	800±/1100± psi
Target reservoir pressure for CO_2 flood	2200± psi
Initial oil formation volume factor	1.312
Solution gas-oil ratio @ original pressure	420 ft^3/bbl
Gas-oil ratio at start of secondary recovery	1060 ft^3/bbl
Gas-oil ratio at current conditions	600±^{3}/bbl
Oil viscosity at 60°F and 1100± psi	1.18 cp
Source of injection water	Produced and Ogallala

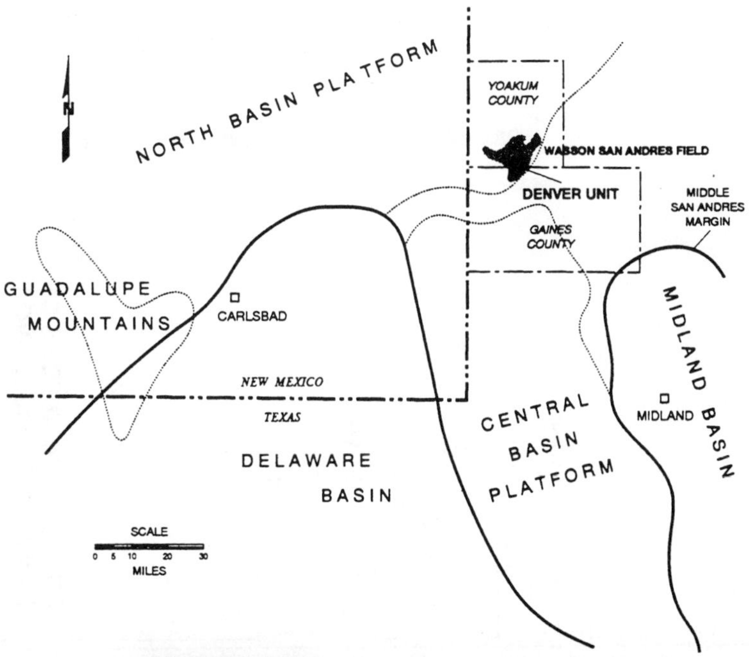

LOCATON MAP
DENVER UNIT
WASSON SAN ANDRES FIELD

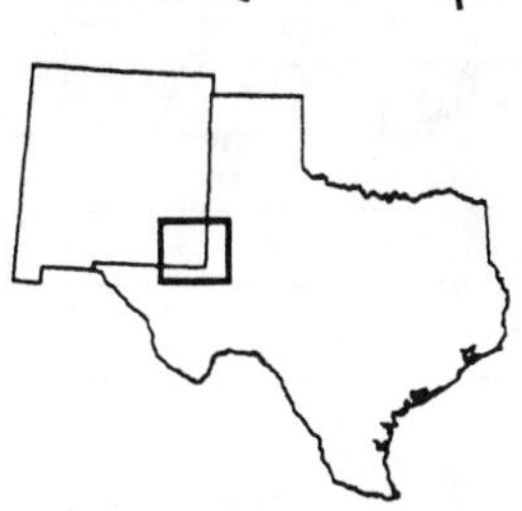

FIGURE 1

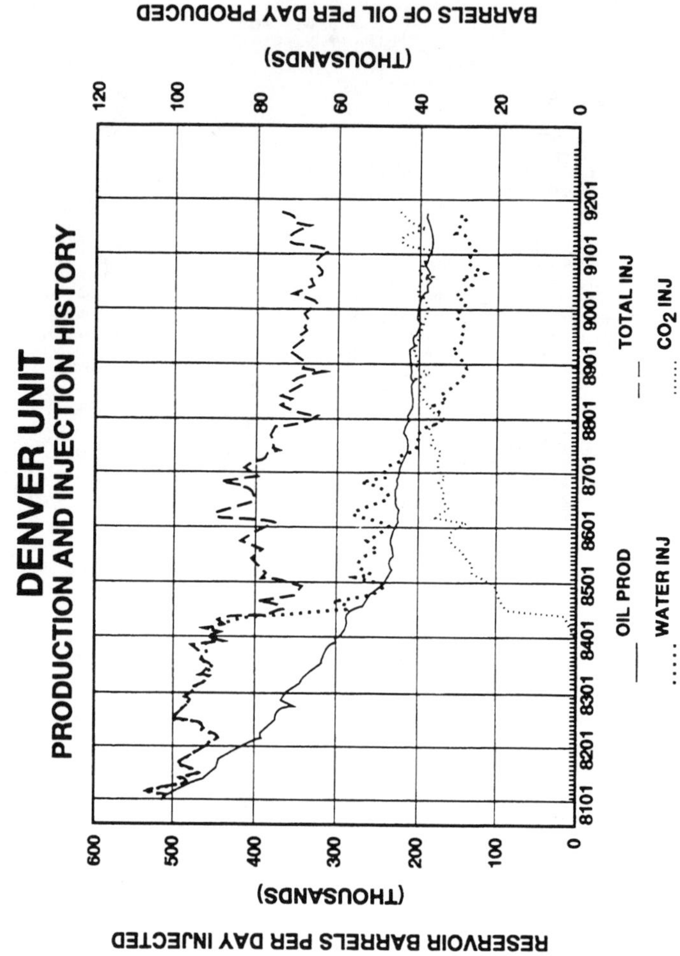

FIGURE 3

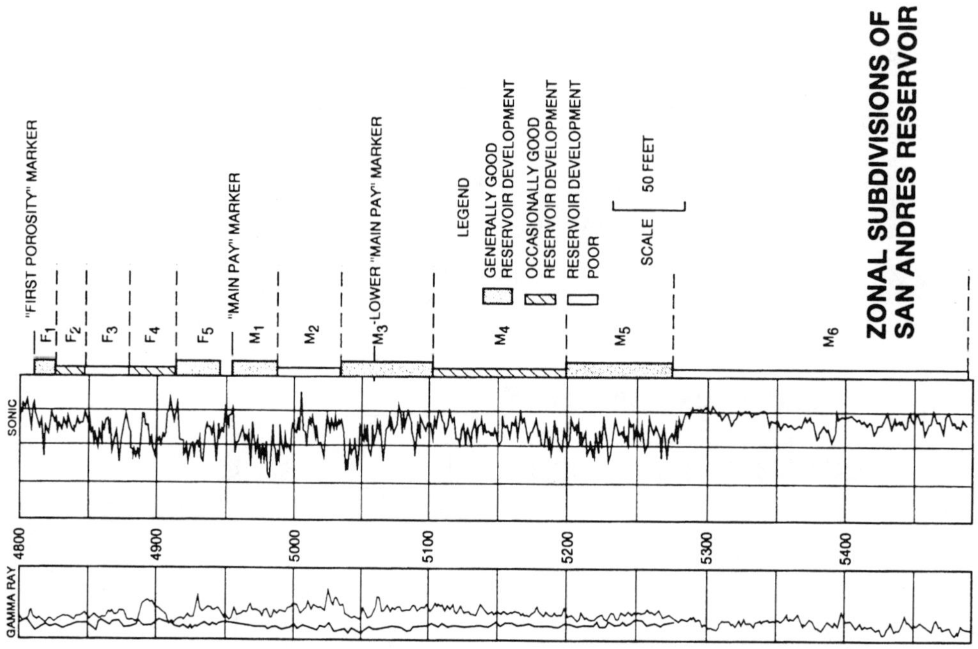

FIGURE 2

DENSITY COMPARISON
WATER, PURE CO_2, AIR

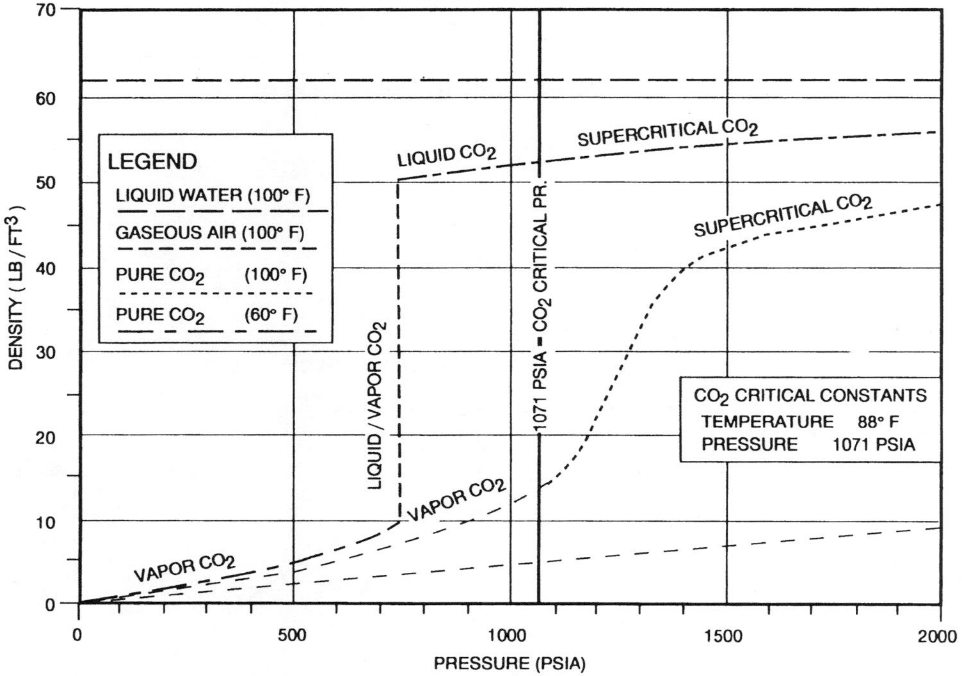

FIGURE 4

SOLUBILITY OF CO_2 IN WATER AT 100° F

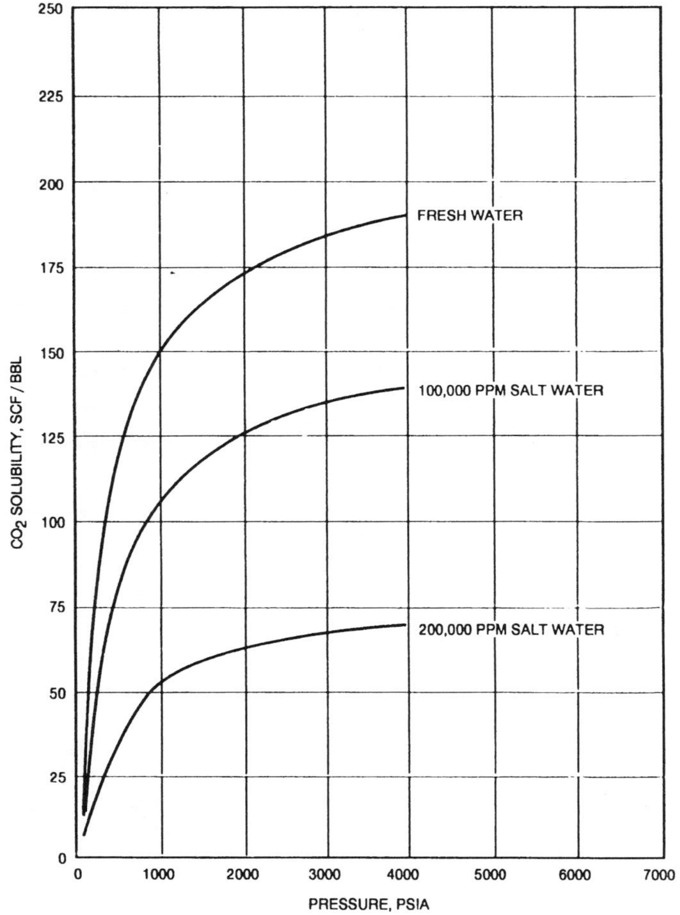

FIGURE 5

SOLUBILITY of CALCITE in PURE WATER

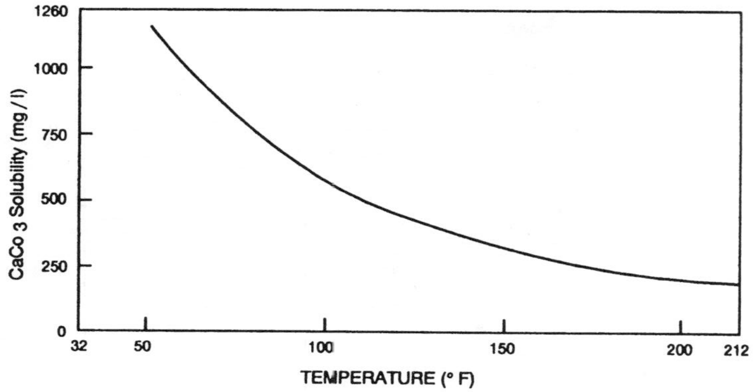

FIGURE 6

SOLUBILITY OF GYPSUM IN SYNTHETIC WEST TEXAS SUPPLY SYSTEM WATER

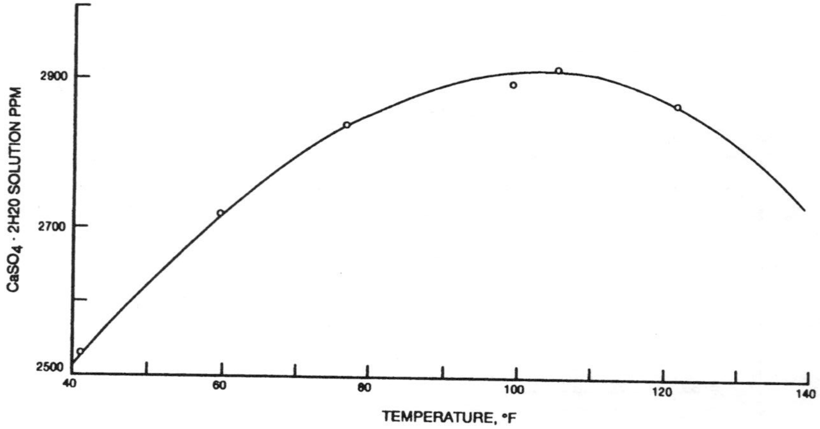

FIGURE 8

BRINE pH in a CO_2 Response Well

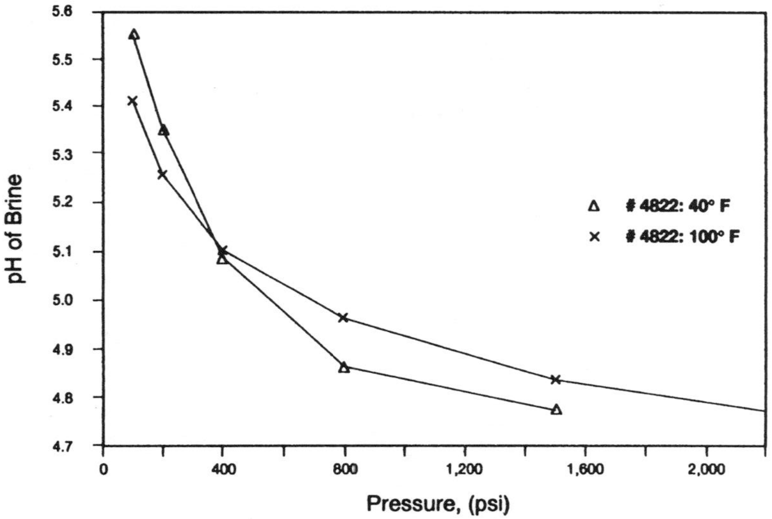

Range of calculated pH.

FIGURE 7

VAPOR - LIQUID PHASE RELATIONSHIP FOR CO_2

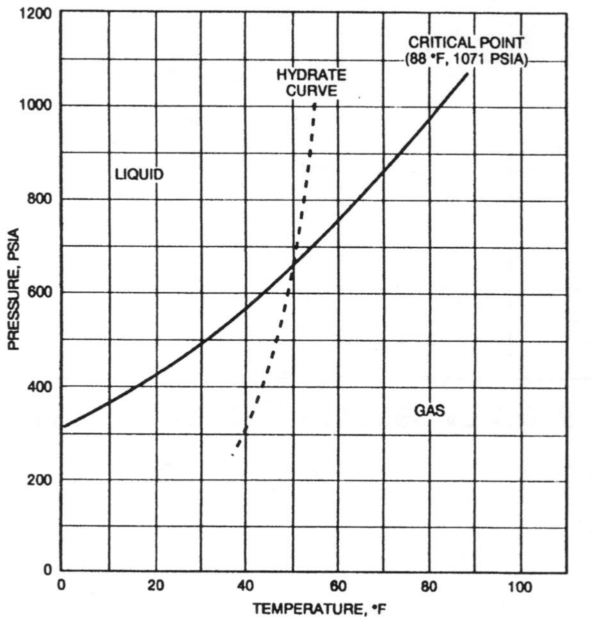

FIGURE 9

SUBMERSIBLE EFT WELLHEAD DETAIL

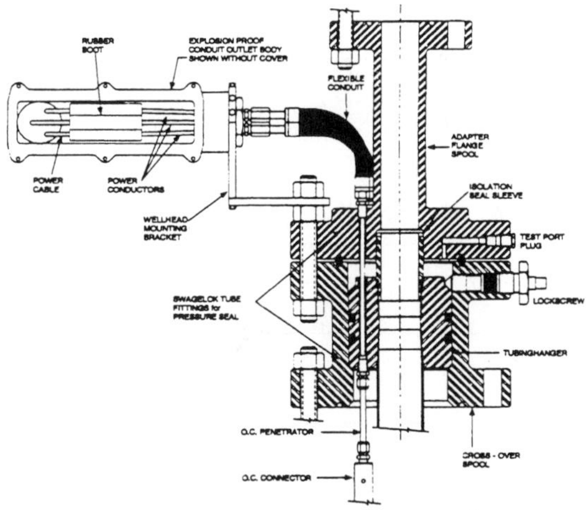

FIGURE 10

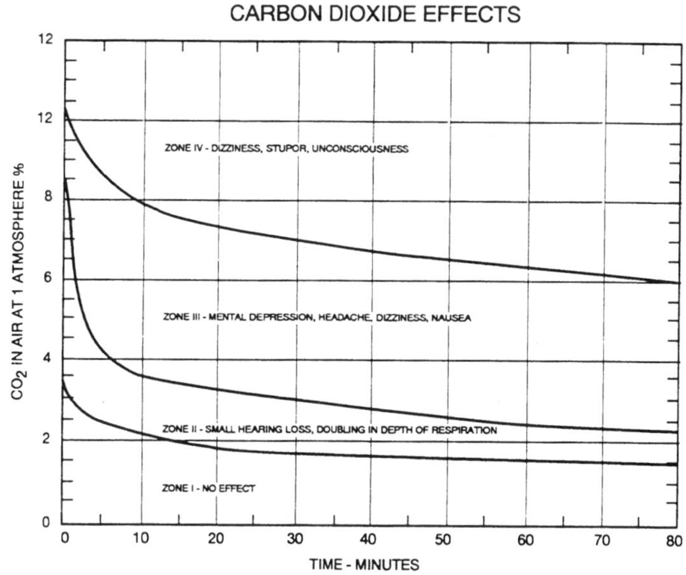

FIGURE 12

HIGH PRESSURE DOUBLE PACK STUFFING BOX

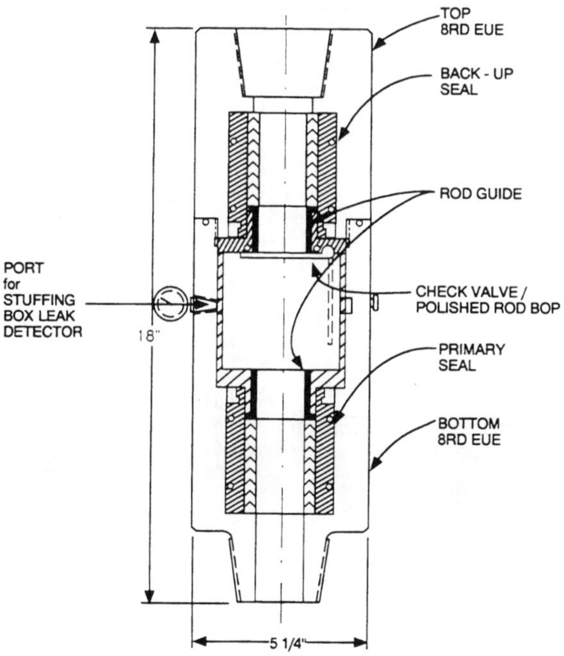

WEIGHT - 70 LBS.

FIGURE 11

RECOMMENDED POPULATED AREA WELLHEAD DESIGN CRITERIA

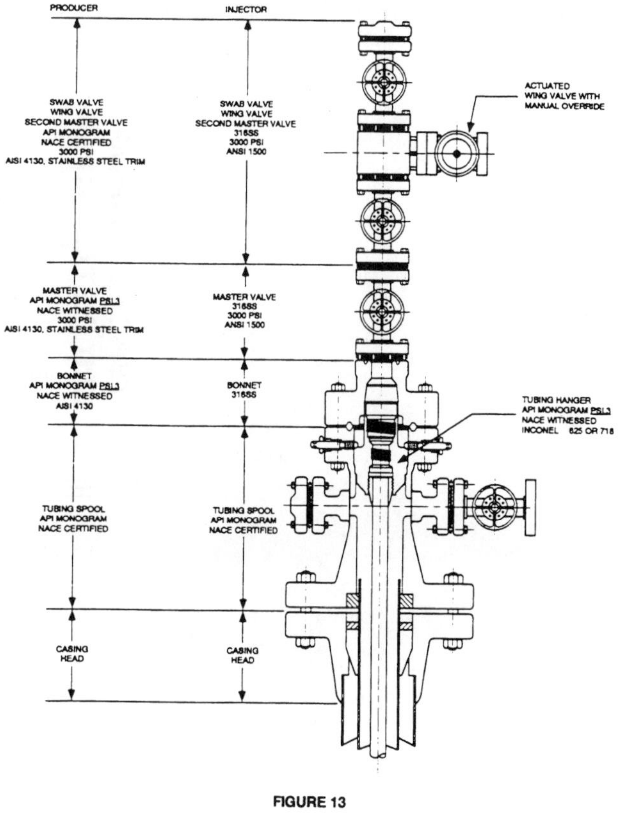

FIGURE 13

POPULATED AREA

TYPICAL PUMPING UNIT INSTALLATION

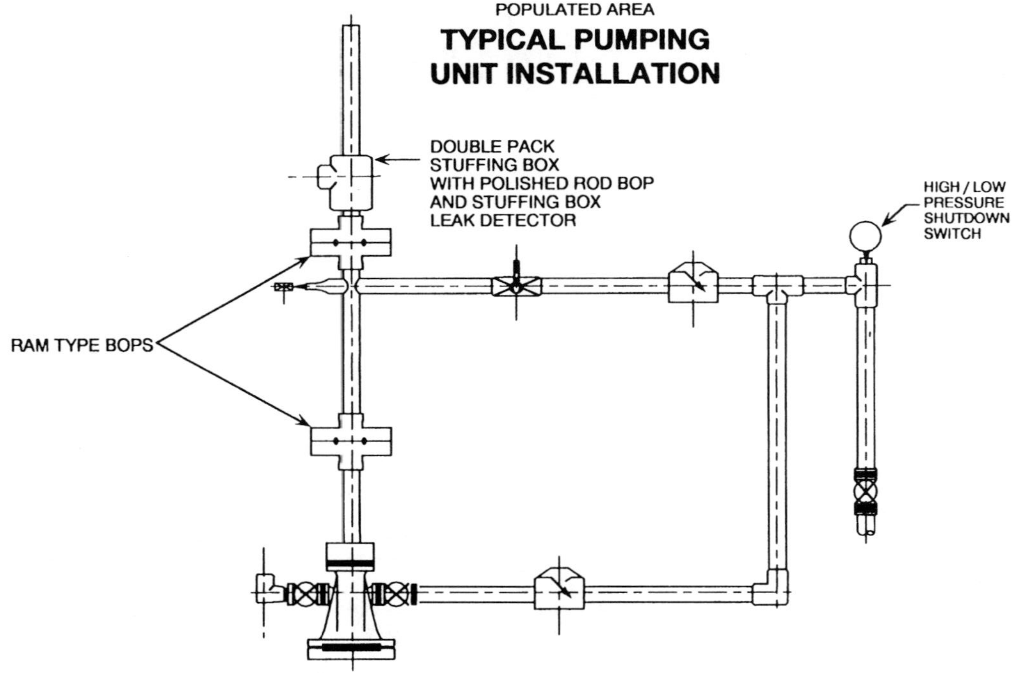

FIGURE 14

RESERVOIR BARREL CONVERSIONS for CO_2

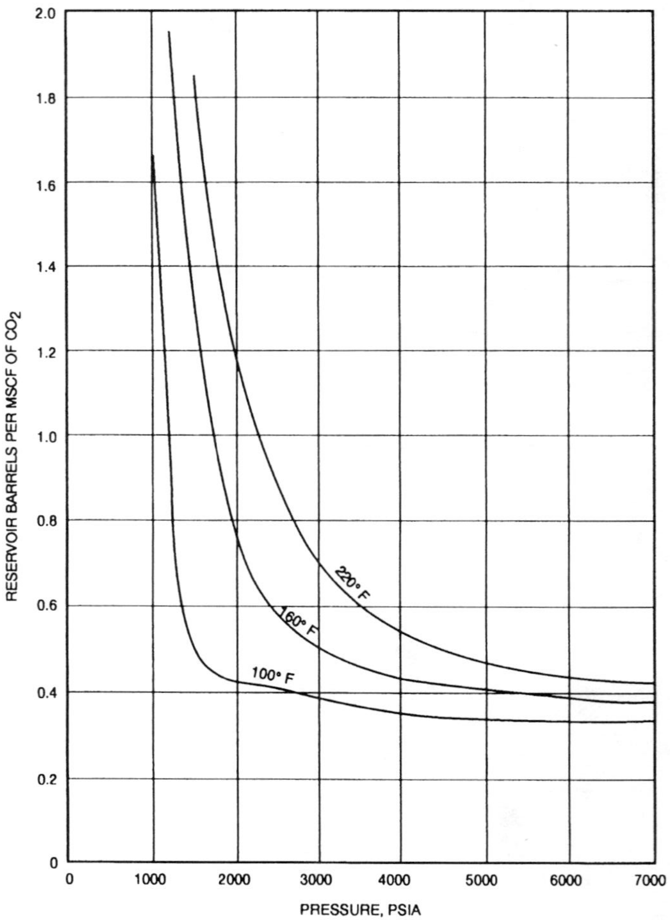

FIGURE 15

EFFECT OF LINER & SELECTIVE STIMULATION ON INJECTION PROFILE OF A DENVER UNIT INJECTION WELL

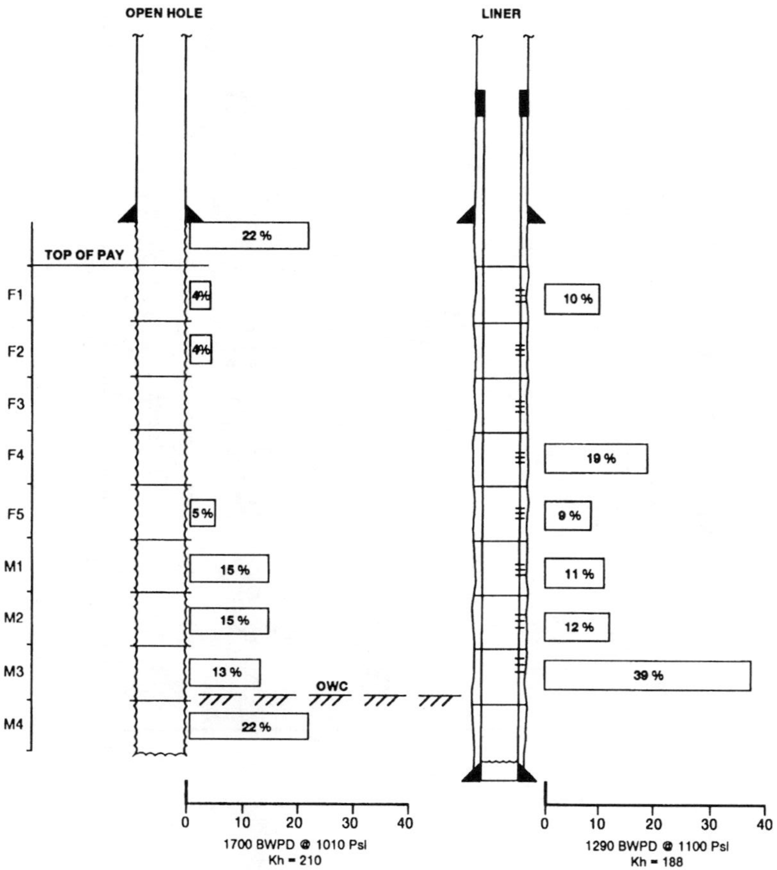

FIGURE 16

EXAMPLE CO_2 DOWNHOLE INJECTION EQUIPMENT

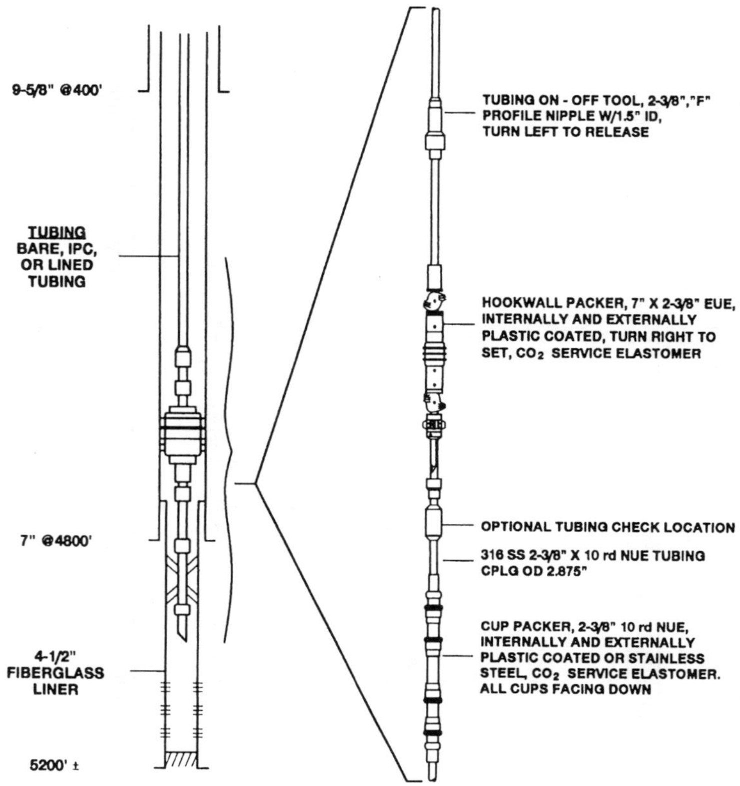

FIGURE 17

WASSON DENVER UNIT

WAG TEST AREA
(THREE WELLS ON SECOND H_2O CYCLE)

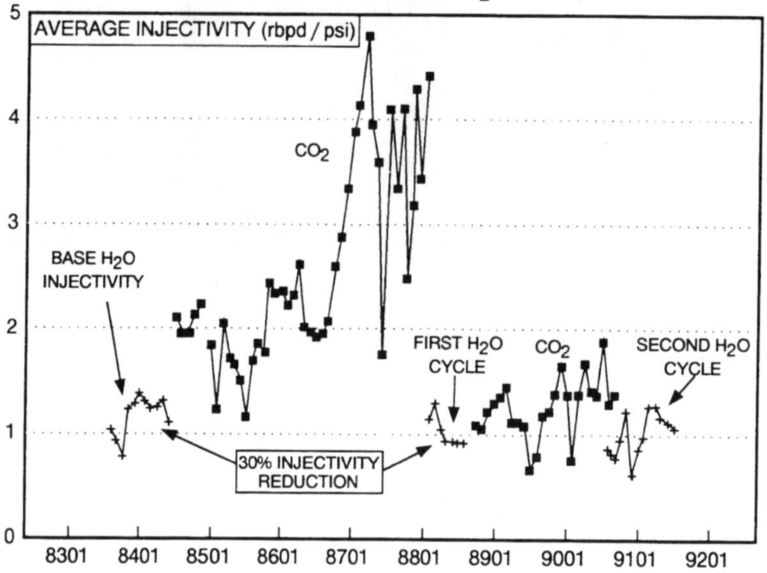

FIGURE 18

WASSON DENVER UNIT

PHASE 1 WAG TEST
(4 WELLS CURRENTLY ON CO_2)

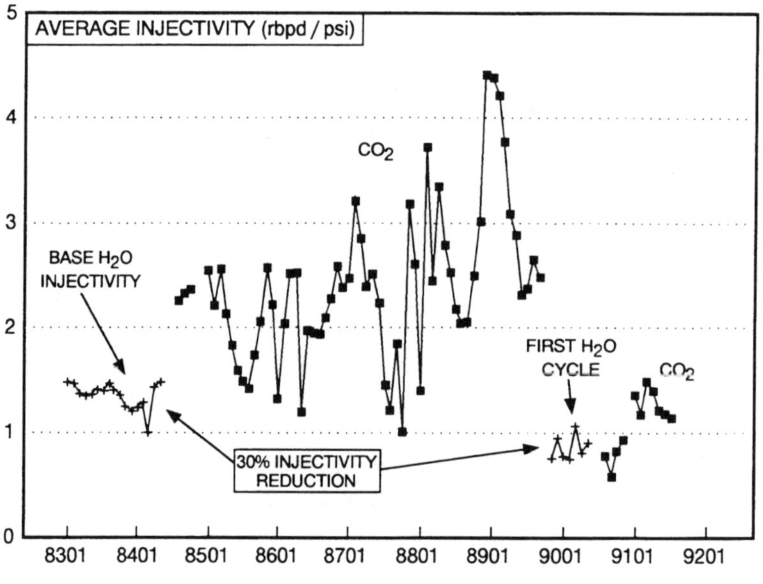

FIGURE 19

SPE
Society of Petroleum Engineers

SPE 16830

CO_2 Injection and Production Field Facilities Design Evaluation and Considerations

by L.R. Linn, Amoco Production Co.

SPE Member

This paper was prepared for presentation at the 62nd Annual Technical Conference and Exhibition of the Society of Petroleum Engineers held in Dallas, TX September 27–30, 1987.

Abstract

Many technical papers have been published on CO_2 flooding from a reservoir standpoint; but, few have ever discussed design considerations of field CO_2 production and injection facilities. This paper will present initial design and installation considerations, design criteria, and initial installation problems associated with Amoco's four West Texas CO_2 projects (Slaughter Estate Unit, Central Mallet Unit, Frazier Unit, Wasson ODC Unit). Additionally, design and operational considerations, based on the experience gained from operating the four CO_2 floods during the last two years, will be discussed for use in future CO_2 facility designs.

Equipment problems which have been experienced during operation, modifications which have been made as a result of experience, and recommended changes for future designs will be discussed in this paper. Actual case histories of operations, equipment design, and equipment problems encountered will be presented. The specific areas to be discussed will include: The CO_2 injection system material selection and layout, CO_2 injection wellhead material selection and configuration, CO_2 injection well downhole equipment, producing wellhead pressure and material requirements, flowline and fluid gathering systems, satellite and central tank battery layout and operation, and the gas collection system layout and material selections.

Special considerations will be given to design details and material specifications that are often overlooked, or not considered. The resulting potential problems and failures of these oversights will be discussed.

Illustrations at end of paper.

Introduction:

In November, 1984, Amoco Production Company commenced CO_2 injection into the Wasson ODC Unit (WODCU), the first of four CO_2 projects in West Texas. One month later, CO_2 injection was initiated in the three remaining projects, the Slaughter Estate Unit (SEU), the Central Mallet Unit (CMU), and the Frazier Unit. Figure No. 1 shows the location of these units, approximately 35 miles southwest of Lubbock, Texas.

The beginning of CO_2 injection marked the completion of over two and one-half years of engineering design and over 18 months of construction. The projects consisted of designing, specifying, and constructing more than 300 miles of 2 inch through 36 inch diameter steel, fiberglass, and polyethylene line pipe; 74 two and three phase production vessel installations; and approximately 400 CO_2 injection wellhead and bottomhole equipment changeouts. Figure No. 2 is a schematic of the CO_2 project facilities.

This paper discusses field facilities problems that were experienced during the initial construction and during the actual operations over the last two years. Furthermore, design and operational considerations and changes which have been made will be discussed for future CO_2 projects.

For the purpose of this paper, four areas of the CO_2 project facilities will be discussed: the CO_2 injection facilities, the field production facilities, the gas collection facilities, and the excess water handling facilities.

CO_2 Injection System

Piping System: CO_2 is provided to all four units from the Bravo Dome Unit through the Bravo Pipeline. The normal pipeline pressure of 1800 to 2000 psi is boosted to 2400 psi for the Slaughter Field units and reduced to approximately 1600 psi for the WODCU. CO_2 for the WODCU can also be provided from the Sheep Mountain and Cortez CO_2 Pipelines.

Each of the four projects utilized a trunkline piping system. Radial systems were investigated and were found to be more costly. The SEU, CMU, and FU systems were designed with a working pressure of 2500 psi, which required ANSI 1500# flanges and valves, and the WODCU system was designed with a 2000 psi working pressure, which required ANSI 900# flanges and valves.

Each project made use of X-grade and grade B internally bare, externally coated, carbon steel pipe, which was installed at a depth of four ft. All of the injected CO_2 is dry and is therefore noncorrosive in these bare carbon steel injection systems.

Valves used throughout the system are trunion mounted ball valves with bare carbon steel bodies and NACE trim. The design of the piping system was such as to allow all of the valves to normally operate in the open position. In places where a positive shutoff was required, a line blind was installed. This was done to minimize seat damage within the vales. Valves in the piping system which were four inches in diameter and larger were buried. Three inch and smaller valves were installed above ground. This was done based on the cost of installing each size valve. The larger valves were less expensive to bury, and the smaller valves were less expensive to install above ground.

Several minor problems have been noted in the piping system since the start-up of CO_2 injection. First, many of the below ground flanges were found to have small leaks due to the bolts not being sufficiently tight. As a result, dry ice blocks formed and pushed the connection and the surrounding earth upwards. In one instance, this phenomenon caused a weld on a four inch diameter, 45 degree ell to break. The solution to this problem is to have as many (if not all) flanged valves and tie-ins above ground as possible. This would allow for early leak detection and quick remedial action. This will cost a little more than burying the valves, but will more than pay for itself if any leaks do occur.

Another minor problem with the existing system is that many of the X-grade fittings are not readily available and can cause unreasonable delays for otherwise routine repairs. Consequently, it was necessary to order spares for those long delivery items. For future designs, it may be more economical to order pipe that would be compatible with "off the shelf" Sch. 80, Grade B fittings. The use of transition pipe spools may be another economical way of using X-grade pipe with Sch. 80 fittings. Both alternatives should be evaluated before ordering pipe.

Valving: The valves used in the CO_2 injection system were trunnion mounted ball valves with ring joint flanged connections. Generally speaking, these valves have worked very well, with only a few exceptions.

Many of the valves were ordered with standard EPDM O-rings in the bonnets and trunnions, which have failed. The failures appear to have been caused by both explosive decompression and material degradation. The decompression problem is due to the relative low density, or a low degree of crosslinking, of the elastomer (70-80 durometer), and the degradation problem is thought to be caused by the incompatibility between the oil based sealant and the EPDM material used to make the O-ring. Future valve orders for high pressure CO_2 service should specify either Buna-N or Viton 90 to 95 durometer O-rings or any other proven elastomers that are compatible with both CO_2 and oil based sealants.

Another minor problem with the CO_2 injection valves has been that of the sealant injection ports. CO_2 tends to dry the sealant in the ports, which causes the internal check to unseat and leak. To resolve this problem, sealant ports with threaded caps have been installed. With the new ports, CO_2 can not leak past the metal to metal seal formed by the caps.

Blowdown: The blowdowns installed in the CO_2 injection system piping basically consisted of a buried, flanged blowdown valve (ball valve), a blowdown stack; and a line blind (see Figure No. 3). In the normal operation of the blowdown, CO_2 pressure is exerted against the blind, and the flanged blowdown valve is open. When a blowdown is required, the blowdown valve is closed, the piping is depressurized through the bleeder valve (initially a 1/2" needle valve), the line blind is rotated open, and the blowdown valve is then opened.

Two minor design changes have been made to the CO_2 blowdowns. First, the buried, flanged valve should be changed to a weld neck valve to eliminate the buried flange connection. A flange leak on this valve at the blowdown on the 12 inch main CO_2 supply line caused the Slaughter Field projects to be shut-in while repairs were being made.

The second change that we would make in future CO_2 injection piping designs is the installation of a full port, 1/2 inch ball valve as a bleeder to depressure the blowdown stack in place of the originally designed 1/2 inch needle valve. With the needle valve, the time required to depressure the piping is about 45 minutes. With the full port ball valve installed, the time required for depressurizing the piping is reduced to 2-3 minutes. Additionally, the needle valves, in general, tend to freeze closed when venting high pressure CO_2.

Injection Wellheads: The CO_2 injection wellheads consist of two major components: the injection control skid and the Christmas tree (see

Figure No. 4). The injection control skid is composed of a bare, carbon steel, junior orifice type meter run; a solar powered control valve; and solar powered instrumentation and microprocessor.

The Christmas trees were totally reconfigured from the master valve upward. All new, 9D aluminum bronze (AlBz) and 316 stainless steel ring joint flanged components were installed to minimize leaks. Because the WAG (water-alternating-gas) method of CO_2 injection is used in these projects, the exotic materials selected were required to be suitable for resistance to corrosion resulting from the occasional mixing of CO_2 and water in the wellhead when making well switches or during a flow back period. The high pressure CO_2 injected during the CO_2 half cycle is dry and is therefore noncorrosive unless mixed with water. The use of either AlBz or 316 SS was determined for each component based on component pricing.

The original CO_2 Christmas tree configuration included installing two AlBz tees above the master valve. One tee is used for CO_2 injection, and the other is used for water injection. Each branch consisted of a check valve, a line blind, and a ball valve. To allow for wireline work without interrupting either CO_2 or water injection, an AlBz gate valve was installed on top of the second tee.

Several repeated problems with both the injection control skid and Christmas tree have been encountered. Consequently, many design changes have been made.

The only major problem with the CO_2 injection control skid has been with low power pressure and temperature transducer failures caused by lightning. To eliminate this problem, isolating flange kits were installed between the control skids and the injection system piping, which was conducting transients into the skid. Additionally, all of the wellhead instrumentation and microprocessor components were commonly grounded to avoid creating a potential between instruments.

One problem common to both the control skid and Christmas tree, as with the CO_2 injection system valves, was that the 70 durometer EPDM and buna elastomers initially installed in the check valves, ball valves, and orifice fittings were improper for high pressure CO_2 service. As a replacement, and for future designs, 90 durometer Buna or Viton seals would be considered.

The problems encountered with the Christmas trees have required more extensive changes than those required for the control skid. First, the double tee arrangement resulted in a very tall and unstable wellhead. Special tie-down braces were later installed to provide a more rigid structure. Newer wellheads have incorporated a cross in place of the two tees. As a result of the lower profile, the wellheads are considerably more stable.

Early in the project, the AlBz tees, along with the AlBz check valves and the 316 SS line blinds, were found to be very susceptible to casting defects. The components passed a standard hydrotest but failed in high pressure gas service. The initial specifications called for an industry standard casting (a level 3 casting), but were later changed to a more stringent level 2 casting, which requires radiographic inspection.

Another problem that has developed over the last several years involved the AlBz clappers in the check valves. These clappers have worn to the point of being inoperable. In general, the AlBz in this service has shown very poor wear resistance. These checks are being replaced with 316SS lift disc valves.

Finally, new casing valves were found to be required. Most of the pre-CO_2 valves were equipped with improper elastomers and were unable to withstand the high pressures associated with CO_2. Additionally, many of the 400 CO_2 injection wells were only equipped with one casing valve, and as discovered on the first blowout, two valves are required for proper well control. One valve is normally used as a blowdown valve while the other valve is being tied onto a kill truck. Additionally, if either of the valves become inoperable, the remaining valve can be used.

<u>Bottomhole CO_2 Injection Well Equipment</u>: The only new bottomhole equipment installed in each CO_2 injection well consisted of plastic coated tubing and either a new or a reconditioned tension set type packer with 85-90 durometer perioxide cured nitrile CO_2 compatible elastomers.

The new tubing installed incorporated several design changes to the pin ends. First, the standard API sharp end, as shown in Figure No. 5, was rounded to allow for a smooth surface which could easily be plastic coated. The second change incorporated the extension of the standard upset length from four inches to six inches. This was done to allow the tubing to be rethreaded in the future if corrosion attack of the pin end persists.

Both thick film and thin film phenolic "plastic" coatings were used in the 350 miles of tubing installed. However, the thick film generally provides more consistently acceptable coating jobs, which were required to be 100 percent holiday free. With both coatings, quality control is essential. Our experience has shown that having a third party inspector or company inspector at the coating company to ensure an acceptable coating job is money well spent.

Several problems have occurred with the injection well downhole equipment since CO_2 injection commenced. First, the tension type packers that were initially installed have been found to be less than satisfactory. The tension packers have a depth limitation of about 5000 ft., and at the depth of these wells (4900 ft.), the packers are very sensitive to additional stresses caused by rapid temperature changes, as experienced during a water to CO_2 switch. In several instances, the packer shear pins sheared, which caused the packers to unseat and created blowouts. Additionally, on/off tools, which have proven to be very beneficial, can not be used with this packer. Currently, tension packers are being replaced with neutral set packers as wells are worked over, or at any other time the tubing is removed from the wellbore.

In conjunction with changing the packers, profile nipples and on/off tools are being installed. With this new equipment, the wells are much easier to control, since in most cases, it is possible to control the well by setting a (wireline) profile plug before pulling or unlatching the tubing. Also, with neutral set packers, the circulation of the packer fluids and displacement of trapped, corrosive fluids is much more effective. Corrosive wellbore fluids, which flow into the annular space above the packer while a packer is being set, can be displaced by setting a profile plug, unlatching the on/off tool, and circulating inhibited water down the tubing. However, since on/off tools could not be used with tension packers, this procedure of displacing the trapped, corrosive fluids was not performed. As a result, several tubing failures due to external corrosion on the bottom tubing joints have occurred in wells equipped with these tension-set packers. With tension packers, costly, weighted mud must also be used in killing the well.

Field Production Facilities

Producing Well: No changes were made to either the artificial lift equipment or the wellhead equipment during the implementation of the CO_2 project facilities. Although modifications were anticipated to be required, none were initially made due to the potentially large investment and the uncertainty of design parameters for shut-in pressures and pumping conditions.

Modifications to the wellheads are presently being implemented on CO_2 breakthrough wells. They include: installing Schedule 80 nipples, 2000 psi rated ball valves on the tubing head, and new elastomers in both the secondary seal and tubing slips seal. For beam pumped wells, the blowout preventer elastomers are also changed. For electrical submersible pumps, the tubing valve is being replaced with a 2000 psi rated gate valve. Figure No. 6 illustrates all of the above changes.

Once completed, the modifications will allow for a maximum shut-in pressure of approximately 2000 psi. It should be noted that the majority of the production wells in the four CO_2 projects have casing which is more than 40 years of age and may not be capable of withstanding pressures in excess of 1000 psi. Consequently, the potential for casing repairs is expected to increase during the mature stages of the CO_2 floods.

Many equipment changes and tests are being conducted with respect to all artificial lift equipment. No significant conclusions about optimum artificial lift for CO_2 flooding can be made at this point in time. Undoubtedly, artificial lift equipment requirements for CO_2 floods will be the subject of numerous technical papers in the future.

Production Flowlines and Fluid Gathering Lines: The majority of the pre-CO_2 flood fluid gathering lines and individual well production flowlines were replaced with 150 psi cyclic pressure rated fiberglass pipe. The design operating pressure for the production flowlines is 50-55 psi, which is also the maximum desired back pressure at the wellheads. The fluid gathering lines, on the other hand, are sized to have a maximum operating pressure of 125 psi (with the exception of the remote header fluid gathering lines, which will operate at a maximum of 30-35 psi). In all instances, the replacement lines were installed to increase line capacity, and not for corrosion resistance. All totalled, more than 91 miles of 3, 4, and 6 inch diameter fiberglass production flowlines and more than 25 miles of 4 inch through 14 inch diameter fluid gathering lines were replaced.

No equipment failures have occurred in the fiberglass piping since the original pipeline construction was completed. However, due to the poor quality control that was discovered during the initial installations, stringent manufacturing specifications and inspections should always be included in major construction projects of this nature.

Remote Production Headers, Satellite Batteries, and Central Tank Batteries: The remote header, satellite battery, and central tank battery (CTB) installations are among the most important of the CO_2 flood production facilities.

The function of the remote headers and satellite battery headers is to collect, or centralize, production from numerous wells and to provide individual well test facilities for oil, water, hydrocarbon gas, and CO_2 gas production. Operating pressures for the remote headers are in the range of 30 to 35 psi, while the satellite batteries' headers operate from 25 to 30 psi.

All of the remote and satellite battery headers in the projects (with the exception of the WODCU) existed prior to the initiation of the CO_2 flood. The WODCU facilities are all new, and larger, facilities. These were required in order to handle the expected high gas production rates (1 to 5 MMSCFD per well).

The only modifications made to the testing facilities were the inclusion of an automated, multiple orifice tube metering skid and an infrared CO_2 analyzer, which measures the percentage of CO_2 gas in the total volume of gas.

The main function of the satellite batteries is to remove the majority of the hydrocarbon and CO_2 gas from the produced fluids and to pump the remaining produced fluids through the fluid gathering system to the CTB. As a backup to the pumps, an automatic low pressure (70 psi) dump system was incorporated into each of the 46 new satellite battery production separators, which would allow for operation during a power outage or a pump malfunction. Additionally, a total production separator bypass system was installed at each satellite battery.

All of the production stream, dump system, bypass system, and gas piping was bare carbon steel. The theory used in this original design was that the corrosion inhibitor flushed into the production wells would be produced and would create a protective film in all of these surface facilities. Due to the reduced fluid velocity and increased residence time in the production separators, sludge and solids can accumulate and impair the creation and regeneration of a protective film. An internal plastic coating was therefore applied.

The only modification to any of the eight central tank batteries in the four units was the addition of a 10 ft. x 30 ft. three phase production separator, which serves to provide for a second stage of gas separation during normal operation and as a backup if one of the satellite batteries is operating in a bypass or pressurized dump mode.

Many minor problems have occurred with the CTBs, the satellite batteries, and the gas testing facilities since the production facilities were put into service more than three years ago. First, numerous piping failures caused by corrosion have been reported, especially in the Wasson ODC Unit satellite and central tank battery facilities. All of these failures have been in locations where stagnant produced fluid has lain dormant in the bare steel for several months. These locations include the test side of each well header leg, the production separator bypasses, and the emergency dump valves and piping.

The below ground steel piping has been replaced with either fiberglass or coated steel piping, and the above ground valves and piping have been replaced with coated steel components. Future designs would include fiberglass piping and coated steel components wherever possible. The cost to repair several failures would more than offset the initial cost of installing fiberglass pipe or coating the steel components.

Finally, with respect to the gas measurement skids, only one failure and one operational change have occurred. The only failure occurred when a pinhole leak, due to corrosion, developed in a weld. The one major operational change made was due to the relatively high (and unexpected) volume of condensate that was found to accumulate in the measurement skids. To eliminate this problem, drip pots were installed on each skid. New skids being installed are fitted with a plastic coated drip header and an automatic blowdown system to drain the gas skid header (see Drawing No. 7).

Gas Collection System

Piping System: The gas collection systems begin at the Mallet and Wasson CO_2 Removal Plants at an inlet pressure of 5 psig. The gas gathering systems are polyethylene trunkline systems, which range in size from 10 inches to 36 inches in diameter and span outward to the satellite and central tank batteries to collect the produced gas. The systems were designed to have the satellites and the CTBs operate at a maximum of 25 psig during peak CO_2 gas production periods. Due to the burial load requirements, an SDR 21 pipe was used, which yields a 48 psi working pressure (based on rating factors). It should be noted that fiberglass pipe was bid and evaluated for installation, but was found to be cost prohibitive. The evaluation included installation cost, pipe and fitting cost, and the cost of gas loss associated with the permeability of the polyethylene pipe. Gas loss, although possible, is generally insignificant.

The fittings used in the piping system were all prefabricated, SDR 17, polyethylene. With the use of these fittings, special chamfers had to be made on fittings adjacent to the high performance butterfly valves in order for the valve disks to have ample clearance.

The valves used were buried, carbon steel valves with 316 SS disks and stems. As a special design feature, the valves were oriented so that the stems and disks were mounted in the horizonal position. This was done to allow debris to flow through the valve without getting into the bearing areas (see Figure No. 8).

One major problem that has persisted during construction and during initial operations has been with the prefabricated polyethylene fittings. Seams on fittings 24 inches and larger in diameter have frequently split. Repairs thus far have been performed by using plastic coated steel fittings. Upon investigation, it was discovered that the installed cost of steel fittings was actually less than the polyethylene counterparts. Additionally, the delivery of the steel fittings is generally a fraction of the time required for the polyethylene fittings. Consequently, steel fittings, especially in the larger sizes, will be evaluated for use in future projects.

Drips: The drips installed were fabricated from 3 ft. x 30 ft. fiberglass pipe sections and were strategically placed on the upslope sides of the elevation changes directly below the polyethylene pipe. At this point in the line the slug is at its lowest velocity, and is therefore less likely to flow over the drip without being caught.

Several failures have been experienced in the drips. Two of the failures were due to settling of the polyethylene gas gathering line pipe, which caused the fiberglass flanged nozzle (located on the top of the drip) to be pushed through the surrounding fiberglass drip body. The third failure occurred in a fiberglass drip body seam. An inspection of the failure showed that the seam had a wall thickness which was less than specified. However, the settling or improper bedding of the drip also contributed to the failure. The drip appeared to be in a bind at the location of the seam.

One alternative to the current design, which will reduce these problems, would be to install an internally and externally coated steel drip located two to three feet off center from, but parallel to and deeper than, the polyethylene pipeline. The connection between the top of the drip and the bottom of the pipe could be made using a non-rigid connection (i.e. hose). Although this design has not been tested, we feel that it would work satisfactorily.

Lease Custody Meter: The new lease metering facilities incorporate the use of multiple orifice type meter runs. They were sized to work individually at lower produced gas rates initially and eventually in unison at higher gas rates.

No failures have occurred in any of the installations; however, one major operational change has occurred. Gas samples prior to CO_2 flooding were taken quarterly and showed very little, if any, compositional changes from period to period. But with the breakthrough of CO_2 and heavier hydrocarbon gas, compositional changes are occurring more frequently. Consequently, gas samples may be required more often. Additionally, several continuous gas samplers are being tested for future use.

Saltwater Disposal (SWD) System

The SWD systems constructed in the four CO_2 floods utilize either low pressure transfer of produced water (at 100 psi) or high pressure disposal (at 1000 psi with a design maximum of 1800 psi) into disposal wells. In the CMU, a low pressure system was used to transfer excess produced water to another Amoco operated property requiring makeup injection water for waterflooding, while the remaining three units made use of high pressure systems which transported pressurized water from the waterflood injection stations to deep (12,000 ft.) disposal wells by means of plastic coated high pressure trunkline systems.

Numerous operational problems, equipment failures, and design changes have occurred in the high pressure WODCU, SEU, and Frazier Unit SWD systems. However, the CMU low pressure transfer system has worked extremely well. Therefore, the following will apply only to the high pressure systems.

Piping: The piping system functions to transport excess produced water from the existing waterflood injection stations to the disposal wells. In all cases, a trunkline layout was less expensive than a radial layout. Consequently, trunkline systems were installed.

The piping used was Sch. 60 and Sch. 80, Grade B, internally plastic coated and was connected with a proprietary interference fit coupling. Additionally, ANSI 900# ring joint flanges were used at the tie-ins and valve locations.

Numerous corrosion caused failures have occurred in the high pressure piping. Two of these failures were caused when tightening of RTJ flange bolts caused the metal ring gasket to crack the internal plastic coating. We believe that the remaining failures were due to a set screw being loose in the machine used to hydraulically press the six inch diameter pipe into the six inch interference fit couplings. Our conclusion as to the cause of these failures is based on the failure locations, which were 18 inches from the couplings and concentrated in one area of the piping system. Also, the manufacturer had one of the failures analyzed and found a partial imprint of the set screw.

Since these original SWD systems were installed, new, tougher coatings and new internal pipe liners have become available. Future high pressure SWD system design evaluations will consider these new products. Additionally, ANSI 900# or 1500# raised face flanges, instead of the RTJ flanges, should be used for internally plastic coated systems.

As with the CO_2 injection system flanged connections, the SWD system flanged connections were buried and have been found to leak. In several of the units, the valves have been excavated and raised above ground level. Future systems will be designed with above ground valves and flanged connections.

The valves used in the SWD system were remanufactured, carbon steel, trunnion mounted ball valves with 316 SS trim. The remanufactured valves were used due to the lack of availability of new valves at the time the systems were designed and constructed. Many of these valves were found to have improper trim components and undersized gear reducers. As a result, many of them were sent back to the supplier for repairs. Most of the reported problems are associated with the failure of the valves to produce a positive seal when closed. Other problems with the valves, such as gear reducers breaking, grease fittings breaking, and lubrication line leaks, have also been reported.

It is recommended that reconditioned valves not be purchased for critical service projects. Inspecting reconditioned valves, such as these, before purchase is not of appreciable benefit unless the inspection is performed by an original manufacturer's representative who knows the acceptable tolerances and other performance and manufacturing criteria of the valves.

System Operations: Two major design changes have been instituted on the three high pressure disposal systems. The first, which applied to all three systems, is a new method of maintaining suction tank levels at the water injection stations (see Figure No. 9). The original design incorporated the use of a low tank level switch with a SWD control valve which regulated the volume of water pumped from the water station header to the SWD system. When the low level switch was activated, the SWD control valve would close. By regulating the volume of water transferred to the SWD system, this valve also controlled the injection header pressure. When this valve closed, as a result of low suction tank level, it caused the water injection pump bypass valves to open (in order to maintain the set header pressure) and to circulate water back to the suction tanks. This design did function, but the header pressure fluctuated, because the movement of the bypass valves from the closed to the open position was appreciably slower than the movement of the SWD system control valve from the open to the closed position. As a result, it was decided that the SWD system control valve should only control header pressure and not tank level. A bypass loop from the SWD line back to the suction tanks was installed downstream of the control valve, which, in this location, would not affect the header pressure. An actuated ball valve, which would open when the low level switch was energized, was installed on this throwback line. With this new design, the suction tanks can be filled using the water in the disposal system when required.

The second operational change, which was not part of the initial design, is that the SEU and Frazier Units' SWD Systems have been modified to dispose water at low pressure (150 psi). The modifications included the installation of at least one centrifugal pump at each water injection station, new taps into the SWD system, and a high level switch (located on the suction tanks) which activates the pumps.

This option of water disposal at low pressure was not a part of the original design, since wellhead pressures of the disposal wells in the SEU and Frazier Units were expected to be about 1600 psi for peak disposal rates. However, due to a more phased implementation schedule, the actual SWD rates and pressures required are somewhat lower than the original design criteria.

The result of disposing at low pressure, rather than at high pressure, is that significantly less electrical horsepower is required. Additionally, it is a safer system to operate. Future high pressure saltwater disposal systems which are connected to the water injection stations will be designed to operate on low pressure when it is possible.

Conclusion

In conclusion, no insurmountable operational problems or equipment failures have occurred in the four CO_2 projects since the commencement of CO_2 injection in late 1984. Problems have been encountered, but none which could not be addressed quickly with prompt, cooperative engineering and operations efforts. Furthermore, design modifications and recommendations have been implemented that will help reduce future problems.

Overall, the original designs have worked exceptionally well. This is due to the excellent coordination and communication that has existed between all offices and groups involved during the initial design stages, construction phases, and operational stages of the CO_2 floods.

Acknowledgements

I am grateful to Amoco Production Company management for allowing me to be involved in the original design, implementation, and operation of the West Texas CO_2 Floods and for permission to publish this paper. Additionally, many thanks to all of the engineers involved in the original design and implementation of the CO_2 floods.

Amoco's CO_2 Projects

FIGURE NO. 1

AMOCO'S CO_2 PROJECTS

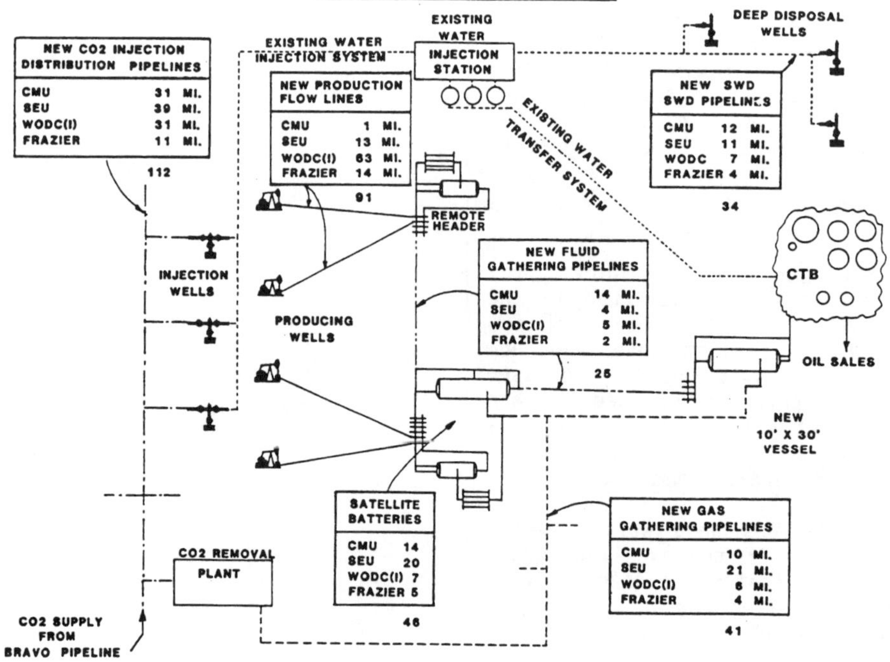

FIGURE NO. 2

CO2 TRUNKLINE BLOWDOWN STATION

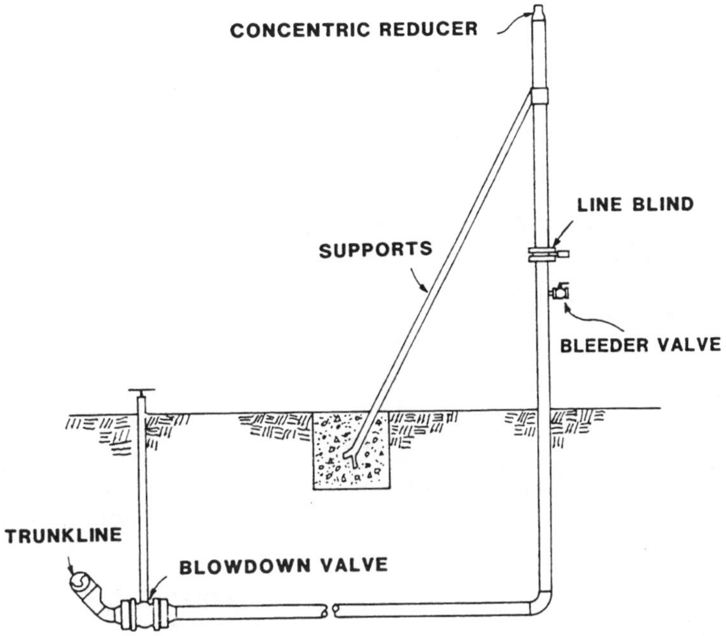

FIGURE NO. 3

CO2 INJECTION WELLHEAD

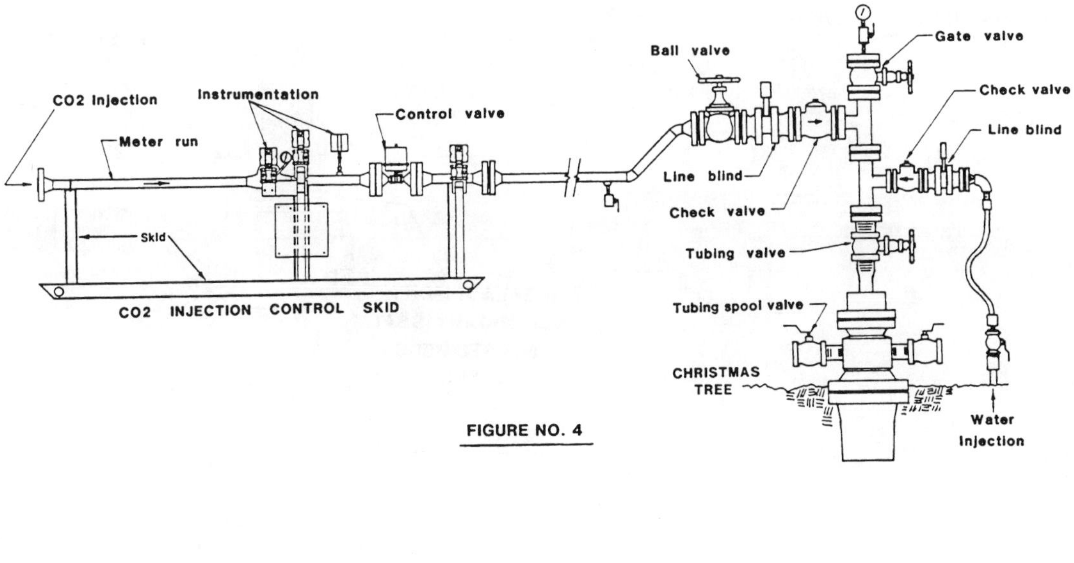

FIGURE NO. 4

MODIFIED API TUBING ENDS

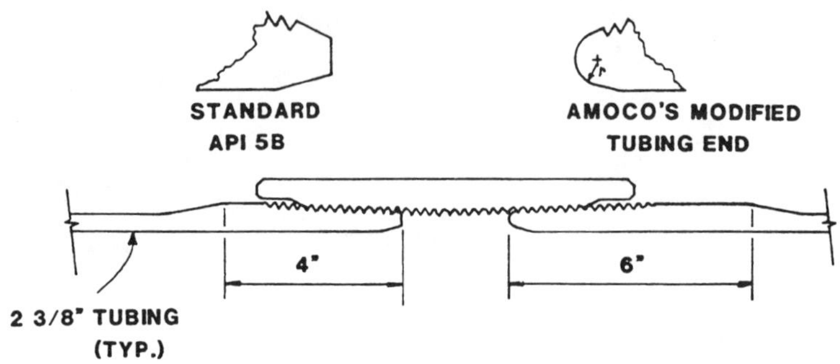

FIGURE NO. 5

CO2 PRODUCTION WELLHEAD MODIFICATIONS

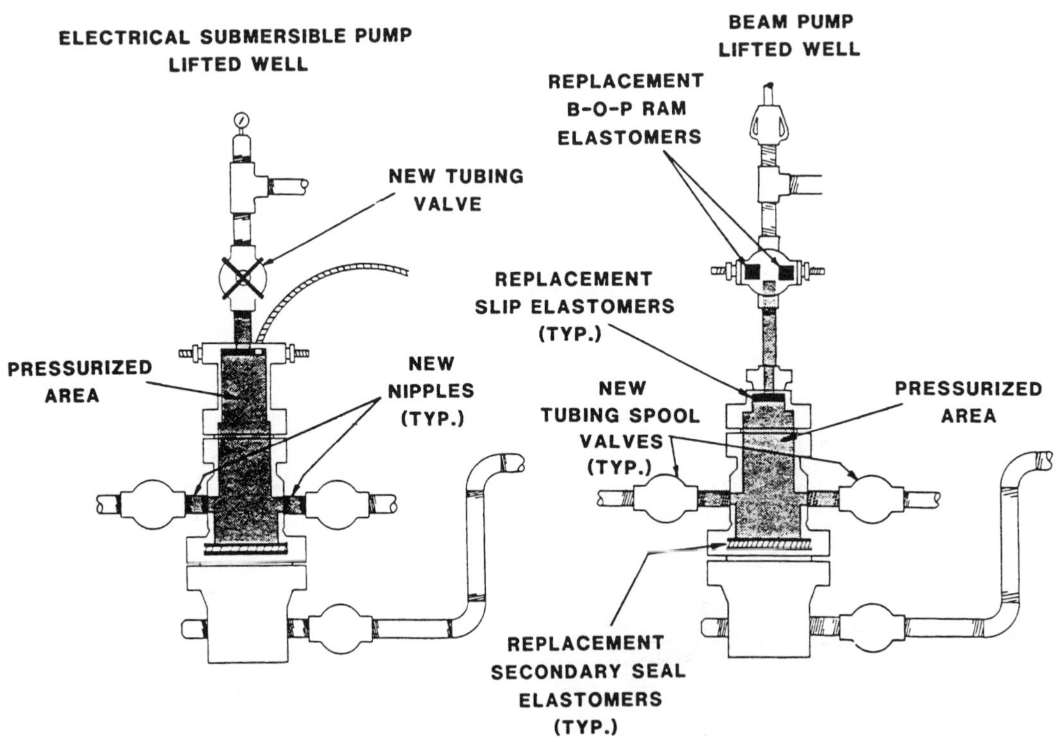

FIGURE NO. 6

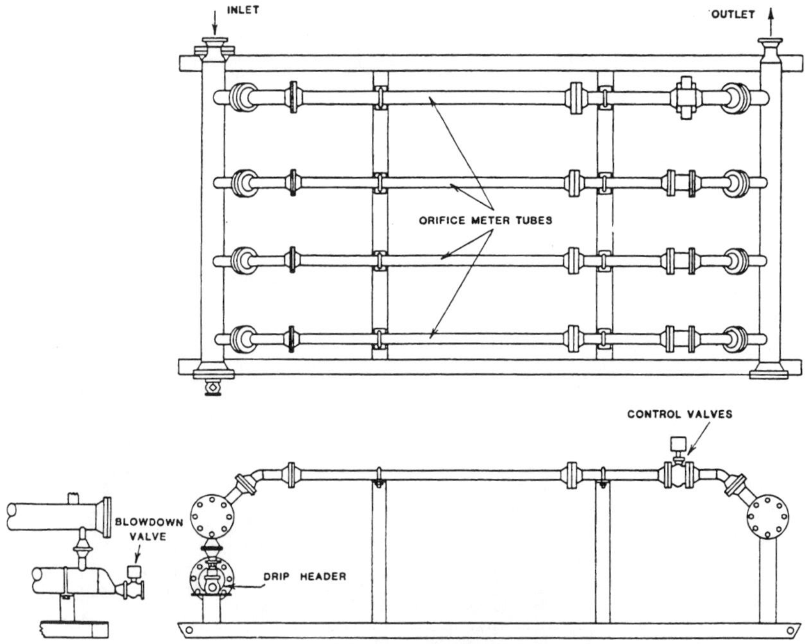

GAS MEASUREMENT SKID

FIGURE NO. 7

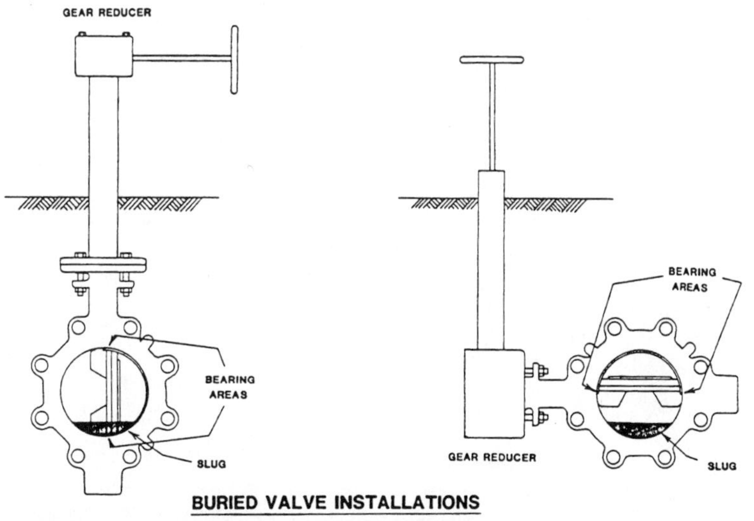

BURIED VALVE INSTALLATIONS

FIGURE NO. 8

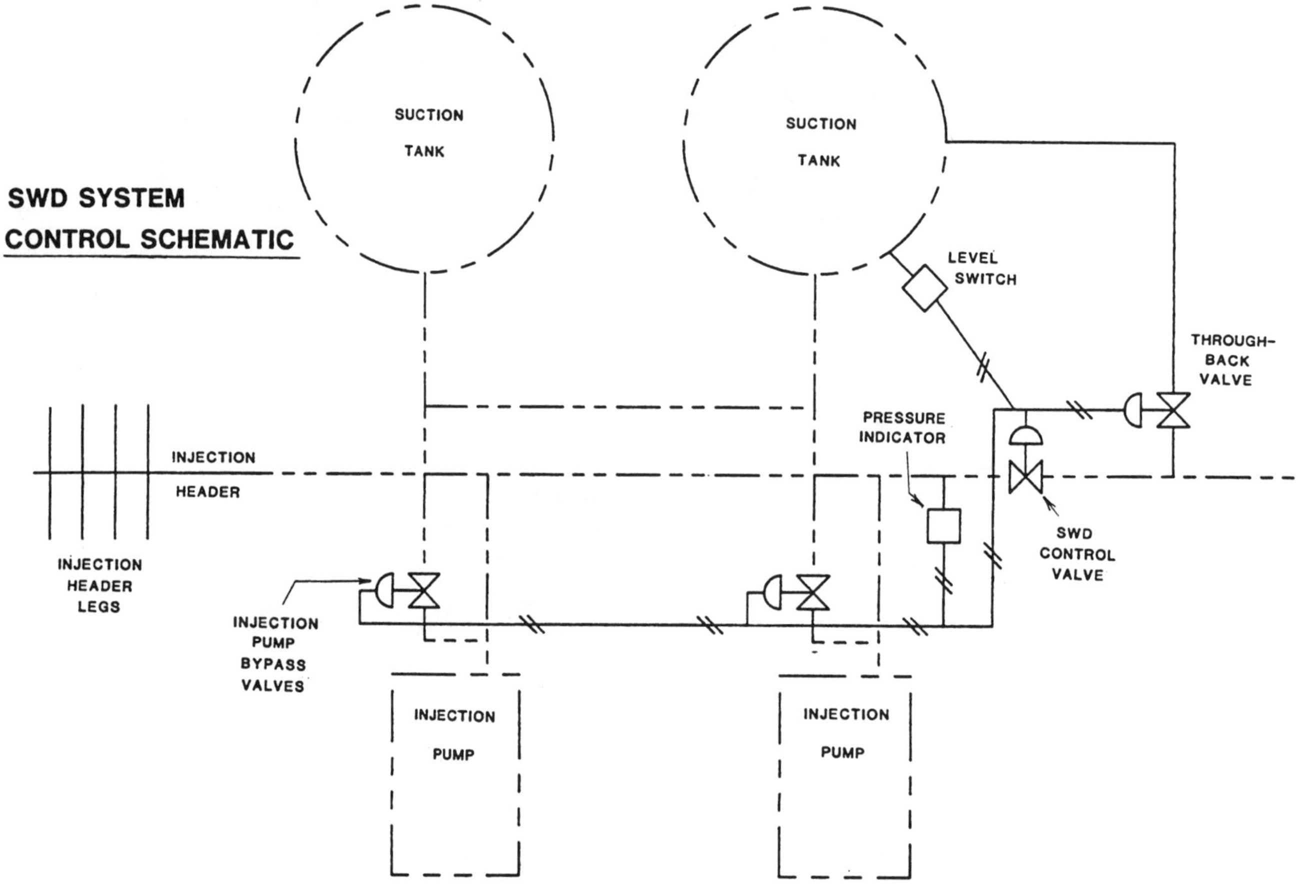
SWD SYSTEM
CONTROL SCHEMATIC
SUCTION
TANK
SUCTION
TANK
LEVEL
SWITCH
THROUGH-
BACK
VALVE
PRESSURE
INDICATOR
INJECTION
HEADER
INJECTION
HEADER
LEGS
INJECTION
PUMP
BYPASS
VALVES
SWD
CONTROL
VALVE
INJECTION
PUMP
INJECTION
PUMP

FIGURE NO. 9

Field-Scale CO_2-Flood Simulations and Their Impact on the Performance of the Wasson Denver Unit

C-F. Hsu,* SPE, Shell Western E&P Inc.; **J.I. Morell,*** SPE, Shell Development Co.; and **A.H. Falls,**** SPE, Shell Western E&P Inc.

Summary

Recent advances in software and hardware technology have made possible the development of field-scale, fully compositional CO_2-flood simulations capable of capturing areal variations in performance on an individual well basis. This paper first describes the general methodology developed, including key elements used to construct such large models, followed by the validation of this approach. Two modeling studies conducted on the Wasson Denver Unit are then presented to highlight several of the significant outcomes realized so far: high-grading the profitability of new CO_2 projects, pinpointing best-well candidates to return to production, identifying infill and horizontal drilling locations, and identifying and quantifying injectant losses.

Introduction

The Denver Unit is located in Yoakum and Gaines County, Texas. The ongoing CO_2 project in this unit is one of the world's largest. Each day, more than 400 MMscf of CO_2 are injected into the 1,500 wells within the 21,000-acre project area, while 38,000 bbl of oil is produced. Since mid-1984, incremental oil recovery attributable to this tertiary process has exceeded 60 million bbl. As of January 1994, CO_2 developments of the oil column (OC), the historical producing interval of the San Andres formation, have been relatively complete; the largest untapped reserves within the Unit are the 650 million bbl of paleo residual oil accumulated below the OC.

As part of the flood-management effort, field-scale simulations have been used to provide well-specific forecasts (a key piece of information that can help uncover flood-improvement opportunities and high-grade new CO_2 projects). An equation-of-state (EOS), compositional simulator has been chosen in the study because it can provide more reliable predictions of CO_2/oil phase behavior than a black-oil-type model.[1] Recent advances in simulation technology made this effort possible.[2] A study by Hill *et al.*[3] has indicated that the simulator used in the present study can provide reasonable predictions of the production response of a CO_2 pilot in the South Welch Unit.

In this paper, basic building elements of the modeling methodology and the process of validating the models are described first. Subsequently, key results of two field-scale simulation studies are summarized to highlight the impact of simulations on the field performance. The first case illustrates how we used a "transition-zone (TZ) model" to design and high-grade CO_2-flood expansions to recover residual oil in the TZ. The second case demonstrates the use of a "Battery 1 model" to improve the ongoing CO_2 developments.

Background

The Denver Unit is the largest unit in the Wasson field. The unit produces oil from the San Andres formation, a mid-Permian-aged dolomite located at subsea depths ranging from approximately −1,250 to −2,050 ft (see **Fig. 1** for the top structure of the San Andres formation). The Denver Unit initially contained more than 2 billion bbl of oil in the OC, which is the interval of the San Andres hydrocarbon (HC) accumulation above the producing oil/water contact (OWC). In addition, the San Andres formation contains more than 650 million bbl of oil in a TZ (sometimes referred to as a paleo residual oil zone), which is the interval between the OWC and the base of zone (BOZO). Historically, the OC was the only target for the primary, waterflood, and CO_2 developments. Results of a TZ pilot, however, indicated that the residual oil in the TZ can be recovered by a CO_2-miscible process.

The Wasson San Andres field contains a primary gas cap. The subsea depth of the initial gas/oil contact (GOC) was estimated to be −1,325 ft when the field was unitized in 1964. Because the San Andres formation in the Denver Unit is stratigraphically highest among all units operating the Wasson field (see Fig. 1), more than 90% of the gas cap resides within the western portion of the Denver Unit. The presence of gas cap has a significant impact on the CO_2-flood performance.

Modeling Methodology

Reservoir Characterization and Geologic Model. As shown in **Fig. 2,** the San Andres formation can be stratigraphically divided into two major intervals: "first porosity" and "main pay." The deeper main-pay interval consists of dolomitized open marine packstones and grainstones and possesses better reservoir properties than the first-porosity interval. The first-porosity interval was deposited in a more restricted marine intertidal environment. It has a finer crystalline matrix and is less continuous than the main pay. Supratidal mudstones and anhydrites that lie above the first-porosity interval form the seal for the hydrocarbon accumulation. On the basis of the character of gamma ray and sonic logs, we subdivide the first porosity interval into five intervals (F_1 through F_5) and the main pay into six zones (M_1 through M_6).

Simulation layers were constructed to represent reservoir structure and zonation because they generally resemble actual flow units. In addition, some geological zones were subdivided into more simulation layers because of the large variation in reservoir properties within them. For example, the upper part of Zone M3 in Fig. 2 was modeled as a separate simulation layer in the TZ model because it was much more permeable than the rest of the zone. The M1 zone was also subdivided into two layers because of the high-permeability contrast.

Well logs of both the San Andres and deeper Clearfork wells were included in the gridding process to ensure good descriptions of the TZ. For each of these wells, layer depths and thicknesses were determined from marker tops, while the average porosity in each layer was calculated by averaging actual sonic log porosities between markers. Permeabilities were calculated from ϕ/k correlations, derived *a priori* from core measurements and then averaged between markers. Areal variations of thickness, porosity, and permeability in the simulation layers were generated by a $1/r^2$ weighted average of the values at several closest control wells. (However, no interpolations were made across the layers.) Wells outside the simulated area were also used in this process to ensure that properties did not change inappropriately at the boundaries of the model. The gridded permeability distribution, as shown in **Fig. 3,** revealed the presence of an extremely high-permeability region near the center of the TZ sweetspot area where a TZ pilot was conducted. Pilot core data confirmed the presence of very permeable, highly dolomitized grainstones in the upper M3 zone. This special rock type was nicknamed "Bertha."

Effective vertical permeability in a reservoir simulation model depends on gridblock size. Because the gridblocks in the model are relatively thick, typically more than 10 ft, a relatively small k_V/k_H ratio of 0.025 was used (over the entire area) to represent the small

*Now with Shell E&P Technology Co.
**Now with BDM Petroleum Technologies

Original SPE manuscript received for review 15 February 1995. Revised manuscript received 2 December 1996. Paper peer approved 9 December 1996. Paper (SPE 29116) first presented at the 1995 SPE Symposium on Reservoir Simulation held in San Antonio, Texas, 12–15 February.

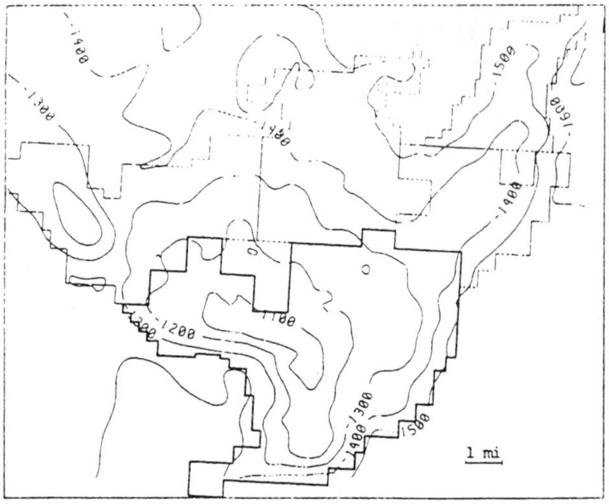

Fig. 1—Top structure of the Wasson San Andres formation and boundaries of the Denver Unit.

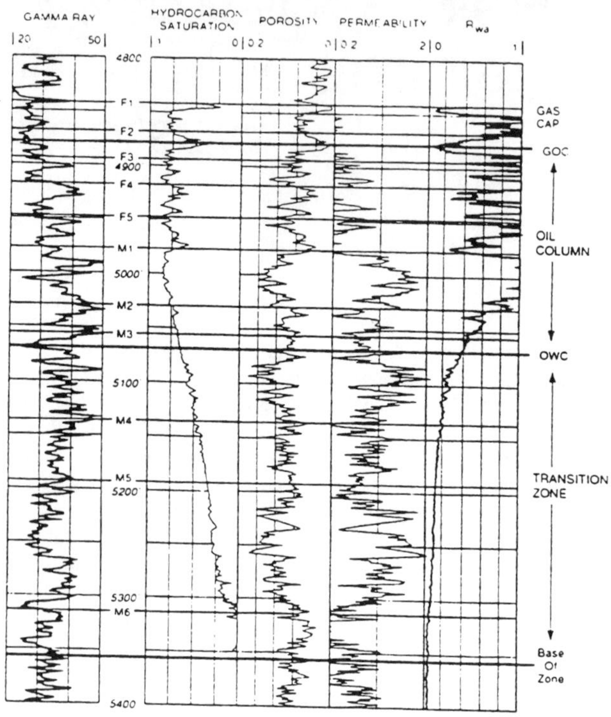

Fig. 2—Typical well logs through the Denver Unit San Andres formation (depth shown is the measured depth from the surface).

mud layers not explicitly represented. In addition, vertical transmissibilities between some simulation layers were further reduced by an additional factor of 0.25 to capture the continuous, thicker mudstones present within the TZ model area. Because there was no field evidence indicating gravity segregation of CO_2, the effective k_V should be very small. A recent simulation study of the Seminole field also indicated that a low k_V/k_H was needed to match the GOR during the primary production although the actual core data indicated a much higher vertical permeability.[4]

Initial Oil Saturation. Initial oil saturations within the gridblocks of the model were assigned according to their porosity and structural position following an empirical porosity/height correlation derived from oil-based core measurements of water saturation vs. the height above BOZO. This approach was used because traditional Archie saturations cannot be determined accurately from resistivity logs. (Injection of fresh water has made the salinity of the reservoir brine an unknown.) Typically, initial oil saturations grade from more than 85% in the OC, to about 65% near the OWC, and to less than 20% at the BOZO. Below BOZO, oil saturations are very low. Although capillary pressure was neglected in most simulation runs (for computing speed), selected runs including the capillary pressure indicated little impact owing to high viscous force of the process and low vertical permeability.

Phase Behavior and Fluid Properties. The Soave-Redlich-Kwong EOS[5] was chosen to describe the CO_2/oil phase behavior. In the model, the oil was approximated as five pseudocomponents: C_1, C_2 through C_5, C_6 through C_{14}, C_{15} through C_{25}, and C_{26+}. Parameters in the EOS model were tuned to match constant-composition-expansion, multiple-contact, and slim-tube laboratory experiments.[6,7] Thereafter, interaction parameters were altered slightly to produce a smaller three-phase region at the expense of a slightly less accurate match to the constant-composition experiments. **Fig. 4** illustrates the comparison of the *p-x* diagram predicted by the EOS and the experimental data. Simulations incorporating this adjusted model ran much faster and predictions were essentially the same as those made when the original EOS model was used. Ref. 8 details how we circumvented the three-phase flash calculation to speed up the simulation.

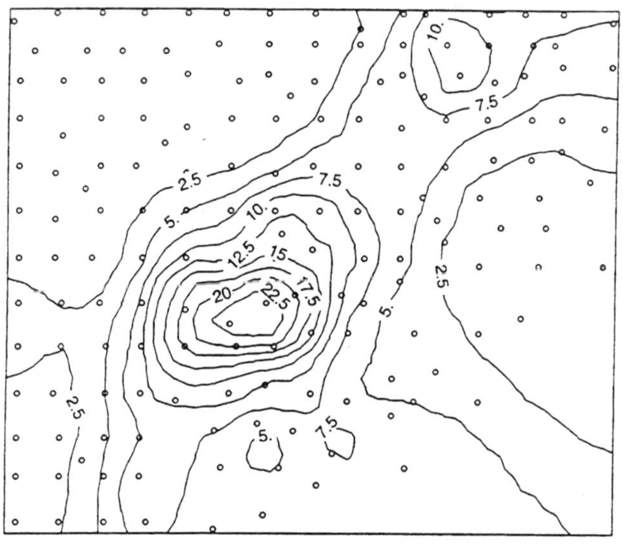

Fig. 3—Variation in the permeability of the upper M3 Layer in the TZ model (circles represent San Andres Wells; permeability in millidarcies).

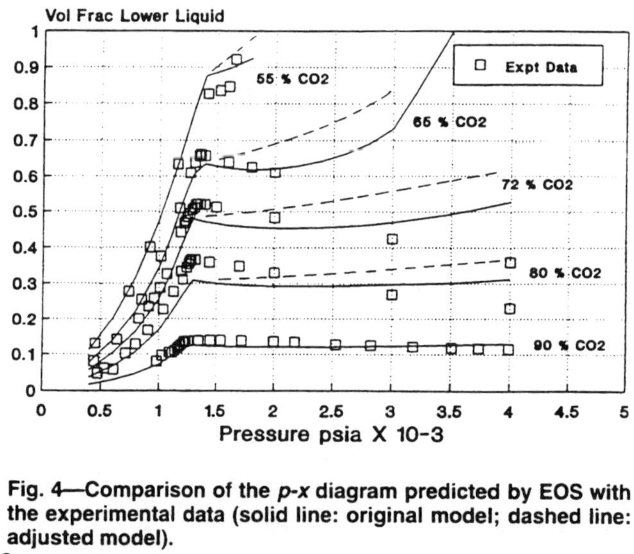

Fig. 4—Comparison of the *p-x* diagram predicted by EOS with the experimental data (solid line: original model; dashed line: adjusted model).

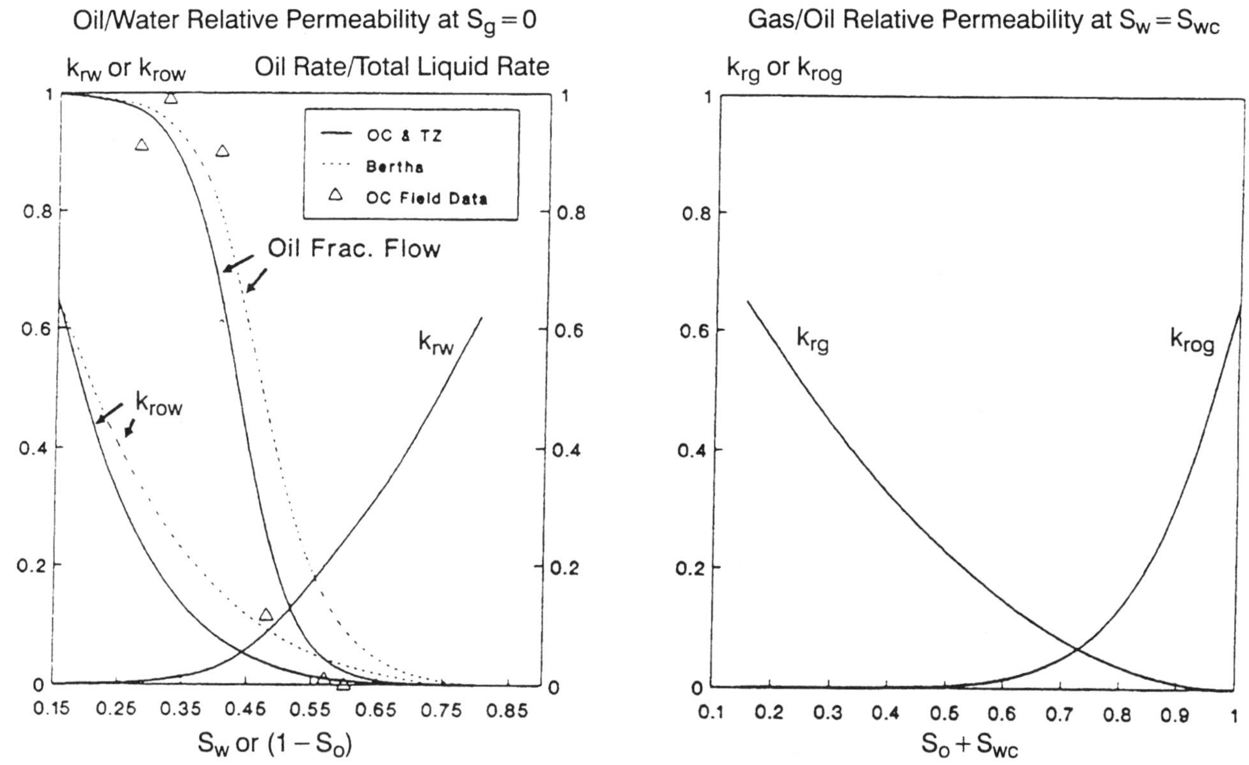

Fig. 5—Relative permeability curves used in the TZ model.

Relative Permeabilities. ***Waterflood.*** **Fig.** 5 displays the two-phase oil/water at $S_g = 0$ and gas/oil at S_{wc} relative permeability curves used in the model. The oil/water curves were drawn through the scatter of laboratory data.[9,10] Also plotted in Fig. 5 is the oil fractional-flow curve derived from the relative permeability curves; it agrees quite well with isolated zone production test and core saturation data measured on the OC interval in several Denver Unit wells. For simplicity, we used the same curves to describe both the OC and most of the TZ. The only exception was that the curves for "Bertha" (see Fig. 3) were stretched to accommodate a lower (20%) endpoint oil saturation as suggested by the pilot core data. The oil relative permeability was interpolated from k_{row} and k_{rog} with the Stone[11] II method.

CO_2 Flood. The waterflood k_{rw}, k_{row}, and k_{rg} curves depicted in Fig. 5 were also used for the CO_2 flood. The tail of the water relative permeability curve, by contrast, was truncated so that the endpoint water saturation incorporated trapped water above connate-water saturation (S_{wc}). An average measured 20% trapped-water saturation[9,10] above the 15% S_{wc} was used for the OC, and 10% trapped-water saturation was used for the TZ, which is generally believed to be more water-wet. The CO_2 flood was modeled with a water-repellant relative permeability model in which the water relative permeability was calculated from k_{rw}, the lower hydrocarbon phase (oil) relative permeability was predicted from k_{row}, and the upper hydrocarbon phase (which can be either the CO_2-rich phase in the OC or the vapor phase in the gas cap) was treated as an intermediate phase with its relative permeability interpolated between k_{row} and k_{rg} by use of the Stone II model. (Stone's model was the most appropriate three-phase option in the MORE simulator when the study was performed. A linear interpolation model would have been selected if available then.)

History Match and Model Validation. Owing to lack of the detailed production history before unitization (1964), we constrained the simulated primary production period to honor cumulative oil production on a well-by-well basis. Next, the waterflood (from 1965 to mid-1984) and CO_2-flood periods (mid-1984 to present) were history matched. For the history match, rate (as opposed to pressure) constraints were chosen because well-specific kh and skin are unknown. (Most wells are extensively fractured.) Therefore, we specified the actual individual well gross production rates (in reservoir barrels) or fluid (water or CO_2) injection rates as the well constraints. With these constraints, the quality of the match was judged from how well simulated oil, water, and gas production and reservoir pressure tracked actual data.

Tuning of the model was accomplished primarily by enlarging well-completion intervals to account for the effect of well stimulation. Initially, the completion interval of each well was set to be the mechanical completion interval. With the interval so limited, however, the model predicted insufficient water injectivity. This is to be expected because of historical well stimulation treatments (e.g., nitroglycerine jobs and acid fractures) and injection above formation parting pressure. Consequently, for injectors, the first 50 to 75 ft of the TZ interval was opened up at the start of the waterflood so that actual injection volumes could be introduced without exceeding the formation parting pressure. To match CO_2-flood response, all simulation layers were opened to injection at the start of the CO_2 flood; otherwise, the CO_2 response predicted by the model occurred much too quickly. (This is supported by several neutron logs that indicated the presence of CO_2 $\approx$200 ft above and/or below the perforated interval.) For producers, well files were searched to determine the times when the wells were stimulated. Each time a well was stimulated, the completion interval was enlarged beyond the mechanically completed depth. The amount the well was "deepened" was calculated from the actual water-cut increase after the stimulation assuming that the deepened interval produced only water. To account for fracture-only communication, however, only a fraction (approximately 0.25, for example) of the well kh was assigned to the interval opened below the mechanical completion.

Illustrative Applications

Fig. 6 delineates the areas encompassed by the two field-scale models presented in this section.

TZ Model: A Model for Planning New CO_2 Projects. Although the TZ is present in all Wasson San Andres units, it varies in richness across the trend. The best TZ resource lies under the Denver Unit OC that has been CO_2 flooded for more than a decade. This so-called TZ sweetspot of the unit encompasses a more-than-2,000-acre area in the southeastern portion of the unit where the TZ richness exceeds 60,000 STB/acre and the gross thickness exceeds 300 ft. Stratigraphically,

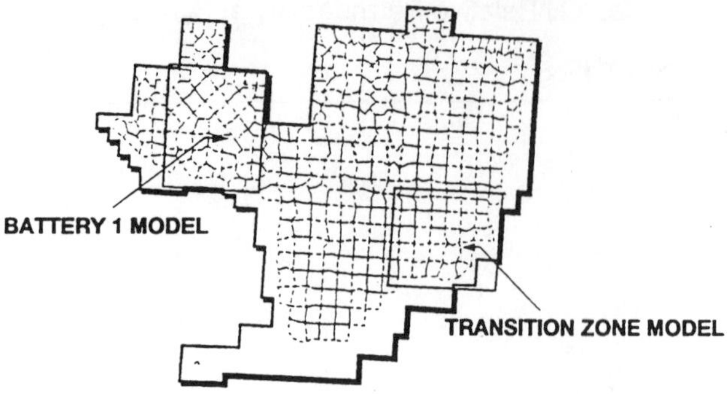

Fig. 6—Areas encompassed by the TZ and Battery 1 models.

the strata of the TZ sweetspot are the downdip equivalents of the Wasson San Andres OC accumulation as displayed in **Fig. 7.**

Screening studies have indicated that the TZ sweetspot can be developed most profitably with the wells being used in the OC flood and commingling injection and production. Because the quality of the reservoir in the TZ is better than that of the OC in this area, however, there is a question as to whether the TZ can be developed in this way without jeopardizing the OC flood. In addition, field data show fresh water and CO_2 to be present in the TZ in some locations. Thus, it is likely that some of the oil in the TZ at discovery may have already been recovered, making a CO_2 development less attractive.

To address the uncertainties surrounding the TZ sweetspot development, a three-dimensional (3D) field-scale simulation study was performed. The objectives were (1) to determine the oil saturations now present in the TZ so that the very best patterns can be selected, (2) to quantify the risks of commingling injection and production, (3) to high grade the project by optimizing completion depth and timing, and (4) to plan facilities and well work.

The model included 170 wells in 29 full (inverted) nine-spot patterns as well as 12 half nine-spot patterns along the northern and western boundaries. The centerlines of the half patterns were assumed to act as no-flow boundaries, and wells situated on these edges were treated as "half" wells. On the eastern and southern boundaries of the model, pore volumes (PV's) of the edge gridblocks were enlarged (as much as 10-fold) to mimic the downdip water zones connected to the simulated area. This was done to repre-

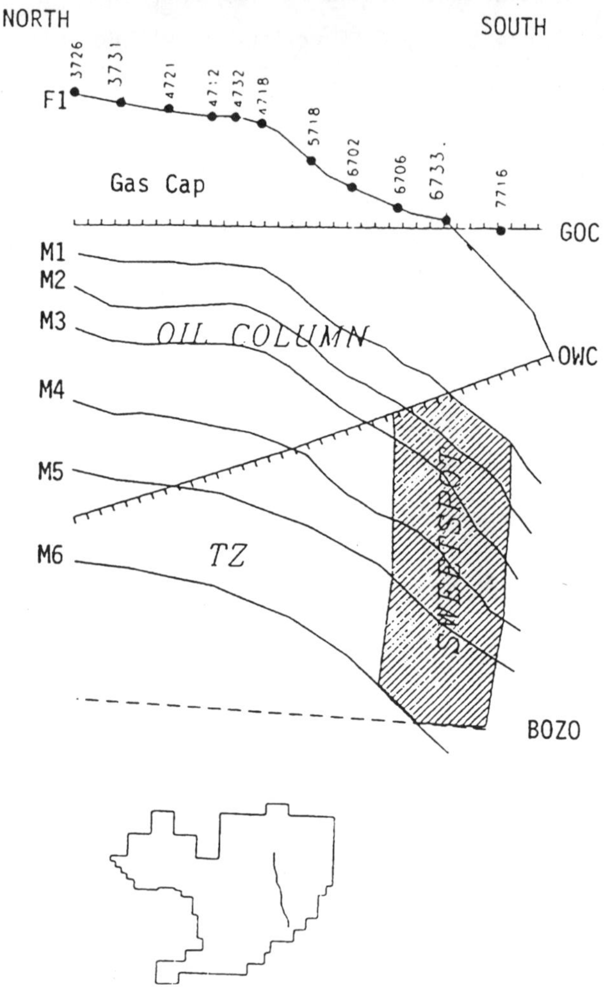

Fig. 7—North-south cross section illustrating the Denver Unit TZ sweetspot strata.

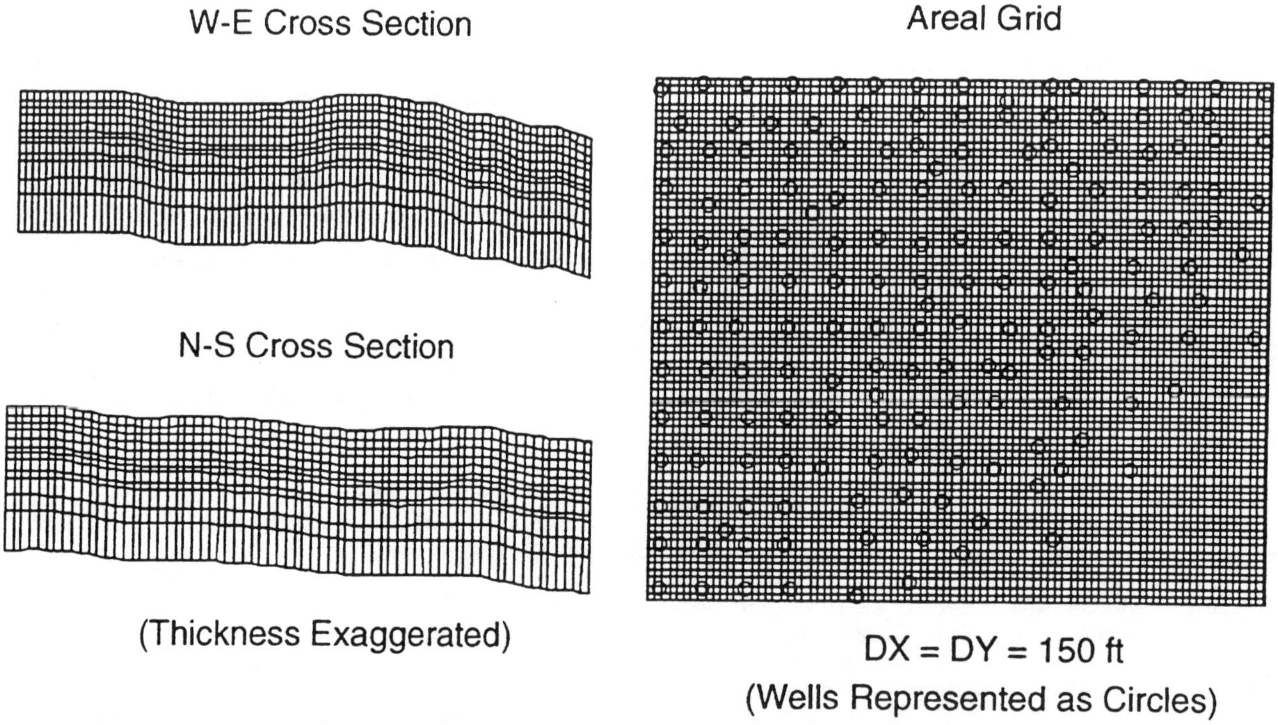

Fig. 8—Three-dimensional 83 × 69 × 12 grid structure used in the TZ model.

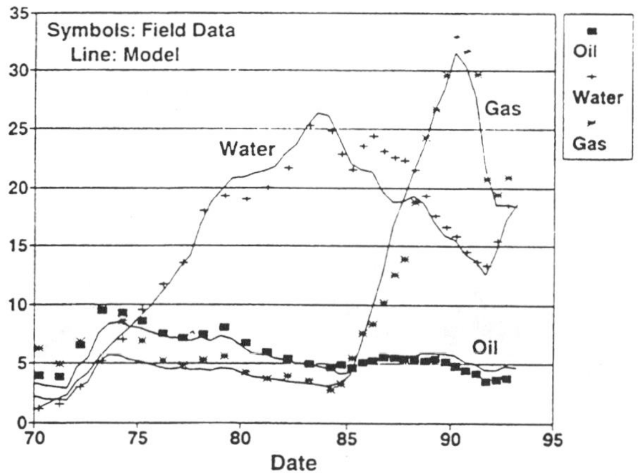

Fig. 9—History match of total production within TZ model area.

sent better actual reservoir pressure changes caused by over-/under-injection because pressure plays a large role in CO_2-flood predictions because of its impact on phase behavior. **Fig. 8** displays the 3D 83 × 69 × 12 grid structure of the TZ model. Other modeling details have been described in the previous sections.

Fig. 9 contrasts actual and simulated production histories of the entire simulated area during the history-match period. Well-by-well history matches are very good during the waterflood but not as good (although still satisfactory) during the CO_2 flood. A more detailed model, incorporating heterogeneity on a smaller scale, would likely be needed to improve the CO_2-flood history match. Because forecasts of various future development options made with the current simulation model differ greatly, discrepancies in the CO_2-flood history match should be less significant when difference cases are analyzed.

In making predictions with the history-match-validated model, the gross reservoir fluid rates from production wells were maintained at their historical levels or, if the well was deepened, limited to 1,000 RB/D of reservoir fluid, the lift capacity of a new pump. When the CO_2 production increased in the forecast, this well constraint automatically reduced the total liquid (water plus oil) production rate so that the total reservoir withdrawal rate remained constant, a convenient method for mimicking the gas interference with the pump.

Injection rates were also specified to give a pattern-level throughput of 0.06 total (OC plus TZ) hydrocarbon PV (HCPV) per year, subject to a bottomhole-pressure limit of 3,600 psi. The injection strategy followed a 1 : 1 water-alternating-gas (WAG) scheme until the total amount of CO_2 injected (since the start of OC flood) reached 60% of total (OC plus TZ) HCPV. Development cases were compared with a base case in which current OC CO_2-flood areas receive an ultimate 60% CO_2 slug.

One of the primary reasons for this simulation study was to quantify the target oil in TZ. **Fig. 10** maps the simulated TZ oil content as of January 1994. As can be seen, much of the oil that was initially present in the TZ along the northern boundary of the model appears to have already been recovered during past operations. This seems to have been the result of an aggressive stimulation program within this area during the late 1980's. In CO_2 development forecasts, these patterns produce far less oil from TZ than the patterns to the south. Consequently, the 12 patterns in this area were left out of the development.

Having established the area that should be developed, the simulation model was then exercised to optimize the design of the project. Completion depth, project staging, and deepening vs. twinning injectors were all studied. A completion depth of ≈ 150 ft below the OWC was chosen because it struck the best balance between oil recovery, CO_2 utilization, and water production.

The purposes of staging the TZ sweetspot project were (1) to keep produced fluid volumes within the capacities of existing facilities, (2) to optimize the use of high-volume lift equipment, and (3) to expedite the oil production response. In the proposed TZ project, injection wells are deepened within the first 6 months of the project; in contrast, producer deepenings are performed over a 4-year period. This keeps produced volumes within current handling capacities. It also reduces the number of high-volume pumping units that need to be purchased. As CO_2 breaks through and lightens fluid columns, high-volume lift equipment acquired for the producers deepened in the first 2 years of the project will be moved to wells deepened in the last 2 years. To expedite the oil response, producers for which the simulation indicates that (1) mobile oil is still present and (2) CO_2-flood fronts in TZ have advanced an appreciable fraction of the interwell distance are scheduled up front.

A key question in developing the areas of the TZ sweetspot was whether to drill new "twin" injectors or just deepen the existing injectors and commingle the injection. The potential advantage of drilling new injectors is to provide more injection control so that the TZ development can be accelerated without cycling a large amount of CO_2 through the OC. To evaluate these options, two simulation runs were made. In the first case, the injectors were deepened. In the second case, the existing OC injectors were twinned by adding a well in the adjacent gridblocks and completing them in the TZ. The fact that injection intervals have not been limited historically to the mechanical completion was incorporated into the latter case by completing into one geologic layer above and below the targeted completion interval in the newly drilled wells. The newly drilled twins were forecasted to inject a continuous CO_2 slug for 3 years at approximately twice the rate that can be attained with WAG. Injection into the TZ in the deepening case was driven by the ongoing 1 : 1 WAG in the OC.

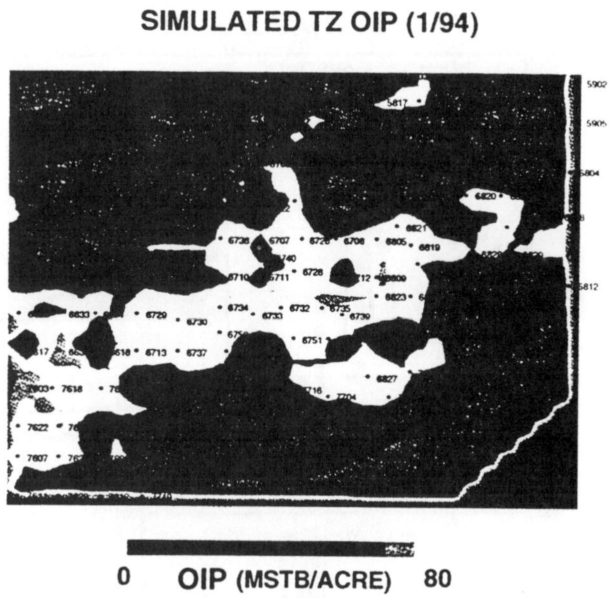

Fig. 10—Simulated oil in place (OIP) in the TZ sweetspot area as of January 1994 (light color represents high oil content).

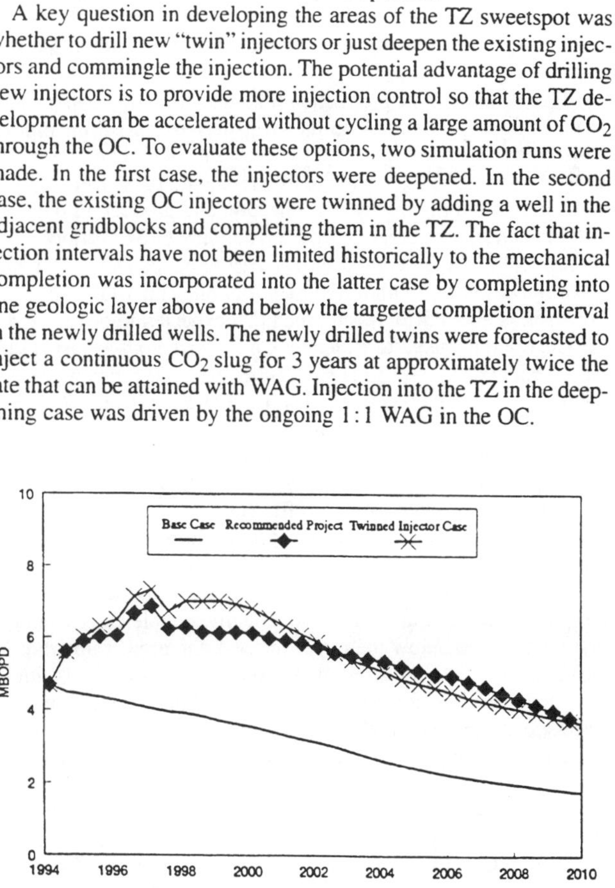

Fig. 11—Oil production forecasts for the no development, recommended project, and project in which injectors are twinned rather than deepened.

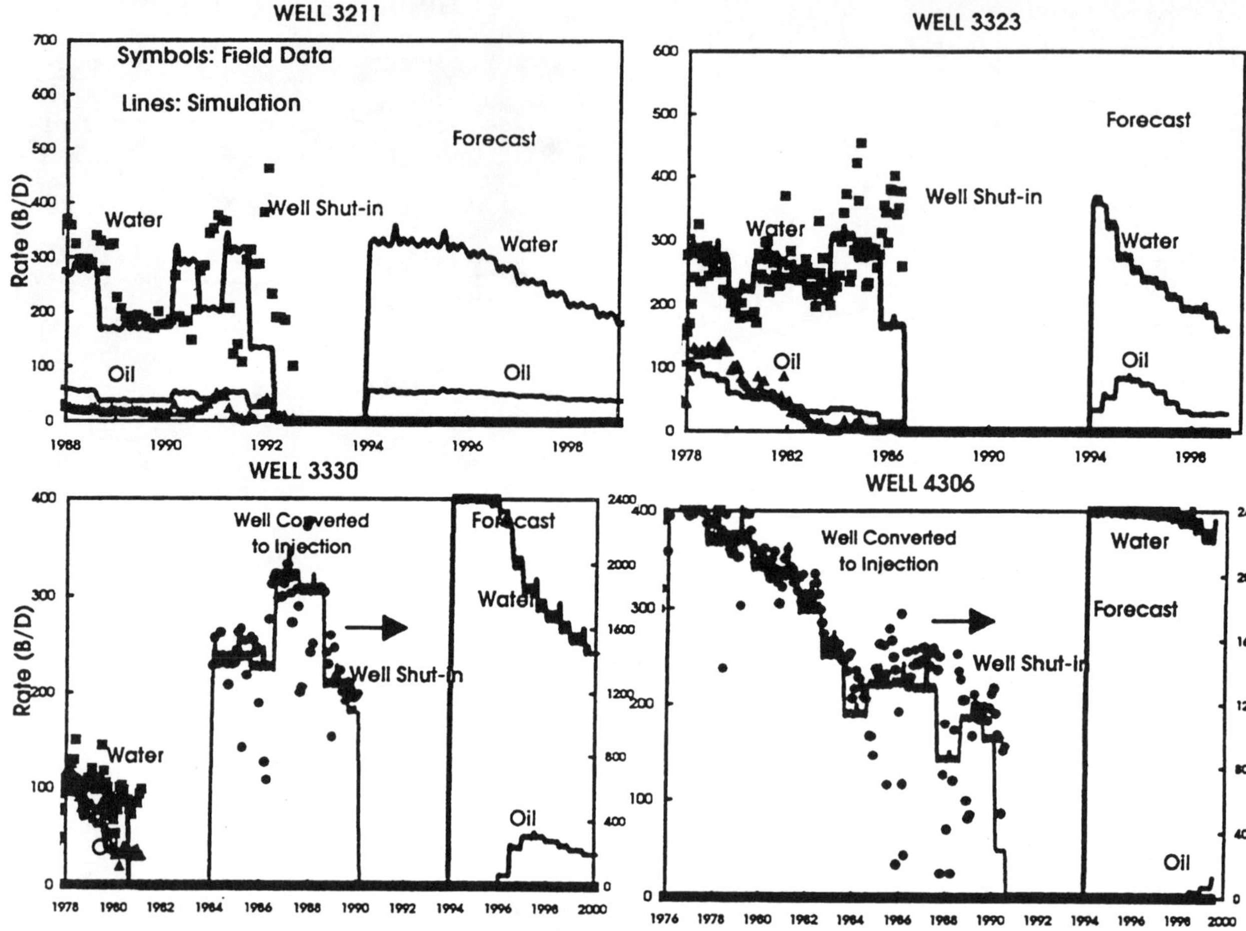

Fig. 12—Example of favorable and unfavorable RTP conversions as predicted by the Battery 1 model.

Fig. 11 contrasts the production functions from the deepened and twinned injection cases. Also shown as the base case is the production forecast from the sweetspot project area without doing the TZ CO_2 development. According to the simulation runs, deepening the wells and commingling the TZ injection and production with that from the OC does not jeopardize the ongoing OC CO_2 flood: 17.2 million bbl of incremental oil is forecast to be recovered at a good CO_2 utilization of 7 purchased Mscf/bbl oil. This CO_2 utilization, however, is about 50% greater than that expected to be realized in the OC flood in the eastern Denver Unit. The degrading target richness, anticipated losses to layers deeper than 150 ft below the producing OWC, and lower trapped-water saturation in the TZ all contribute to the higher CO_2 requirements.

The twin injector case shows significantly greater response than the deepened-injector case (18.9 vs. 17.2 million bbl of incremental oil) but uses purchased CO_2 less efficiently (8.4 Mscf/bbl oil). With the poorer CO_2 utilization and the cost of drilling new injectors, the return on the incremental investment of the twin-well case over the deepening case is fairly low. This, combined with the risk that the new wells may end up being inadvertently completed beyond one layer above and below the target interval, the greater up-front investment, and longer payout, makes the twin-well case less attractive than the injector-deepening option. Consequently, the TZ sweetspot project was implemented by deepening the existing injectors and commingling TZ injection with that going into the OC.

Battery 1 Model: A Model for CO_2-Flood Surveillance. The western portion (Battery 1) of the Denver Unit has not exhibited the fast response to CO_2 injection experienced in the eastern portion of the unit. The muted response to CO_2 is believed attributable to a smaller target (thinner OC) and significant losses of the injected CO_2 to the reservoir's original gas cap. A field-scale simulation model was constructed to evaluate options for improving the performance of this area of the field. The initial goals of the study were (1) to identify and quantify losses of CO_2 to the gas cap, (2) to determine which of more than 30 temporarily abandoned (TA) wells should be returned to production to capture oil mobilized by CO_2, (3) to evaluate CO_2-flood potential of the south flank of the battery, and (4) to determine the optimum injection strategy for the remainder of the CO_2 flood. Although not originally planned, the model has also identified drilling locations for short-radius horizontal wells in the simulated area.

The Battery 1 model covers a 5-sq-mile area consisting of Denver Unit Secs. 32, 33, 42, 43, 52, and 53, as shown in Fig. 7. Eleven reservoir layers were used with 130-ft gridblock spacing, resulting in more than 89,000 gridblocks. Sonic logs from 90 of the 186 wells in the model were used to characterize the porosity and permeability of each reservoir layer. Following the methodology presented in the previous sections, an excellent history match was obtained without additional tuning.

Of the 30 TA wells that are available for return to production, simulation results suggested that only five would show favorable response [defined as 50 BOPD production response within 6 months of return-to-production (RTP) or injector-to-producer (ITP) conversion]. Two of these wells (Wells 3211 and 3323) were recently completed. Initial tests on these wells indicate 10 to 20 BOPD and an increasing oil cut. Examples of favorable and unfavorable RTP/ITP candidates are shown in **Fig. 12.**

Of the 36 CO_2 injectors in the model, 25 were determined to have significant losses to the gas cap. Because the model does not report production and injection by layer, it is not possible to quantify the fraction of the injectant lost; however, comparative runs indicate that CO_2 losses have reduced current oil production by 900 BOPD (33%) and will reduce ultimate recovery by approximately 5 million bbl oil from the area modeled. Corrective actions, including options for remedial well work and/or conversion to water injection, are currently being evaluated.

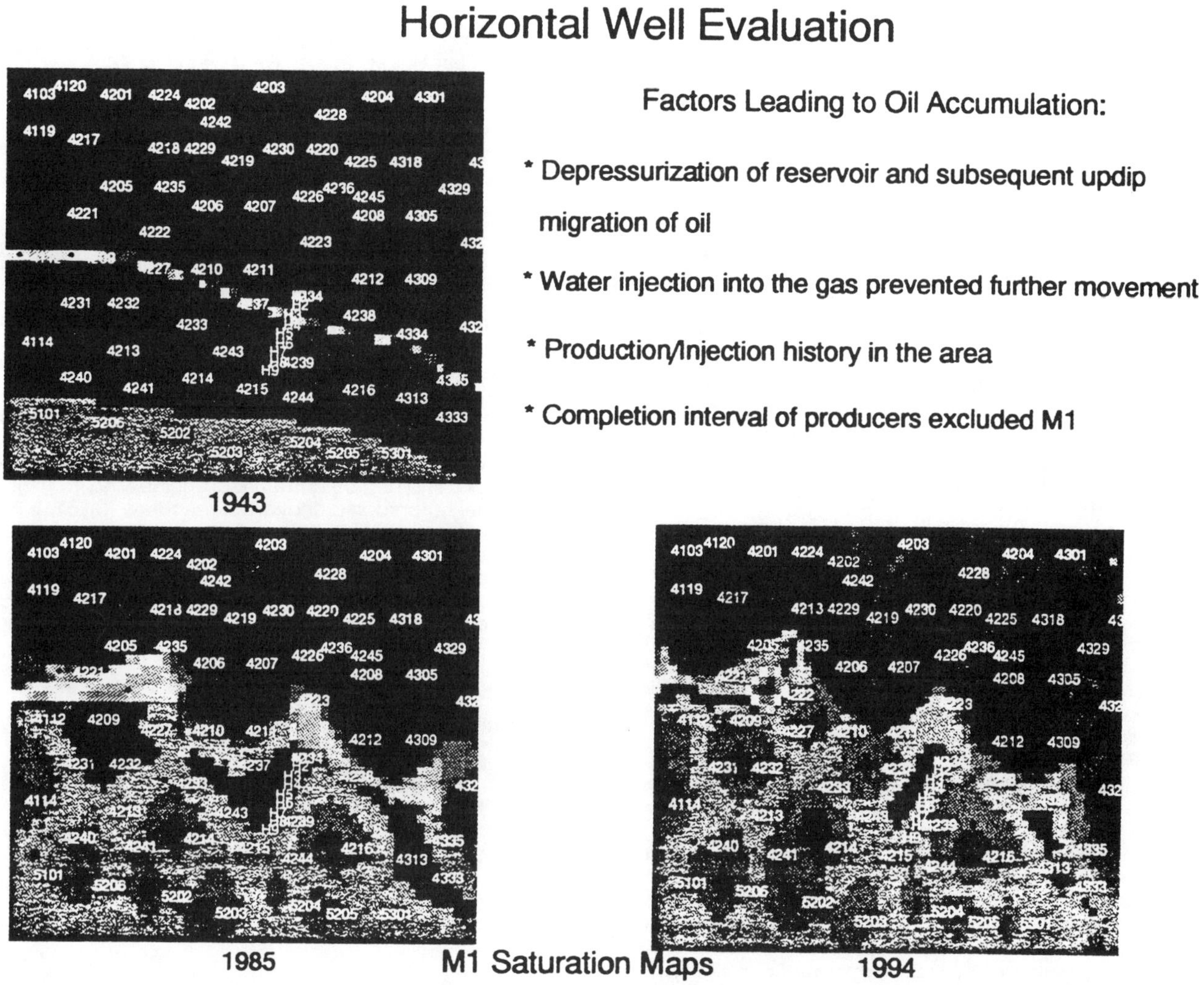

Fig. 13—Layer M1 saturation maps of simulated oil saturations in evaluation of Horizontal Well DU 4234 showing location of the well for 1943, 1985, and 1994.

This model has also proved effective in locating regions of unswept oil along the south flank of Battery 1. These locations have become targets for a recently initiated horizontal-drilling program in the Denver Unit. Simulation results suggested that oil has been swept updip along the original GOC. The traditional practice of avoiding well completion above the GOC has left much of this oil unrecovered (see **Fig. 13**). Two horizontal wells have been drilled on the basis of the results of this model (Wells 4234 and 4313).

The first horizontal well, Well 4234, clearly confirmed the presence of oil above the original GOC. This well was completed with 36 ft of open hole exposed above the GOC and produced 100 to 200 BOPD at 90% oil cut with a solution-gas GOR. The second well, Well 4313, also left open a significant interval above the GOC and has not experienced excessive gas production. However, directional control problems on Well 4313 resulted in the well path placed between the targeted F_5 and wet M_1 layers. This interval is indicated as a tight zone on well logs. Production results from the well are lower than predicted; remedial options are being evaluated.

Conclusions

1. A compositional reservoir model, when appropriately constructed following the methodology described in this paper, can provide reasonable predictions of the CO_2-flood performance of the Denver Unit. No *ad hoc* mixing parameters were used in this study.
2. Field-scale CO_2-flood simulations can help plan new CO_2-flood projects. Incorporating the TZ model as an integral part of the re-engineering process, a marginal project for recovering the residual oil in the Denver Unit TZ was revised and high-graded to a profitable venture.
3. Field-scale CO_2-flood simulations can be very valuable to CO_2-flood surveillance. The Battery 1 model pinpointed the best shut-in wells to return to production, identified infill and horizontal drilling locations, and identified and quantified the injectant losses.
4. A compositional model can be quite useful for many practical applications despite the fact that it cannot predict miscible flood residual. After injecting 60% HCPV of CO_2 (a typical slug size of a CO_2 project), the model did predict oil saturations less than those observed in pilots in a region near the injector. Because this region was so small (less than 5% of the total volume of the flood pattern), the discrepancy has little impact on oil production forecasts; however, its impact on the injectivity predictions can be significant.

Acknowledgments

We thank the management, operations staff, and engineering staff of Shell Western E&P Inc. and Shell Development Co. who lent their enthusiastic support to this project and to the preparation of this paper. We are also indebted to J.B. Ward and J.C. Cooper for their reservoir characterization work, to J. Ratulowski for his phase-behavior description, and to J.F. Simmons for his support for this simulation study and coordination of the horizontal drilling program.

Nomenclature

h = height, L, ft
k = permeability, L^2, md
k_{rg} = gas relative permeability

k_{ro} = oil relative permeability
k_{rog} = oil relative permeability in the oil/gas system
k_{row} = oil relative permeability in the oil/water system
k_{rw} = water relative permeability
k_V/k_H = vertical to horizontal permeability ratio
p = pressure, m/Lt^2, psi
r = radius, L
R_{wa} = apparent resistivity
S_g = gas saturation
S_{wc} = connate-water saturation
x = composition
ϕ = porosity

References

1. Todd, M.R. and Longstaff, W.J.: "The Development, Testing, and Applications of a Numerical Simulator for Predicting Miscible Flood Performance," paper SPE 3834 presented at the 1971 SPE Annual Meeting, New Orleans, 3–6 October.
2. Young, L.C.: "Full-Field Compositional Modeling on Vector-Processors," *SPERE* (February 1991) 107.
3. Hill, W.J., Tinney, T.J., and Young, L.C.: "CO_2 Operation Plan, South Welch Unit, Dawson County, Texas," paper SPE 27676 presented at the 1994 SPE Permian Basin Oil and Gas Recovery Conference, Midland, Texas, 16–18 March.
4. Wang, F.P., Lucia, F.J., and Kerans, C.: "Critical Scales, Upscaling and Modeling of Shallow-Water Carbonate Reservoirs," paper SPE 27715 presented at the 1994 SPE Permian Basin Oil and Gas Recovery Conference, Midland, Texas, 16–18 March.
5. Soave, G. Equilibrium Constants from a Modified Redlich-Kwong Equation of State," *Chem. Eng. Sci.* (1977) **27,** 1197
6. Orr, F.M. and Jensen C.M.: "Interpretation of Pressure-Composition Phase Diagram for CO_2/Crude Oil Systems," paper SPE 11125 presented at 1982 SPE Annual Technical Conference and Exhibition, New Orleans, 26–29 September.
7. Gardner, J.W., Orr, F.M., and Patel, P.D.: "The Effect of Phase Behavior on CO_2 Flood Efficiency," paper SPE 8367 presented at 1979 SPE Annual Technical Conference and Exhibition, Dallas, 23–26 September.
8. Hsu, C-F., Morell, J.I., and Falls, A.H.: "Field-Scale CO_2 Flood Simulations and Their Impact on the Performance of the Wasson Denver Unit," SPE 29116 presented at 1995 SPE Symposium on Reservoir Simulation, San Antonio, Texas, 12–15 February.
9. Potter, G.F.: "The Effect of CO_2 Flooding On Wettability of West Texas Dolomite Formations," paper SPE 16716 presented at the 1987 SPE Annual Technical Conference and Exhibition, Dallas, 27–30 September.
10. Schneider, F.N. and Owens, W.W.: "Relative Permeability Studies of Gas/Water Flow Following Solvent Injection in Carbonate Rocks," *SPEJ* (February 1976) 23.
11. Stone, H.L.: "Estimate of Three-Phase Relative Permeability and Residual Oil Data," *J. Cdn. Pet. Tech.* (October–December 1973) 54.

SI Metric Conversion Factors

acre	× 1.233 489	E+03	= ha
bbl	× 1.589 873	E−01	= m^3
ft	× 3.048*	E−01	= m
ft^3	× 2.831 685	E−02	= m^3
md	× 9.869 233	E−04	= μm^2
mile	× 1.609 344*	E+00	= km
psi	× 6.894 757	E+00	= kPa
sq mile	× 2.589 988	E+00	= km^2

*Conversion factor is exact.

SPERE

Chia-Fu Hsu is a senior staff reservoir engineer with Shell E&P Technology Co. in Houston, responsible for the Denver Unit TZ project design. He has worked for Shell since 1981 in various aspects of reservoir management, reservoir simulation, and simulator development. Hsu holds a PhD degree in mechanical engineering from the U. of Minnesota. **Jose I. Morell** is a research engineer with Shell E&P Technology Co. He joined Shell 1992. Morell holds a PhD degree in chemical engineering from the U. of Houston. Photograph is unavailable. **Andrew H. Falls** is Business Unit Leader for BDM Petroleum Technologies. He previously was a reservoir engineering specialist for the Continental Div. of Shell Western E&P Inc. He has nearly 20 years' experience developing and implementing new technology to business advantage in the areas of thermal- and CO_2-recovery methods and unconventional-resource development.

Hsu

Falls

An Approach To Optimize Economics in a West Texas CO_2 Flood

G.J. Pariani, SPE, **K.A. McColloch,** SPE, **S.L. Warden,** SPE, and **D.R. Edens,** SPE, Amoco Production Co.

Summary. EOR projects, notably CO_2 floods, are the next generation of recovery methods in the more mature west Texas waterfloods. Installing and operating a CO_2 flood can be extremely expensive. In this paper, we discuss methods used to make several active CO_2 floods more profitable by reducing operating costs and deferring investments.

Introduction

Our goal in studying several active west Texas CO_2 floods was to determine the optimum near-term cash flow, overall project economics (rate of return, present worth, etc.), and oil recoveries. We developed various CO_2 flood designs with a reservoir simulator by altering specific operating parameters, including the half-cycle slug size, gas/water ratio (GWR), injection schemes, and total CO_2 slug sizes. The resulting injection and production rates were then entered into an economic simulator to determine the most economical set of operating conditions. We also assessed the impact of various economic conditions—oil prices, CO_2 prices, gas processing fees, and operating costs—on the optimization process. This was done to ensure that the optimum set of operating conditions did not change drastically should the economic environment change.

This study shows how optimizing operations can significantly improve the economics of existing CO_2 floods. An additional benefit of the study was insight gained on plant sizing and adjusting injection schemes to minimize or defer investments while maximizing the economics of new projects.

Background

During the late 1970's and early 1980's, Amoco Production Co. committed significant manpower to evaluate the feasibility of full-scale CO_2 flooding in west Texas. Before the feasibility work, Amoco undertook numerous CO_2 pilots in various fields.[1] Because of high crude prices, optimistic price forecasts (**Fig. 1**), and successful pilot recoveries, the future of CO_2 flooding looked promising.

The approach taken at the time was first to history match waterflood performance with an in-house black-oil simulator.[2] The resulting reservoir description then was transferred to a streamline generating model and finally to a miscible streamtube model. The CO_2 prediction was done at a single GWR with the CO_2 injection slug size changed to find an optimum based on present worth. The reservoir simulation provided oil, gas, and CO_2 production rates and CO_2 injection rates for input into an economic model. Because of the man-hours and computer time involved in simulating the CO_2 process, only a few GWR's were tried, with slug size as the main variable.

In these early CO_2-flood designs, CO_2 reinjection was planned and a plant design was needed to remove H_2S and hydrocarbons and to handle the peak predicted total gas production. Because of the single GWR being used, the gas rate was expected to increase to some peak and then decline, leaving excess plant capacity. This type of injection scheme impaired the economics because a large plant investment was needed at the beginning of the project. Unless there was an offset property feasible for CO_2 flooding, the excess plant capacity went unused once the peak was past.

The costs of plants and other investments were relatively high compared with today's environment because of high oil prices and high inflation rates. Most projects were still economical, however, because oil price forecasts were optimistic and tax relief from the Windfall Profits Tax was granted for EOR projects.

One shortcoming of this early approach was the optimistic nature of the miscible streamtube model. Although this model adequately handled loss of miscibility, it was unable to regain it should the pressure rise above the miscible pressure. The model also was too efficient in areal sweep: the displacement was piston-like through the streamtube and all streamtubes were eventually swept for nearly 100% areal sweep efficiency. Injectivity losses also were handled inadequately in the streamtube model. Water hysteresis was accounted for, but solvent relative permeability effects[3,4] were not. The miscible streamtube model also had limited ability to make changes in the pattern configuration and injection schemes. If such changes were to be made, the streamline generator had to be run again for entry into the miscible streamtube model. This was both time-consuming and costly. As a result of these model shortcomings, many economic sensitivities were run with various rate and reserve scenarios to become comfortable with the final CO_2-flood design.

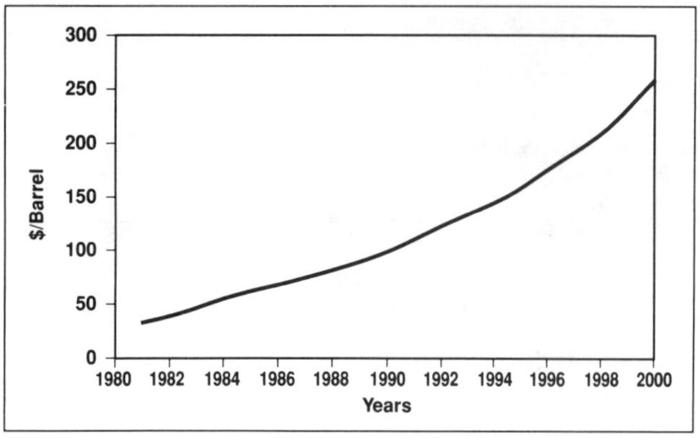

Fig. 1—Example oil price forecast.

"In CO_2-flood design technology, an understanding of the technical and economic parameters is essential."

After many model studies and countless economic runs, four west Texas properties, one in the Wasson field and three in the Slaughter field, were authorized for full-scale CO_2 flooding. CO_2 injection began in Dec. 1984 using the designs from the miscible streamtube model, which called for a 2:1 GWR based on a reservoir-barrel basis. Concurrent with the field-scale CO_2-flood design work, our research department was developing and calibrating a simulator to handle the waterflooding and CO_2 miscible portions of the flood. This model was verified against many of the Amoco CO_2 pilots in west Texas.[1,3] By mid-1986, the new model was ready for use in the full-scale floods.[5]

With tertiary operations based on the then extremely favorable economic conditions initiated, the 1986 oil price collapse radically changed both the overall economic environment and the profitability of CO_2 floods. The goal in studying the active west Texas CO_2 floods in the late 1980's was to use technical advances and operational insight to design tertiary operating schemes to optimize near-term cash flow, point-forward discounted present worth, and oil recovery.

Optimization Design

In CO_2-flood design technology, an understanding of the technical and economic parameters is essential. These parameters and their interrelation are discussed in the following paragraphs. The four major economic relationships considered in the design of a CO_2 flood or in optimizing an existing CO_2 flood are oil price, gas processing costs, CO_2 purchase costs, and operating costs. The most important is the oil price forecast, which, unfortunately, is also the component on which the CO_2-flood design has no effect. History has shown that this component is volatile at best; however, a CO_2 flood can be designed to respond to variances in the economic climate.

The second substantial economic component is the gas processing costs. This assumes that the gas is processed as opposed to reinjection of full well-stream production. The gas is normally processed by a unit-owned facility or by contractual agreement at an outside facility. Costs at a unit facility are controlled by plant operating costs and therefore do not vary with oil price. Processing contracts often have a base processing cost that escalates with inflation and significant increases in oil price. Processing costs have a significant effect on the profitability and design of a CO_2 flood.

The third financial consideration is the CO_2 purchase costs. These costs typically have two components: the actual commodity price and the transportation cost (or pipeline tariff). While the commodity price is tied to oil prices, the supply contract may specify a floor price that is not affected by oil prices. The transportation cost does not usually vary with the price of oil. Therefore, purchase costs are important to the design of a CO_2 flood, especially with projections of lower oil prices where purchase costs will remain high compared with oil prices and will weigh heavily on the profitability of the flood.

The final costs affected by CO_2 design are the base operating costs; however, this component is affected only minimally by the CO_2 flood. Differences in design may affect power consumption (both injection and lift), additional subsurface lift-related costs, corrosion protection, treatment costs, etc.

In the assessment of an ongoing CO_2 flood, three major design parameters have the most significant effect on both the operation and economics of a CO_2 flood: (1) the CO_2 and water half-cycle slug size, which is commonly referred to as a percent of the HCPV of the affected area; (2) the GWR scheme, which usually is represented as a varying ratio over the life of the flood; and (3) the ultimate injected CO_2 slug size, which is represented as a percent of the HCPV.

Several technical, operational, and economic factors affect the design of CO_2 and water half-cycle sizes in a CO_2 flood. Variances in half-cycle sizes will affect mobility control and consequently CO_2 and oil production profiles. Small half-cycle sizes that approach simultaneous injection of water and gas will minimize viscous fingering.[6] With projections of modest oil prices, minimizing the half-cycle slug sizes will also maximize profitability by limiting CO_2 production rates and therefore gas processing costs. In stratified reservoirs, especially where thin, high-velocity (permeability/porosity) layers are present, mobility control and areal sweep can be influenced drastically. Also, near-miscible conditions and floods that consist of large irregular patterns may benefit from minimizing half-cycle slug sizes. Conversely, because the injection wells are switched manually between gas and water, the demands on field personnel need to be recognized in determining the half-cycle sizes (smaller sizes will necessitate more frequent switches). In our assessment of west Texas CO_2 floods, an optimum half-cycle slug design was determined by weighing benefits in cash flow and present worth against operational and manpower considerations.

The second and most important parameter is the GWR design.[6-8] In the early 1980's, CO_2-flood designs typically called for one specified GWR to be maintained over the life of the flood. Through technical advances and operational and financial experiences, a tapered-GWR design has been implemented. This design has two major portions: the GWR design until an "optimum" maximum gas production rate is reached and the GWR scenario designed to maintain the optimum gas production rate. The first part of the design is determined through a near-term cash flow evaluation. The factors dictating the resulting cash flow are oil price, CO_2 purchase and processing costs, reservoir heterogeneity, and pattern geometry. CO_2 flood response can be accelerated through high initial GWR's. A combination of high GWR's, favorable economic conditions, and a symmetrically patterned homogeneous reservoir would very likely produce a very favorable cash flow projection. However, the introduction of a highly stratified reservoir, large unsymmetrical injection patterns, or an unfavorable economic climate could easily alter the cash flow and call for lower initial GWR's. Many times, the profile with the highest near-term oil rates does not result in maximum cash flow projections. Each flood must be evaluated individually.

Determination of an "optimum" gas production rate will be affected by gas processing costs and may be affected by plant processing capacity or contractual agreements. The optimum gas production rates for Amoco's CO_2 floods were determined

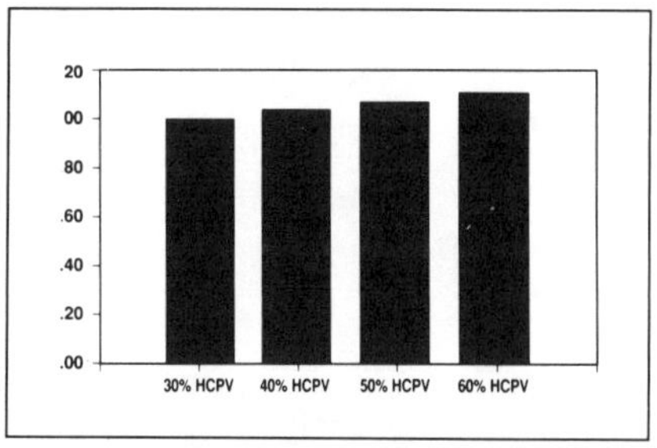

Fig. 2—Example west Texas CO_2-flood pattern—normalized oil recoveries vs. percent HCPV CO_2 injected.

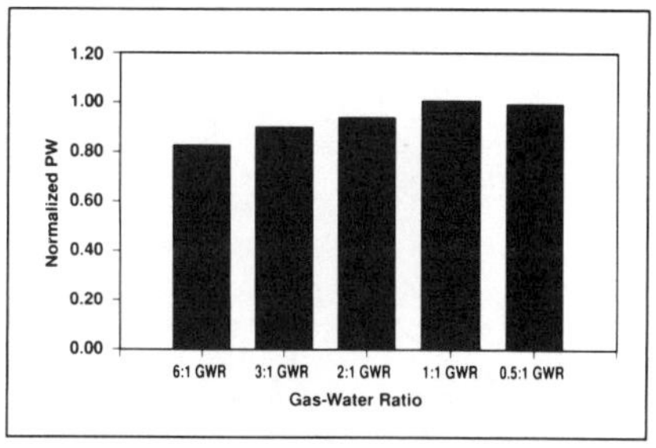

Fig. 3—Example west Texas CO_2 pattern—normalized present worth vs. GWR, 700-Mcf/D gas production and 50% HCPV CO_2 injected.

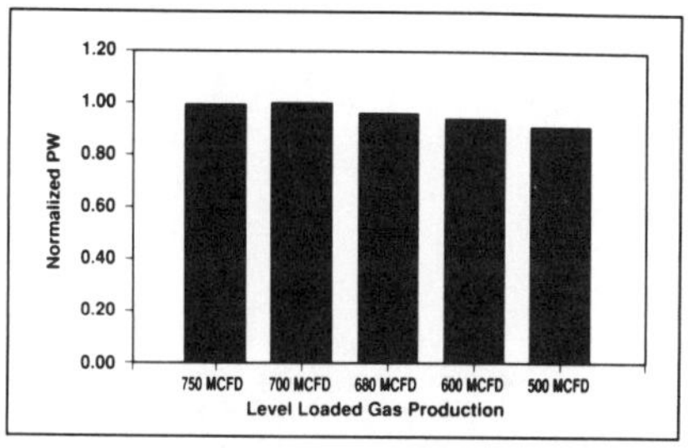

Fig. 4—Example west Texas CO_2 pattern—normalized present worth vs. level loaded gas production rates, 1:1 GWR and 50% HCPV CO_2 injected.

through a present worth evaluation after technical constraints were taken into account (the methodology is discussed later). In the tapered-GWR design, the ratio of injected water to gas is increased gradually after the "optimum" gas production rate is reached through the life of the flood. The gradual increase of injected water in relation to gas (a decrease in the GWR) results in increased mobility control and a constant produced-gas profile.

The tapered-GWR design has several advantages over a constant-GWR design. Plants in the early 1980's were designed for peak gas production rates, which generally left the plants operating well below capacity for most of the life of the flood. With the tapered-GWR design, smaller plants with lower capital investments can be installed and operated at capacity (with higher efficiency) through the life of the flood. In ongoing floods, the use of existing plant capacity can be optimized by implementing different GWR designs areally based on variances in pattern geometry and reservoir heterogeneity. Also, the tapered-GWR design will allow lower facility (gas gathering, flowlines, etc.) costs and fewer operational problems associated with downhole gas separation and high bottomhole producing pressures.

Finally, it was found that even in a point-forward economic evaluation (no additional capital investment required), the tapered-GWR design greatly enhanced the projects' present worths by optimizing tertiary oil production with respect to the cost of purchasing and processing CO_2.

The final design parameter is the ultimate size of the CO_2 slug injected. This parameter is affected most by oil price forecasts, CO_2 processing costs, and reservoir stratification. This parameter has the most significant effect on oil recoveries. In the simulation work on Amoco's west Texas floods, a comparison of water-alternating-gas (WAG) flood designs illustrates that ultimate incremental tertiary recovery was controlled predominantly by the total slug size (**Fig. 2**). Also, in our economic evaluations, the previously discussed tapered-GWR design allowed economical injection of larger total CO_2 slug volumes, thus increasing ultimate oil recovery. Because this design parameter is, by definition, a long-term parameter, it should be re-evaluated with significant changes in the economic climate.

Methodology

The purpose of these studies was to develop operating schemes to maximize the economic performance of the four existing CO_2 floods in an environment of relatively low oil prices. The primary economic optimization criterion was present worth. Near-term cash flow was used as a secondary criterion. Present worth and near-term cash flow were reviewed in each case to ensure that both were in agreement as to which operating scenario would yield the optimum economics.

The approach used in these studies was to develop production/injection profiles for different operational schemes by using a numerical simulator and running economics on each of the profiles to determine the best point-forward economics. A generalized compositional model was used in a black-oil mode with three nonaqueous components (oil, hydrocarbon gas, and CO_2) and solvent relative permeability curves.[3,5] The use of solvent relative permeability curves allowed the model to account for the injectivity losses experienced in pilot and full-scale WAG injection. A three-nonaqueous-component black-oil model was used to save computing time. Previous work on the Slaughter Estate Unit pilot[5] showed that a satisfactory history match could be obtained with this method vs. a fully compositional model as long as the pressure of the flood was sufficiently above the minimum miscibility pressure. Each fieldwide CO_2 flood was first history matched from the beginning of primary through about 4 years of CO_2-flood history. The smallest unit was modeled with a full-field model; the other three were modeled with average pattern models. All the models used major geological zonation with some statistical zonation within the major geological zones.

After satisfactory history matches were obtained, predictions were made with different combinations of half-cycle slug sizes, gas-alternating-water injection schemes, level loaded (constant) gas production rates, and ultimate CO_2 injected slug sizes to determine which scenario would yield the best economics.

The first step in optimizing the floods was to select an optimum half-cycle slug size. Half-cycle slug sizes between 1% and 3% HCPV were analyzed to determine the optimum volume of CO_2 for an injection cycle. The analysis revealed that the 1% HCPV slug size resulted in a present worth 6% greater than the 3% HCPV slug because the smaller half-cycle slug gave better sweep

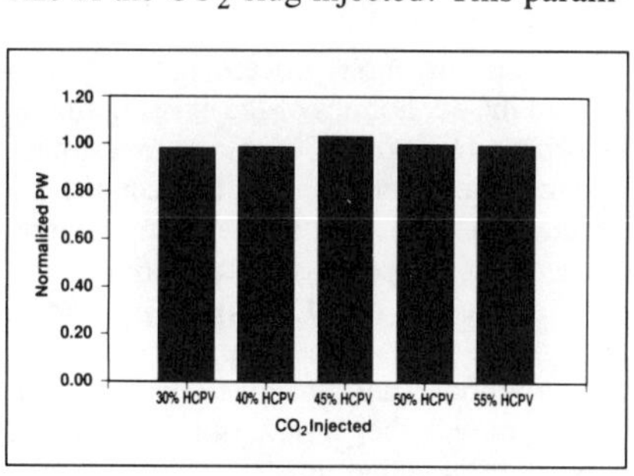

Fig. 5—Example west Texas CO_2 pattern—normalized present worth vs. percent HCPV CO_2 injected, 1:1 GWR and 700 Mcf/D gas production rate.

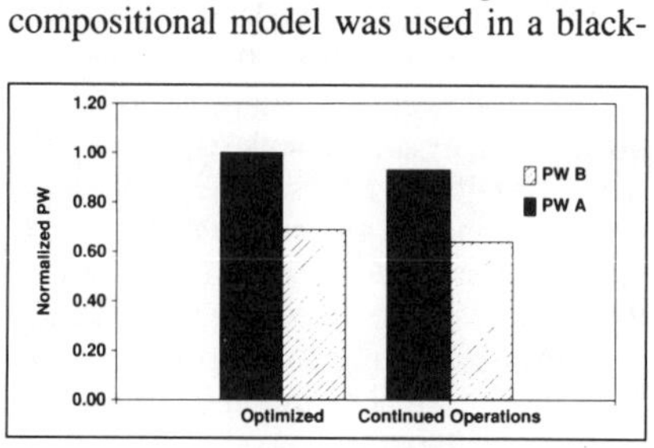

Fig. 6—Four west Texas CO_2 floods—normalized discounted present worth, optimized vs. continued operations.

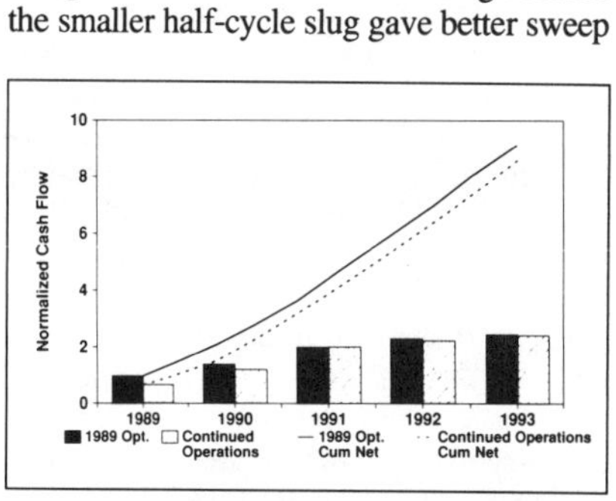

Fig. 7—Four west Texas CO_2 floods—normalized annual cash flows and cumulative net cash flow, optimized vs. continued operations.

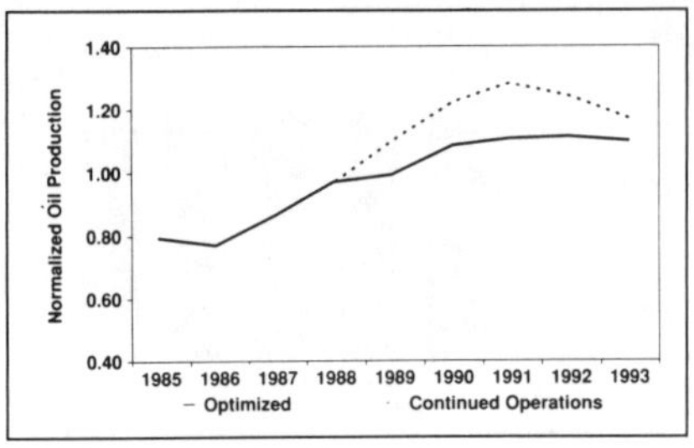

Fig. 8—Four west Texas CO_2 floods—yearly average normalized oil production.

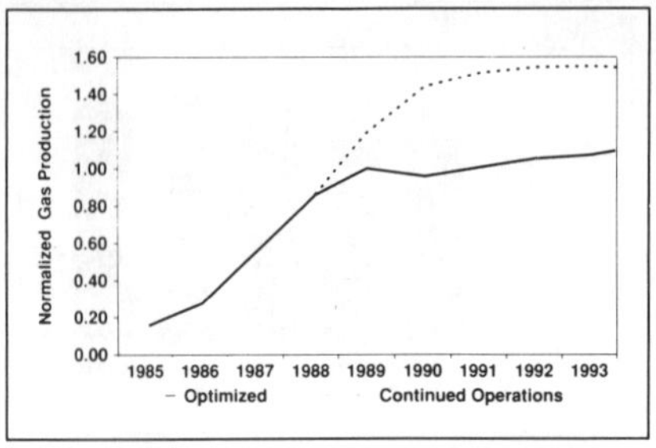

Fig. 9—Four west Texas CO_2 floods—yearly average normalized gas production.

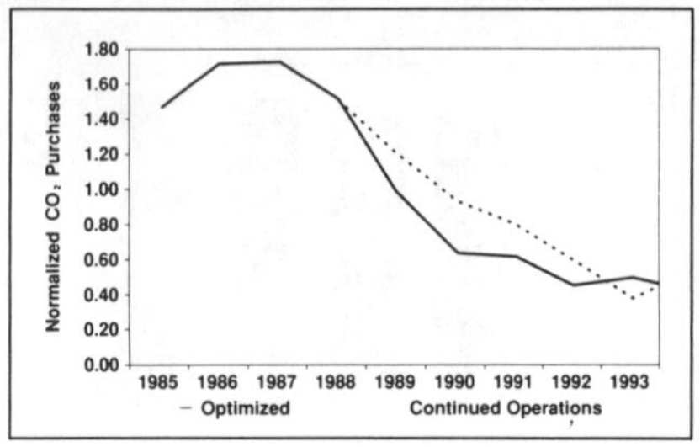

Fig. 10—Four west Texas CO_2 floods—yearly average normalized CO_2 purchases.

control and did not allow CO_2 to cycle through high-permeability, low-PV zones in the reservoir. Half-cycle slug sizes greater than 3% HCPV were not evaluated because sweep would have deteriorated even further with a large CO_2 slug. In some cases, a >3% half-cycle slug size may be appropriate, depending on reservoir characteristics and existing surface facilities (i.e., homogeneous reservoir, sufficient plant capacity, and no CO_2-flood expansion options). Slug sizes less than 1% also were not evaluated because, in most floods, any smaller half-cycle slug would cause operational problems created by the large number of injector switches between CO_2 and water. However, when a pattern has an abnormally large HCPV and/or low permeabilities, a half-cycle slug less than 1% might be appropriate because the length of time required to get the 1% HCPV slug into the reservoir could result in poor mobility control. Therefore, a 1% HCPV half-cycle slug size of CO_2 was selected as optimum for this model and was used in the remaining economic evaluations.

The second step in optimizing the floods was to select an optimum GWR by making several predictions at different near-term GWR's while maintaining a constant-level loaded gas rate and ultimate slug size for each. It was assumed that the optimum GWR selected by this method would be optimum at any other reasonably close gas rate and slug size. The GWR evaluations concentrated in the range of 0.5:1 to 6:1. **Fig. 3** shows a bar graph of present worths vs. the initial GWR's for four point-forward GWR schemes. In each scenario, the GWR's were reduced with time to as low as 0.5:1 to maintain the same constant gas production rate. The present worths for this figure and those associated with the next two steps in the optimization process were normalized on the optimum GWR scheme selected by this first step. An initial 1:1 GWR was optimum for this particular model.

The next step was to determine the optimum loaded gas rate. Several predictions were made at the optimum GWR scheme, but injection was varied later in each prediction to reach and maintain a different gas production rate. **Fig. 4** shows a typical bar graph of the normalized present worths for the optimum GWR scheme at the different level loaded gas rates. The gas rates evaluated generally were ±25% of the optimum level loaded rate. This sample has an economical level loaded rate of 700 Mcf/D per pattern.

The final step was to determine an optimum total slug size. With the previous established parameters, the model was run for various injected slug sizes. **Fig. 5** gives typical results of the analysis. Slug sizes evaluated generally were ±35% of the optimum. In this case, the optimum ultimate CO_2 slug size was at a 45% HCPV. Note that extremely small (10% or less) total slug sizes may not be effective because of the injectivity losses seen in west Texas.

Results

As mentioned previously, the two economic criteria used to determine the optimum CO_2 scheme were near-term cash flow and present worth. These factors and the production/injection profiles were used to compare the new optimization scheme with a continued-operations scheme. The new optimization scheme realized near-term GWR reductions of at least 33% and in some cases 88%, level loaded gas rate reductions of about 40%, ultimate CO_2 slug size reductions between 7% and 63%, and 50% reductions in half-cycle slug sizes.

Fig. 6 shows the present worth at two different discount rates, Rates A and B (Rate A is lower than Rate B). Two discount rates were used to determine whether the economic results were sensitive to the discount rate. The graph was normalized on the Rate A value of the optimized case. The optimized case had a present worth approximately 7% higher than the continued-operations case at each of the discount rates.

Fig. 7 compares annualized yearly cash flows and cumulative cash flows for the first 6 years following the study period. This graph was normalized on the first-year cash flow of the optimized case. The graph shows the importance of the benefits of reducing the near-term GWR's. Yearly cash flows for the first 2 years are greater for the optimized case than the continued-operations case, with the differences diminishing significantly in the next few years. The most significant difference is in the first year when the optimized cash flow is 30% greater.

Figs. 8 through 10 illustrate the changes in the production/CO_2 purchase profiles during the early years of the two operating schemes. Examining the graphs together (which again have been normalized on the first year of the optimized case) shows that the oil and gas production and the CO_2 purchase rates for the optimized case deviate dramatically from the continued-operations case in the first year.

Fig. 8 illustrates that a significant percentage of oil production (as high as 15%) is being deferred during the early years of the new optimized operating scheme; however, even larger reductions are being experienced in gas production and CO_2 purchases (Figs. 9 and 10, respectively). The decreases in gas production (processing costs) and CO_2 purchases (60% and 15%, respectively) more than offset the deferred oil production (revenues).

To ensure that the new optimization would remain valid under significantly different price forecasts, multiple economic sensitivities were run with widely varying scenarios of oil prices, gas processing costs, and CO_2 purchase prices. The first-year values of the alternative price forecasts had variations as great as 30% in oil prices, 35% in gas processing costs, and 60% in CO_2 purchase prices. As discussed, the cost of gas processing and CO_2 purchases, regardless of the beginning price, was tied to the price of oil. The economic sensitivities confirmed that the new optimization was the proper course of action to take on these four floods.

As a result of implementing the new operating scheme, controllable operating expenses (COE) and capital investments were reduced. The areas of COE that were affected the most were gas processing costs and CO_2 purchases. In addition, several capital investments originally planned for these existing floods for both field facilities and plant expansions were either deferred or canceled because of the optimization process.

The optimization process not only improved the economics of the existing floods but opened up other CO_2-flood investment opportunities. For example, the reduced gas production rates in the existing floods opened up plant processing capacity for possible expansion of CO_2 floods into other parts of the fields without large investments for additional plant capacity. This alone can determine whether a new flood or a flood expansion is economical.

Future CO_2-Flood Design

With the development of this CO_2-flood optimization process, a great deal of insight

"The purpose of these studies was to develop operating schemes to maximize the economic performance of the four existing CO_2 floods in an environment of relatively low oil prices."

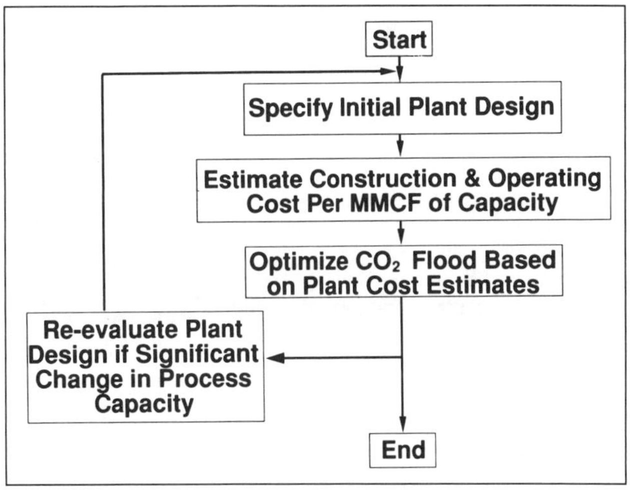

Fig. 11—CO_2-flood plant size optimization process.

was gained on the design of new CO_2-flood projects. Many parameters must be considered in the optimization of any CO_2 flood; however, with the design of a new flood, several additional parameters must be taken into account: plant design, pattern configuration and size, reservoir quality, surface-facilities design, and CO_2 purchase contracts.

Plant design is one of the most critical factors in the design of a new CO_2-flood project. Several variables must be considered that are not options in the optimization of an existing CO_2 flood. These include plant type, plant size, liquid hydrocarbon recoveries, CO_2 product purity, plant location, plant construction cost, and plant operating cost. To handle these multiple design considerations, a three-step design process was developed (**Fig. 11**). The first step is to determine a basic plant type and process design on the bases of plant location, capital and operating costs, desired hydrocarbon recoveries, and desired CO_2 product purity. Second, a cost per million cubic feet of process capacity for plant construction and an operating cost per million cubic feet of gas processed are calculated. Finally, a CO_2-flood production profile is designed with a reservoir simulator and an economics program that minimizes plant investments and maximizes the discounted present worth of the CO_2-flood project. Once an optimum plant size is determined, the plant design is re-evaluated to determine whether any economies can be realized through plant type or process revisions. The process is then repeated if there is a significant design revision.

If gas processing at an outside facility is an option, a similar design process can be used, substituting the contract processing costs and rate limitations for the plant operating costs and process capacity. When evaluating an outside-facility contract, one must be sure to consider not only the process fee but also the hydrocarbon recovery split and CO_2 product purity. Pipeline investments also will be required for produced gas and possibly CO_2 product transport.

Pattern development can significantly affect the design of a new CO_2 flood. Large or irregularly shaped patterns can greatly diminish the effectiveness and consequently the value of a CO_2-flood project. Large patterns or low injector/producer ratios often necessitate the use of high GWR's or continuous CO_2 injection early in the life of the flood to maintain reservoir pressure and to get tertiary oil response early enough to keep the discounted return on investment at a reasonable level. Unfortunately, using a high GWR for a long time can diminish the ultimate incremental recovery and consequently the project value. In some situations, reservoir pressure is insufficient for a miscible CO_2 flood and additional injection support must be installed to allow the project to be successfully CO_2 flooded. An infill drilling program, which has marginal waterflood economics, may increase the value of a future CO_2 flood and significantly enhance the overall project economics. On the other hand, having too dense a pattern development can affect the value of a CO_2 flood negatively owing to increased surface facility requirements. As a result of these conflicting optimums, one must consider future CO_2-flood implications of any infill drilling decisions.

Reservoir quality differences substantially affect the design and value of a CO_2-flood installation. Highly stratified reservoirs are very difficult to CO_2 flood economically as a result of early CO_2 breakthrough and high recycle rates. Any problems noted during waterflood operations with ineffective water injection, early breakthrough, or cycling will be compounded during CO_2-flood operations because of the lower viscosity and higher mobility ratio of CO_2. The most cost-effective way to handle this problem is to use a WAG injection scheme. Decreases in the GWR can then be used to control gas breakthrough rates and recover incremental oil economically. Other options include pattern reorientation and the use of chemical sweep treatments, such as gels or foams.

The magnitude and timing of surface-facility capital investments can be controlled through the CO_2-flood startup plan and the CO_2 injection scheme. More aggressive CO_2 startup schemes will require larger, more expensive field-facility installations, while a more gradual startup will reduce surface-facility size requirements and defer some facility investments. The most effective way to reduce and defer surface-facility investments is to stagger the startup of the CO_2 flood. With this approach, the project is divided into several areas that are then CO_2 flooded in a stepwise manner, starting with the best part of the field and progressing to the poorest. Each area is started on CO_2 injection after oil production has peaked and CO_2 injection is being reduced in the previous area. This minimizes CO_2 supply line investments and defers investments for both CO_2 injection and produced gas handling facilities. Plant size requirements are also reduced because the field's CO_2 production is kept at a low rate for a long time rather than at a very high rate for a short time. One pitfall of this approach is the possibility of deferring startup of a segment beyond its economic limit. Maintenance cost increases resulting from aging wellbores and surface facilities also must be considered in evaluations of tertiary project deferral.

The final design parameter is the CO_2 purchase contract. A thorough understanding of this contract is essential to the design of a successful CO_2-flood project. If the cost of purchased CO_2 is less than the effective cost of recycled CO_2, a flood scheme that increases the use of purchased CO_2 may be optimum. This would be a more aggressive flood scheme using high GWR's or continuous injection and rapid areal coverage of the project area. However, if recycled CO_2 is less expensive, one would want to minimize CO_2 purchases. In this case, the optimum flood scheme would be stepwise, with several flood areas, allowing the recycled gas from each area to be used for injection in the next area.

The above parameters can be incorporated into a successful CO_2-flood design by including them in the optimization plan discussed previously. Pattern configurations and reservoir-quality variance can be accounted for in the reservoir simulator, while CO_2 purchase contract and gas processing cost sensitivities can be handled in the economics program. Plant and surface-facility requirements are determined by including associated plant and facility investments for each CO_2 injection profile in the economics program. Evaluation of a stepwise areal flood progression should be based on the independent optimization of each flood area, without regard to plant or CO_2 supply facility investments. The optimum CO_2-flood design incremental (to waterflood) injection and production profiles for each area can then be shifted or slid several years to produce a rough estimate of an optimized sequential flood by including the associated plant and CO_2 supply investments in the economics program. Once the optimum areal startup timing is determined, it should be confirmed by rerunning each model with the revised CO_2 injection start date.

Conclusions

1. Early CO_2-flood designs, based on a streamtube model, resulted in optimistic economic forecasts. A new simulator improved our ability to evaluate the CO_2-flood process and economics.

2. Four major economic relationships must be considered in the design of a CO_2 flood: oil price forecast, gas processing (or reinjection) costs, CO_2 purchase costs, and base operating costs.

3. The three design parameters that most significantly affect performance, cash flow, and present worth of a CO_2 flood are the CO_2 and water half-cycle slug sizes, the GWR profile, and the ultimate injected CO_2 slug size.

4. Technical variances, such as reservoir heterogeneity, near-miscible conditions, and pattern geometry and size, will significantly affect a CO_2-flood design.

5. Use of a tapered-GWR design should be considered in both ongoing and future CO_2 floods. This design has both technical and economic advantages over a constant-GWR design.

6. Of the CO_2-flood design parameters, the ultimate CO_2 slug has the most significant effect on oil recovery.

7. Optimization of existing floods has the potential of improving project profitability, deferring or eliminating additional facility investments, and opening up other CO_2-flood expansion opportunities.

8. In the design of a new CO_2 flood, the parameters that need to be assessed are plant and surface-facilities design, pattern size and configuration, reservoir quality, and CO_2 purchase contracts.

References

1. Ader, J.C. and Stein, M.H.: "Slaughter Estate Unit Tertiary Miscible Gas Pilot Reservoir Description," *JPT* (May 1984) 837–45.
2. Drennon, M.D., Kelm, C.H., and Whitington, H.M.: "A Method for Evaluating the Feasibility of Tertiary Recovery Using Carbon Dioxide in Several Major Carbonate Reservoirs in West Texas," paper SPE 9323 presented at the 1980 SPE Annual Technical Conference and Exhibition, Dallas, Sept. 21–24.
3. Chopra, A.K., Stein, M.H., and Dismuke, C.T.: "Prediction of Performance of Miscible-Gas Pilots," *JPT* (Dec. 1990) 1564–72.
4. Schneider, F.N. and Owens, W.W.: "Relative Permeability Studies of Gas-Water Flow Following Solvent Injection in Carbonate Rocks," *SPEJ* (Feb. 1976) 23–30; *Trans.*, AIME, **261**.
5. Young, L.C. and Stephenson, R.E.: "A Generalized Compositional Approach for Reservoir Simulation," *SPEJ* (Oct. 1983) 727–42.
6. Christie, M.A. and Bond, D.J.: "Detailed Simulation of Unstable Process in Miscible Flooding," *SPERE* (Nov. 1987) 514–22; *Trans.*, AIME, **283**.
7. Crump, J.G.: "Detailed Simulations of the Effects of Process Parameters on Adverse Mobility Ratio Displacements," paper SPE 17337 presented at the 1988 SPE/DOE Symposium on Enhanced Oil Recovery, Tulsa, April 17–20.
8. Gorell, S.B.: "Modeling the Effects of Trapping and Water Alternate Gas (WAG) Injection on Tertiary Miscible Displacements," paper SPE 17340 presented at the 1988 SPE/DOE Symposium on Enhanced Oil Recovery, Tulsa, April 17–20.

SI Metric Conversion Factors

bbl	× 1.589 873	E−01	= m^3
ft^3	× 2.831 685	E−02	= m^3

Provenance

Original SPE manuscript, **An Approach To Optimize Economics in a West Texas CO_2 Flood,** received for review April 11, 1991. Revised manuscript received July 9, 1992. Paper accepted for publication July 15, 1992. Paper (SPE 22022) first presented at the 1991 SPE Hydrocarbon Economics and Evaluation Symposium held in Dallas, April 11–12.

Authors

Parlani

McColloch

Warden

Edens

Gary Parlani is a staff petroleum engineer with Amoco Production Co. in Houston. He has more than 13 years' reservoir engineering experience, including 6 years of reservoir modeling CO_2 floods, waterfloods, and primary gas reservoirs. He holds a BS degree in petroleum engineering from the U. of Missouri. **Kevin A. McColloch** is an engineering manager with Amoco Production Co. in Houston. Since joining Amoco in 1982, he has worked in production operations, reservoir management, and reservoir simulation. His experience includes both waterflood and CO_2-flood operations, primarily in the Permian Basin area. He holds a BS degree in chemical engineering from Texas Tech U. **Stephanie Warden** is a staff petroleum engineer for Amoco Production Co. in Houston. In 10 years there, she has worked primarily with waterflooding and CO_2 flooding in both operations and reservoir engineering. She holds a BS degree in chemical engineering from Lamar U. **Dennis Edens** is a senior staff petroleum engineer for Amoco Production Co. in Houston. In 18 years at Amoco, he has worked as a reservoir engineer in the U.S. gulf coast, the U.K. North Sea, and west Texas waterfloods and CO_2 floods. He holds a BS degree in mining engineering from Virginia Polytechnic Inst.

Development and Validation of Equation-of-State Fluid Descriptions for CO_2/Reservoir-Oil Systems

Shahin Negahban and **V.J. Kremesec Jr.,*** SPE, Amoco Production Co.

Summary. This paper discusses the required data, methodology, and validation for developing equation-of-state (EOS) fluid descriptions for the vaporizing-CO_2/reservoir-oil systems. Validation of the EOS description is shown by an ability to predict additional first-contact data and slim-tube performance over a wide range of pressures covering both immiscible and multiple-contact-miscible (MCM) conditions. Slim-tube performance is predicted with a fully compositional simulator that uses a physical dispersion formulation. The relative permeability curves and dispersivity of the slim tube were measured and honored during all simulations.

Introduction

CO_2 miscible flooding, one of the principal EOR methods, is being conducted on a commercial scale in many oil reservoirs. One of the key variables of this process is the complex CO_2/reservoir-oil phase behavior at reservoir conditions.

Phase-behavior measurements and slim-tube displacements to measure minimum miscibility pressure (MMP)[1] are the most widely used methods to characterize CO_2/reservoir-oil systems. These data usually are represented with an EOS, which generally uses lumped pseudocomponents to characterize the C_{7+} fraction and binary interaction parameters that must be tuned for specific systems.

Kremesec and Sebastian[2] presented a minimum data set to which the EOS should be tuned. Their procedure resulted in EOS descriptions capable of simulating the performance and the development of miscibility in corefloods above the MMP. Prediction of the performance of these corefloods to validate EOS descriptions was subject to question because the corefloods were affected by adverse viscosity ratios and density differences. The mixing caused by these effects was approximated with numerical dispersivity by adjusting the number of gridblocks to match the CO_2 effluent concentrations. Further, Kremesec and Sebastian did not match displacement results below the MMP and suggested that additional tuning criteria may be necessary to predict immiscible displacements.

In this paper, the tuning criteria of Kremesec and Sebastian are shown to result in EOS descriptions that are capable of predicting both immiscible and miscible displacements over a wide range of pressures. The validation of the EOS used a compositional simulator with a physical dispersion formulation and no numerical dispersion. For immiscible performance predictions, measured slim-tube gas/oil relative permeability data are necessary and use of physical dispersivity eliminates the gridblock number adjustment. There are no adjustable parameters in the simulations.

Previous attempts to simulate slim-tube performance used various empirical formulations to describe corresponding slim-tube parameters.[3-7] This paper shows that an EOS fluid description and measured slim-tube parameters are all that are needed for quantitative predictions of slim-tube MMP and rate performance over a wide range of pressures for the vaporizing-CO_2/reservoir-oil systems. It also is shown that slim-tube displacements are not strongly affected by adverse viscosity ratios and density differences.

This work also presents a method for the development of EOS fluid descriptions for vaporizing-CO_2/reservoir-oil systems. This method was used in developing a new 10-component EOS fluid description for a liquid/liquid CO_2/west Texas reservoir-oil system (Oil A in Ref. 2). This description is superior to the 24-component EOS description of Kremesec and Sebastian because it predicts all the results within a smaller tolerance with fewer components. Minimizing the number of components reduces CPU usage significantly.

Slim Tube

The slim-tube tests were conducted with an apparatus described previously.[1] The slim-tube gas/oil relative permeability curves were measured from an immiscible displacement of dodecane by nitrogen at room temperature and a constant pressure drop of about 800 psia [5.5 MPa]. **Fig. 1** shows the experimental k_g/k_o data determined by the Johnson-Bossler-Naumann[8] method.

The slim-tube displacement is conducted in a long, narrow tube so that hydrodynamic flow instabilities, such as viscous fingering and gravity tonguing, are reduced greatly.[1,9] To test this, the dispersivities of contact-miscible displacements at adverse viscosity ratios and density differences were determined. The first test consisted of a displacement of an octane/butanol mixture by decane. The density difference, $\Delta\rho$, and viscosity ratio, F_μ, were matched closely for these fluids (F_μ=0.84 and $\Delta\rho$=0 g/cm^3). In the second test, dodecane was displaced by hexane (F_μ=4 and $\Delta\rho$=0.11 g/cm^3), and in the third test, a mixture of deodorized mineral spirits and No. 31 mineral oil was displaced by hexane (F_μ=40 and $\Delta\rho$=0.21 g/cm^3). The measured dispersivities from these tests were between 0.01 to 0.03 ft [0.30 to 0.91 cm], which strongly suggests that flow instabilities are suppressed. Because the flow is essentially 1D and dominated by phase behavior, 1D simulations of slim-tube displacements can be made.

Numerical Simulator

A generalized compositional simulator with a physical dispersion formulation was used to simulate the slim-tube tests.[10,11] The Amoco version of the Redlich-Kwong EOS (ARKES)[12] was used to describe fluid properties and phase equilibria in the simulator. Equilibrium phase viscosities were calculated with the Stiel-Thodos correlation,[13] which requires pure component critical volumes and phase densities calculated by ARKES.

All slim-tube simulations were carried out in the 1D mode with measured slim-tube relative permeability data (Fig. 1). To reduce the CPU time requirement, the majority of simulations were conducted with a dispersivity, α, of 0.05 ft [1.52 cm]. The error from using this value is within the experimental error, and CPU time was cut by more than 50%. **Fig. 2** shows the predicted and measured slim-tube displacement results on arithmetic probability paper. The numerical formulation requires that the number of gridblocks be such that the maximum gridblock length is twice the dispersivity and increasing the number of gridblocks does not change the result.[10,11]

Method for Tuning ARKES Parameters

A three-step iterative procedure was followed for tuning ARKES parameters.

1. Starting with an experimental C_{7+} distribution, the first step involves adjustment of molecular weights, mole fractions, and critical pressures of the lumped pseudocomponents to match the experimental base oil density and C_{7+} molecular weight and specific gravity; these properties were varied by about 1% to 2% per iteration.

2. Next, all the binary interaction parameters, k_{ij}, of CO_2/C_{2+} hydrocarbon pairs were set to 0.05 (other interaction parameters

*Now with Amoco Corp.

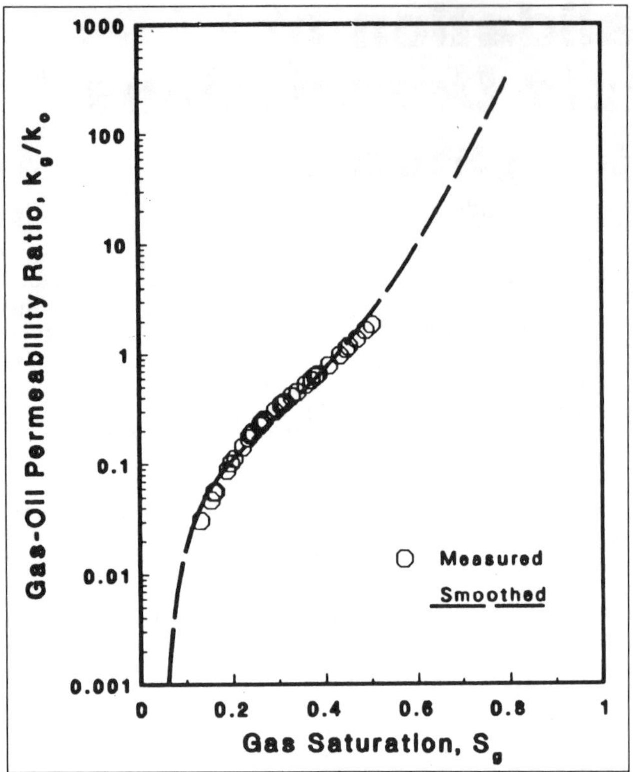

Fig. 1—Slim-tube gas/oil permeability ratio curve.

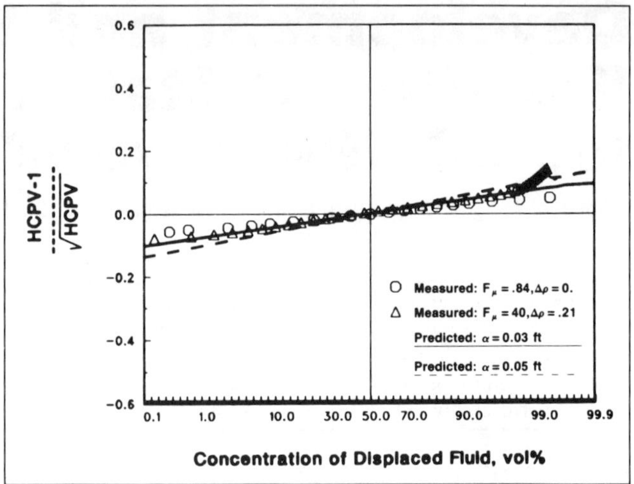

Fig. 2—Error function plot of slim-tube results.

were kept at the default values).[12] Adjustment of the CO_2/C_{7+} binary parameters were continued until a good match of the *p-x* locus was obtained. Minor adjustments were then made on the C_{7+} molar distributions to improve the match of the critical point (if it occurred) and the droplet points (the points at which a second liquid phase evolved). This was followed by checking, with a computer algorithm developed by Turek *et al.*,[14] that the calculated MMP fell within ±100 psia [±0.7 MPa] of the experimentally determined slim-tube MMP. If the slim-tube MMP was over- or underpredicted, the k_{ij} parameters of CO_2/C_{7+} components were adjusted and the entire procedure was repeated.

3. Finally, the predicted and experimental CO_2/reservoir-oil mixture densities and viscosities were compared. If a difference greater than 2% existed, adjustments were made on the critical pressures and molar distributions of C_{7+} components to match densities. If needed, the critical volumes of C_{7+} components were adjusted to match the mixture viscosities (variation of critical volumes will affect only viscosity calculations). Steps 2 and 3 were then repeated.

Although this procedure may not be unique, we were able to develop an excellent set of ARKES parameters for the CO_2/Oil A system.

Reservoir Oils

Two types of reservoir oils were considered in this study. Oil A is a typical west Texas oil from the San Andres formation with a low temperature of 105°F [41°C] and a moderate viscosity of 3 cp [3 mPa·s]. On first contact with CO_2, this oil displays vapor/liquid (V/L), vapor/liquid/liquid (V/L/L), and liquid/liquid (L/L) equilibria.[15] Oil B is from the Rocky Mountains and has a relatively high temperature of 160°F [71°C] and a viscosity of 1.8 cp [1.8 mPa·s]. First-contact mixture of CO_2 and Oil B exhibits only V/L equilibria.[16]

ARKES Fluid Description for CO_2/H_2S/Oil A System

Three sets of experimental PVT data are reported in the literature for the CO_2/Oil A and CO_2/H_2S/Oil A systems at 105°F [41°C].[15,17] **Table 1** presents the Oil A compositions, C_{7+} molecular weights and specific gravities, and the bubblepoint pressures at 105°F [41°C]; differences in Oil A compositions are primarily a result of sampling from different wells.

p-x data for a 70/30 mol% CO_2/H_2S/Oil A system, CO_2/Oil A mixture densities, and the MMP obtained from the 80/20 mol%

TABLE 1—OIL A COMPOSITIONS

Components	80/20 mol%* CO_2/H_2S	70/30 mol%* CO_2/H_2S	100 mol%** CO_2
N_2	1.08	—	—
C_1	9.63	10.87	14.52
CO_2	1.72	1.10	—
C_2	4.60	3.66	3.69
C_3	5.41	4.35	4.66
C_4's	5.40	5.52	3.89
C_5's+C_6	11.10	7.32	5.66
P_1	33.10	38.08	32.61
P_2	10.8	9.70	11.99
P_3	9.75	9.70	11.61
P_4	7.00	9.60	11.37
H_2S	0.41	0.10	—
C_{7+}	60.65	67.08	67.58
C_{7+} molecular weight	232 (232)	223 (235)	232–247 (251)
C_{7+} specific gravity	0.88 (0.86)	0.88 (0.86)	0.88 (0.87)
Bubblepoint at 105°F, psi	600 (578)	500 (478)	626 (628)

*Ref. 17.
**Ref. 15.
P denotes C_{7+} pseudocomponents.
Numbers in parentheses were calculated with ARKES.

TABLE 2—ARKES FLUID DESCRIPTIONS FOR OIL A

Component	Ω_a	Ω_b	Ω_c	Acentric Factor	Molecular Weight	Critical Temperature (°R)	Critical Pressure (psia)	Critical Volume (ft^3/lbm mol)	k_{ij} H_2S	k_{ij} CO_2
N_2	0.42287	0.09389	0.00000	0.03724	28.0200	227.160	493.13	1.4400	0.19850	−0.05500
C_1	0.45268	0.08890	0.00000	0.01047	16.0420	343.000	666.45	1.5900	0.08470	0.16481
CO_2	0.40341	0.07975	0.00000	0.23100	44.0100	547.578	1070.74	1.5100	0.08150	
C_2	0.41017	0.08141	0.00000	0.09916	30.0680	549.594	706.54	2.3700	0.05000	0.05000
C_3	0.41510	0.07987	0.00000	0.15230	44.0940	665.730	616.04	3.2100	0.05000	0.05000
C_4's	0.43076	0.08144	0.00000	0.19660	58.1200	758.743	545.81	4.1100	0.05000	0.05000
C_5's$+C_6$	0.44269	0.08023	0.00000	0.26380	77.4530	865.046	460.00	6.2500	0.05000	0.05000
P_1	0.46448	0.07319	0.00000	0.44960	145.0000	1136.517	338.00	13.2500	0.05000	0.05000
P_2	0.47604	0.06190	0.00000	0.72220	242.0000	1383.140	215.00	13.7500	0.08000	0.10000
P_3	0.47115	0.05479	0.00000	0.90470	331.0000	1481.290	126.00	17.0000	0.08000	0.11000
P_4	0.42585	0.03936	0.00000	1.34310	486.0000	1611.230	103.00	25.0000	0.05030	0.11000
H_2S	0.41145	0.08109	0.00000	0.09110	34.0800	672.354	1306.21	1.5650		0.08150

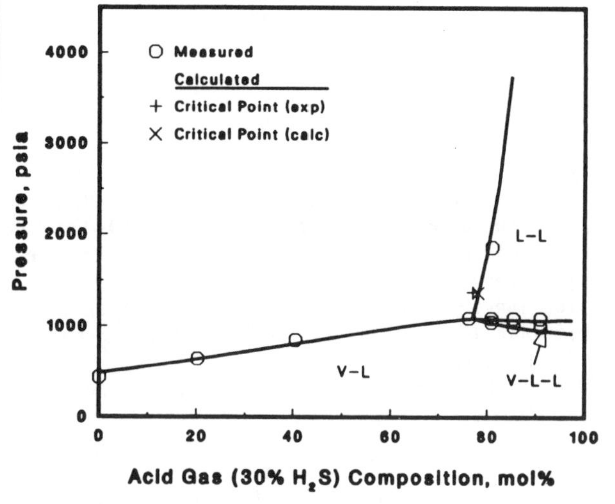

Fig. 3—Calculated and measured *p-x* loci for the acid gas (70/30 mol% CO_2/H_2S/Oil A system.

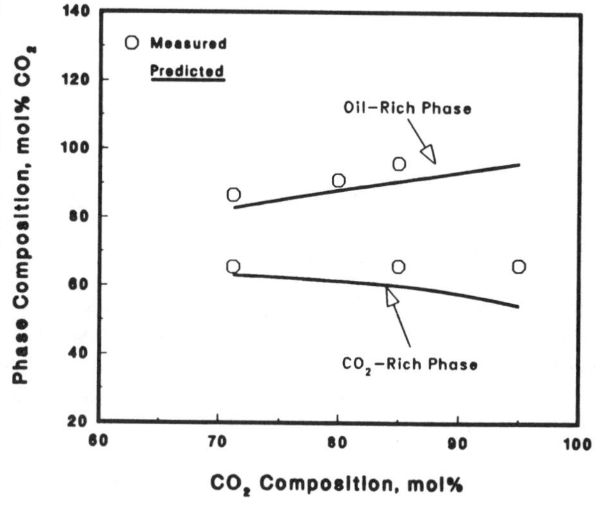

Fig. 5—Predicted and measured phase composition for the CO_2/Oil A system at 2,000 psia.

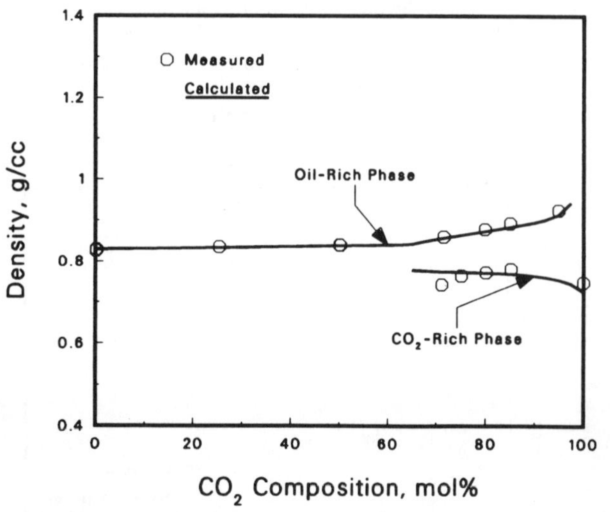

Fig. 4—Calculated and measured densities for the CO_2/Oil A system at 2,000 psia.

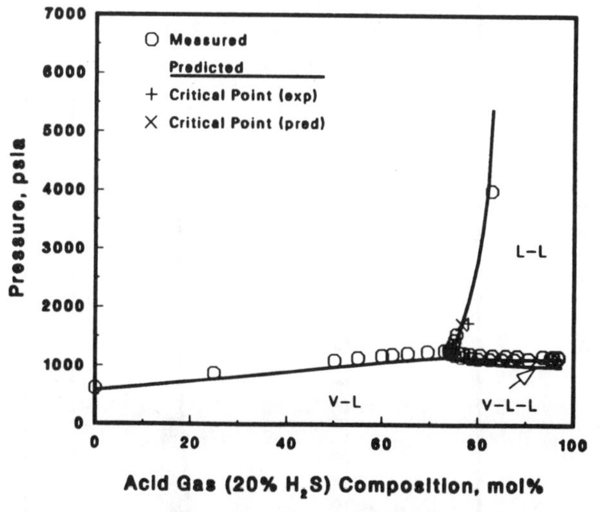

Fig. 6—Predicted and measured *p-x* loci for the acid gas (80/20 mol% CO_2/H_2S)/Oil A system.

CO_2/H_2S displacements of Oil A were used in tuning the ARKES parameters. **Table 2** presents the new 10-component ARKES fluid description. Calculated C_{7+} properties and bubblepoint pressures given in Table 1, are in very good agreement with measured data. The calculated MMP of 1,156 psia [8.0 MPa] is well within the ±100-psi [±0.7-MPa] accepted tolerance of the experimental slim-tube value of 1,100 psia [7.6 MPa].

Fig. 3 compares the calculated and experimental *p-x* data used in tuning ARKES parameters. The match between the calculated and experimental V/L, V/L/L (upper and lower consolute pressure

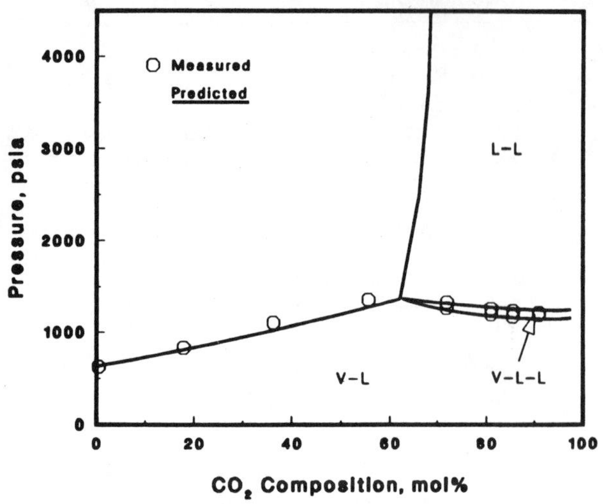

Fig. 7—Predicted and measured *p-x* loci for the CO_2/Oil A system.

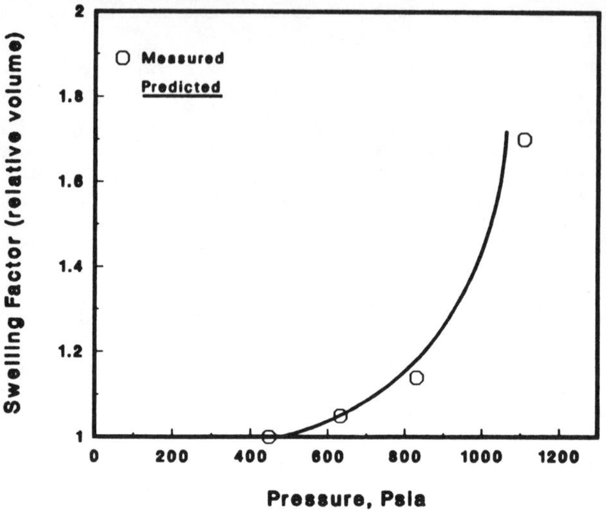

Fig. 8—Predicted and measured swelling factors for the acid gas (70/30 mol% CO_2/H_2S)/Oil A system.

curves), and L/L immiscibility regions is excellent. Note that the match between the calculated and experimental critical point also is excellent.

Fig. 4 compares the calculated and experimental phase densities. Excellent agreement is shown for the oil-rich phase. The inconsistency between the calculated and measured data of the CO_2-rich phase mainly results from two factors: the EOS fluid description underpredicts the pure CO_2 density by about 5% and the calculated CO_2 compositions in the CO_2-rich phase are not in complete agreement with the measured values (see **Fig. 5**).

Validation of Oil A EOS Parameters. Excellent comparisons of the predicted and experimental *p-x* data for the 80/20 mol% CO_2/H_2S/Oil A and CO_2/Oil A systems at 105°F [41°C] are shown in **Figs. 6 and 7**, respectively. Fig. 5 indicates that the predicted and measured phase compositions for the CO_2/Oil A system at 2,000 psia [13.8 MPa] are in good agreement. **Fig. 8** compares the predicted and experimental swelling factors over the pressure range of 500 to 1,200 psia [3.4 to 8.3 MPa]. The predicted volumetric properties are also in excellent agreement with the measured values, as shown in **Fig. 9**.

Nine slim-tube displacements of Oil A with an 80/20 mol% mixture of CO_2/H_2S were conducted at 105°F [41°C] over the pressure range of 912 to 1,250 psia [6.3 to 8.6 MPa]. **Fig. 10** shows the ultimate oil recovery as a function of pressure; ultimate recoveries were determined after 6 to 7 HCPV of solvent injection by subtracting the measured residual oil saturation from 100%. The MMP was determined to be 1,100±100 psia [7.6±0.7 MPa] for this system.

Six simulations were conducted with the new 10-components ARKES parameters. The results, also plotted in Fig. 10 (square symbols), show that the recoveries are in good agreement with experiments. Production rate data are unavailable for this set of slim-tube tests, but production rates from gravity-stable vertical Berea core tests (no viscous fingering) were well matched by simulation with measured relative permeability and dispersivity.

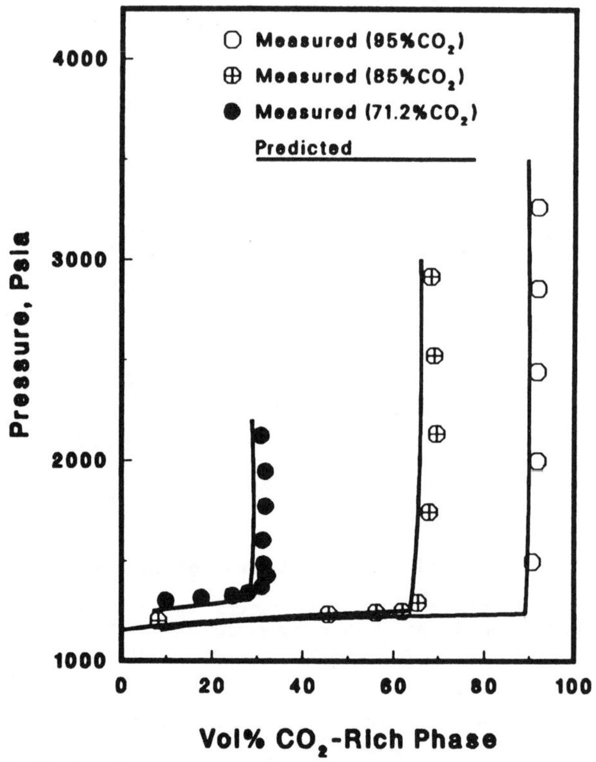

Fig. 9—Predicted and measured phase volume distributions for the CO_2/Oil A system.

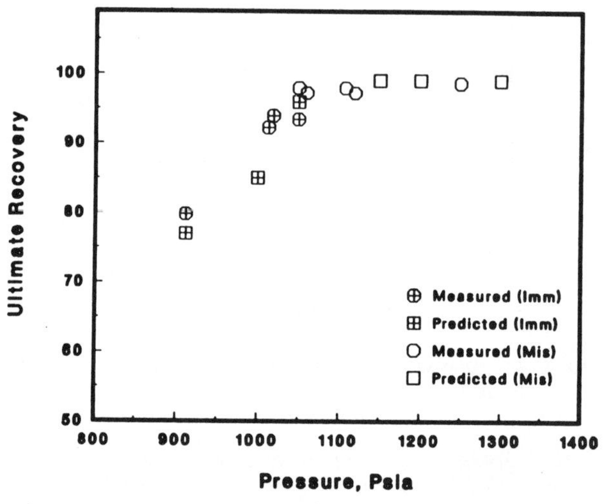

Fig. 10—Predicted and measured slim-tube recoveries for the acid gas (80/20 mol% CO_2/H_2S)/Oil A system.

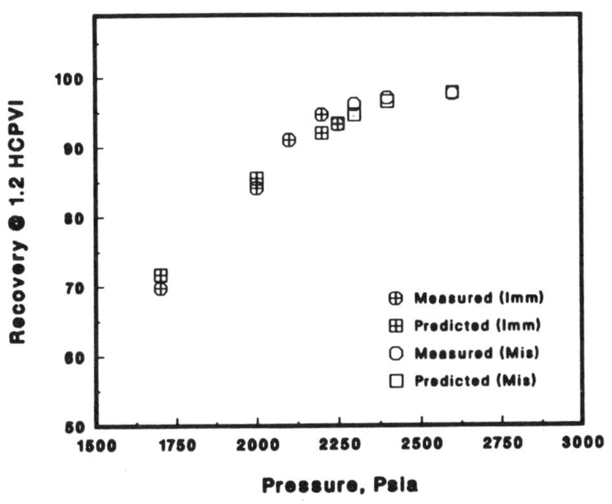

Fig. 11—Predicted and measured slim-tube recoveries for the CO_2/Oil B system.

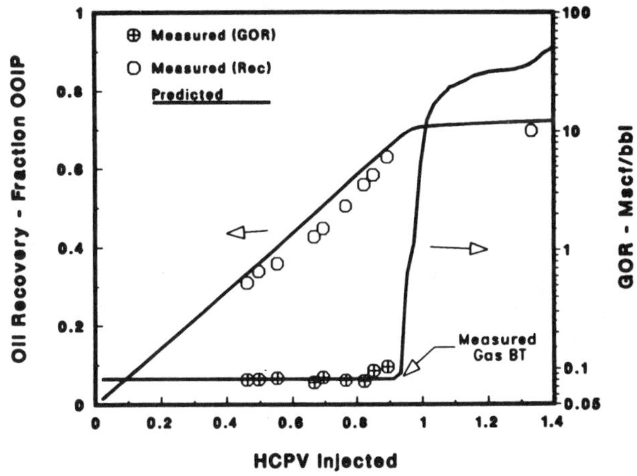

Fig. 12—Predicted and measured oil recovery and GOR for the CO_2/Oil B system at 1,700 psia.

ARKES Fluid Description for the CO_2/Oil B System

The ARKES parameters for the CO_2/Oil B system at 160°F [71°C] are reported by Kremesec and Sebastian.[2] They reported good agreement between the calculated and experimental first-contact data and core tests above the MMP.

Prediction of Slim-Tube Performance. Eight slim-tube CO_2 displacements of Oil B were conducted. The displacements were carried out at 160°F [71°C] over a pressure range of 1,700 to 2,600 psia [11.7 to 17.9 MPa]. Oil rates and GOR were obtained. The MMP was determined to be 2,300±100 psia [15.9±0.7 MPa] for this system.

Six slim-tube compositional predictions were carried out. The predicted oil recoveries at 1.2 HCPV injection over the entire range of test pressures (1,700 to 2,600 psia [11.7 to 17.9 MPa]) are in excellent agreement with the measured values, as shown in **Fig. 11**. The predicted oil recovery and GOR histories match the experimental data very well over the range of pressures, see **Figs. 12 through 14**.

As stated earlier, a dispersivity of 0.05 ft [1.52 cm] was used in all simulations. For comparison, a simulation was conducted with dispersivity of 0.025 ft [0.76 cm] for this system at 2,600 psia [17.9 MPa]. As Fig. 14 shows, the differences between the results (the dotted and solid lines) are well within the experimental error. The CPU time requirement for the run with $\alpha=0.025$ ft [0.76 cm] was more than twice that needed when $\alpha=0.05$ ft [1.52 cm] was used.

When attempts were made to simulate immiscible slim-tube displacements with relative permeability curves reported in the literature, oil rates, breakthrough time, and recovery generally did not match. For example, with Auxiette and Chaperon's[18] curve, which is less efficient than the curve in Fig. 1, lower-than-measured oil rate, earlier breakthrough (~5% PV), and lower ultimate recovery [≈8% original oil in place (OOIP)] were predicted vs. experimental values. Thus, validation of the EOS below the MMP required the use of measured relative permeability curves.

Conclusions

1. Kremesec and Sebastian's[2] tuning criteria yield an EOS fluid description capable of simulating the displacement performance of vaporizing enhanced gasdrive processes over a wide range of pressures, including both immiscible and MCM conditions.
2. Measured slim-tube relative permeability must be used to predict immiscible performance, and the use of measured dispersivity results in no gridblock number adjustment. Thus, slim-tube performance was predicted without adjustable parameters.
3. A tuning method was presented for determining EOS parameters from first-contact data and slim-tube MMP. It can serve as a benchmark for judging future EOS descriptions.
4. A new 10-component EOS description was developed for a typical west Texas oil. The description matches and predicts quantitatively first-contact data and slim-tube MMP.

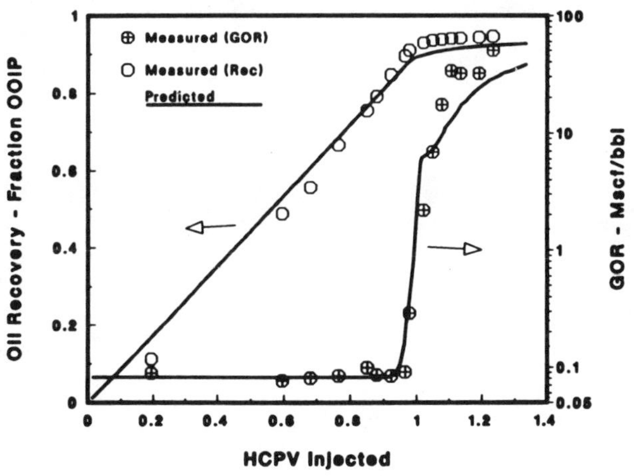

Fig. 13—Predicted and measured oil recovery and GOR for the CO_2/Oil B system at 2,200 psia.

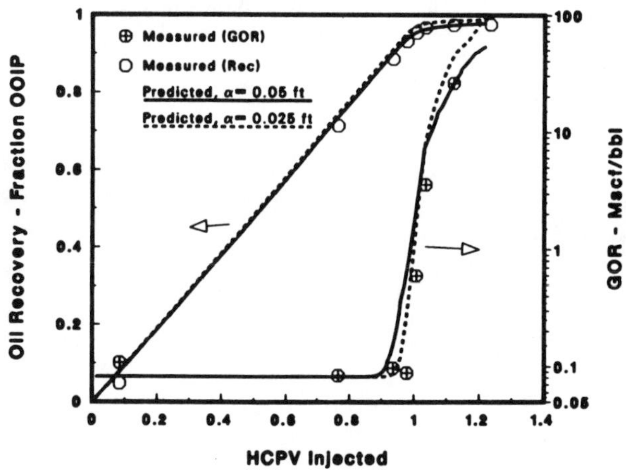

Fig. 14—Predicted and measured oil recovery and GOR for the CO_2/Oil B system at 2,600 psia.

Authors

Negahban

Kremesec

Shahin Negahban is a senior research engineer with Amoco Production Research in Tulsa. He currently is involved in reservoir studies and model development. He holds BS and MS degrees from the U. of Michigan and and a PhD degree from the U. of Kansas, all in chemical engineering. **Vic Kremesec** is an engineering consultant for Amoco Corp. in Naperville, IL, working on remediation of contaminated soils and groundwater. He is chairman of Amoco's Petroleum Environmental Research Forum Underground Discussion Group and a member of the API Soils/Groundwater Technical Task Force. He holds a PhD degree in chemical engineering from Northwestern U.

Nomenclature

F_μ = viscosity ratio
k_g = gas permeability, md [μm^2]
k_{ij} = interaction parameter
k_o = oil permeability, md [μm^2]
p = pressure, psi [MPa]
S_g = gas saturation, fraction
x = composition
α = dispersivity, ft [cm]
$\Delta\rho$ = density difference, g/cm^3
$\Omega_a, \Omega_b, \Omega_c$ = EOS parameters

Acknowledgments

We acknowledge T.A. Renner and G.F. Potter for providing measured slim-tube parameters and L.R. Smith, R.S. Metcalfe, and G.L. Trent for providing slim-tube displacement data.

References

1. Yellig, W.F. and Metcalfe, R.S.: "Determination and Prediction of CO_2 Minimum Miscibility Pressures," *JPT* (Jan. 1980) 160–68; *Trans.*, AIME, **269**.
2. Kremesec, V.J. Jr. and Sebastian, H.M.: "CO_2 Displacements of Reservoir Oils From Long Berea Cores: Laboratory and Simulation Results," *SPERE* (May 1988) 496–504.
3. Hagen, S. and Kossack, C.A.: "Determination of Minimum Miscibility Pressure With a High-Pressure Visual Sapphire Cell," paper SPE 14927 presented at the 1986 SPE/DOE Enhanced Oil Recovery Symposium, Tulsa, April 20–23.
4. Kossak, C.A.: "The Simulation of Phase Behavior and Slim-Tube Displacements With Equation of State," paper SPE 14151 presented at the 1985 SPE Annual Technical Conference and Exhibition, Las Vegas, Sept. 22–25.
5. Sigmund, P.M. *et al.*: "Laboratory CO_2 Floods and Their Computer Simulation," *Proc.*, World Pet. Cong., Bucharest (1979).
6. Gardner, J.W., Orr, F.M. Jr., and Patel, P.D.: "The Effect of Phase Behavior on CO_2-Flood Displacement Efficiency," *JPT* (Nov. 1981) 2067–81.
7. Orr, F.M. Jr., Silva, M.K., and Lien, C.L.: "Equilibrium Phase Compositions of CO_2/Crude Oil Mixtures—Part 2: Comparison of Continuous Multiple-Contact and Slim-Tube Displacement Tests," *SPEJ* (April 1983) 281–91.
8. Johnson, E.F., Bossler, D.P., and Naumann, V.O.: "Calculation of Relative Permeability From Displacement Experiments," *JPT* (Jan. 1959) 61–63; *Trans.*, AIME, **216**.
9. Henry, R.L. and Metcalfe, R.S.: "Multiple-Phase Generation During CO_2 Flooding," *SPEJ* (Aug. 1983) 595–601.
10. Shiralkar, G.S. and Stephenson, R.E.: "A Generalized Formulation for Simulating Physical Dispersion and a New Nine-Point Scheme," *SPERE* (Feb. 1991) 115–20.
11. Negahban, S., Shiralkar, G.S., and Gupta, S.P.: "Simulation of the Effects of Mixing in Gasdrive Core Tests of Reservoir Fluids," *SPERE* (Aug. 1990) 402–08.
12. Turek, E.A. *et al.*: "Phase Equilibria in CO_2-Multicomponent Hydrocarbon Systems: Experimental Data and an Improved Prediction Technique," *SPEJ* (June 1984) 308–24.
13. Jossi, J.A., Stiel, L.I. and Thodos, G.: "The Viscosity of Pure Substances in the Dense Gaseous and Liquid Phases," *AIChE J.* (1962) **8**, 59–63.
14. Turek, E.A., Luks, K.D., and Baker, L.E.: "Calculation of Minimum Miscibility Pressure," *SPERE* (Nov. 1987) 501–06; *Trans.*, AIME, **283**.
15. Turek, E.A., Metcalfe, R.S., and Fishback, R.E.: "Phase Behavior of Several CO_2/West-Texas-Reservoir-Oil Systems," *SPERE* (May 1988) 505–16.
16. Graue, D.J. and Zana, E.T.: "Study of a Possible CO_2 Flood in Rangely Field," *JPT* (July 1981) 1312–18.
17. Chaback, J.J. and Turek, E.A.: "Application of the Redlich-Kwong Equation of State to Phase Behavior of Mixtures of West Texas Oils With Acid Gases (CO_2 and H_2S/CO_2)," paper presented at the 1985 American Chemical Soc. Natl. Meeting, Miami Beach, April 28–May 3.
18. Auxiette, G. and Chaperon, I.: "Linear Gas Drives in High-Pressure Oil Reservoirs: Compositional Simulation and Experimental Analysis," paper SPE 10271 presented at the 1981 SPE Annual Technical Conference and Exhibition, San Antonio, October 5–7.

SI Metric Conversion Factors

ft	× 3.048*	E−01	=	m
ft^3	× 2.831 685	E−02	=	m^3
°F	(°F−32)/1.8		=	°C
lbm mol	× 4.535 924	E−01	=	kmol
psi	× 6.894 757	E+00	=	kPa
°R	°R/1.8		=	K
scf/bbl	× 1.801 175	E−01	=	std m^3/m^3

*Conversion factor is exact.

SPERE

Original SPE manuscript received for review Oct. 9, 1989. Revised manuscript received May 10, 1991. Paper accepted for publication March 5, 1992. Paper (SPE 19637) first presented at the 1989 SPE Annual Technical Conference and Exhibition held in San Antonio, Oct. 8–11.

CO_2 Flow Patterns Under Multiphase Flow: Heterogeneous Field-Scale Conditions

Yih-Bor Chang,* SPE, **M.T. Lim,** SPE, **G.A. Pope,** SPE, and **K. Sepehrnoori,** SPE, U. of Texas

Summary. A finite-difference, equation-of-state (EOS), compositional simulator has been used to study CO_2 flooding. First, unstable first-contact-miscible (FCM) displacements were simulated with a fine mesh to investigate the transition from gravity override to viscous fingering. Next, a direct comparison was made for FCM and multiple-contact-miscible (MCM) displacements under the same conditions to investigate effects of phase behavior on the growth of viscous fingers. Then, the effects of gravity, physical dispersion, capillary pressure, phase behavior, and heterogeneity were combined and simulated for CO_2 flooding on a field scale with stochastic permeability fields.

Introduction

For many years, viscous fingering has been considered one of the dominant flow patterns for displacements at adverse mobility ratios and has been an active research topic both experimentally and theoretically. Most of the studies, however, have been conducted under simplified conditions. For example, Peaceman and Rachford,[1] Christie and Bond,[2] Araktingi and Orr,[3] Ewing *et al.*,[4] Christie,[5] Christie *et al.*,[6] and Waggoner *et al.*[7] simulated viscous fingering for FCM displacements. Remarkably, none of these studies considered multiphase flow, which commonly occurs in MCM floods. One motivation for our studies was to determine the effect of multiphase flow on viscous fingering.

Many factors affect viscous finger growth. Fayers and Newley[8] and Christie *et al.*[9] showed that FCM displacements are dominated by viscous fingering only at low gravity number, N_g (defined as the ratio of gravity to viscous forces). At high N_g, displacements are dominated by gravity override. Fingering is defined as viscous instability. This distinction is important because fingering and gravity override do not scale the same and do not have the same remedies for improving oil recovery. Fingering disappears for mobility ratios $\leqq 1$, whereas gravity override disappears only when fluid densities are equal. Furthermore, water-alternating-gas (WAG) injection is commonly used to stabilize fingering in CO_2 flooding. Christie[5] and Christie *et al.*[6] demonstrated that viscous fingering was reduced for miscible WAG displacements at optimal WAG ratios.

Most fingering studies have been limited to quasihomogeneous porous media with small permeability variations. In typical oil reservoirs, however, a larger heterogeneity with a 0.6 to 0.8 Dykstra-Parsons coefficient is typical. Araktingi and Orr[3] and Waggoner *et al.*[7] included heterogeneities of this magnitude in their studies. Araktingi and Orr applied a random-walk model to simulate unstable flow in statistically generated heterogeneous porous media. Their model was restricted to incompressible FCM fluids. Waggoner *et al.* used a finite-difference simulator to simulate unstable displacements and introduced the notion of constructing flow regime maps that we illustrate for MCM displacements below. Fayers *et al.*[10] investigated the effects of heterogeneity for lean gas, FCM, and surfactant displacements. However, only limited heterogeneity runs were made in their study.

The motivation of this study was to investigate the importance of viscous fingering in MCM flooding, in particular in CO_2 flooding. The phase behavior of CO_2 flooding is always more complicated than that of FCM displacements even when the pressure is far above the minimum miscibility pressure (MMP). The viscosity contrast between swept and unswept zones for CO_2 flooding can be reduced because of extraction of hydrocarbon components from the oil phase to the CO_2-rich phase, or vice versa, and because relative permeabilities for multiphase flow lower the fluid mobilities compared with those for single-phase flow. Both factors will reduce the mobility ratio at the displacement front and thus stabilize the displacement. Another important factor is the permeability distribution. Displacements may channel when reservoirs have permeability fields of large correlation length.[3] The distinction between viscous fingering and channeling is also important because they do not scale the same way and do not have the same effect on oil recovery or the same remedies. Channeling is caused by permeability variations; it will not occur in the presence of uniform permeability or random permeability fields but will occur at unit mobility ratio. These effects plus gravity, capillary pressure, and dispersion reduce the tendency for viscous fingering.

Because all these factors can exist for CO_2 flooding, one may wonder whether fingering is ever important for field applications of CO_2 flooding. If it is not, under what conditions is fingering important? What are the dominant flow patterns under field conditions?

We began with simulating unstable FCM displacements to investigate the conditions under which viscous fingering is the dominant flow pattern when gravity effects are included. In addition, simulations were made to investigate possible 3D effects on these unstable displacements. Then, the effects of phase behavior on viscous fingering were studied by comparing FCM displacements with CO_2 flooding. Quasihomogeneous reservoirs were used in these initial simulations. The complexity of the simulated conditions was further increased in the final study when reservoir heterogeneity was included in addition to gravity and physical dispersion. The effects of heterogeneity, gravity, capillary pressure, and dispersion on the flow patterns were investigated for field-scale CO_2 flooding.

Simulator Description

UTCOMP[11] is a 3D, EOS, compositional simulator. Four phases may flow simultaneously: aqueous, oleic, and gaseous phases and a second, nonaqueous liquid. Water is allowed only in the aqueous phase. The UTCOMP solution scheme is analogous to that of implicit pressure/explicit saturation solutions. First, the gridblock pressure is solved for implicitly by explicit dating of saturation-dependent terms. Then, the material-balance equations are solved explicitly for the total concentration (in moles) of each component. A third-order finite-difference method[12] is used to reduce numerical dispersion and grid-orientation effects when solving the material balance equations. The third-order method is the key to the feasibility of our simulation because it allows accurate simulations with a reasonable number of gridblocks and computer time. Mesh refinement and stability studies[12] have shown this third-order method to be far superior to the first-order finite-difference method (one-point upstream weighting) most commonly used in compositional simulators and significantly superior to the second-order methods, such as two-point upstream weighting, sometimes used in reservoir simulators; however, the computing cost per timestep per gridblock is almost the same as the lower-order methods. Furthermore, it can be used for nonuniform gridblocks and multiphase flow in three dimensions. This third-order method was originally restricted to cell

*Now at Western Atlas Software.

Original SPE manuscript received for review Jan. 15, 1992. Revised manuscript received Oct. 27, 1993. Paper accepted for publication Jan. 11, 1994. Paper (SPE 22654) first presented at the 1991 SPE Annual Technical Conference and Exhibition held in Dallas, Oct. 6–9.

TABLE 1—SUMMARY OF SIMULATION CASES AND DIMENSIONLESS NUMBERS

Case	Description	Grid ($N_x \times N_y \times N_z$)	M^o	N_g	R_L	$N_{Pe,\ell}$	$N_{Pe,t}$	V_{DP}	λ_{Dx}	λ_{Dy}	λ_{Dz}	N_{RL}
1a	2D vertical FCM (van der Poel)	1×150×20	7.9	0.0035, 0.021, 0.04, 0.05	16.4	600	216	*	0	0	0	—
1b	3D FCM (van der Poel)	8×50×20 16×150×20	7.9	0.021	16.4	600	216	*	0	0	0	—
2a	2D areal FCM	80×20×1	10	0	50	200	∞	0.095	0.025	0.1	—	—
2b	2D vertical FCM	80×1×10	10	0	50	200	80	0.63, 0.80	0.025, 0.5	—	0.1	—
3a	2D areal MCM	80×20×1	5.3	0	50	200	∞	0.095	0.025	0.1	—	1408
3b	2D areal MCM	80×20×1	5.3	0	50	200	80	0.095, 0.26, 0.63	0.025, 0.1, 0.5, 1.0, 3.0	0.1	—	1408
					Isotropic							
4a	2D vertical MCM	80×1×20	5.3	0.0055	50	200	80	0.095, 0.26, 0.63, 0.80	0.025, 0.1, 0.5, 1.0, 3.0	—	0.1	1408
4b	2D vertical MCM	80×1×40	5.3	0.0055	50	200	80	0.095, 0.63	0.1, 0.5	—	0.1	1408
4c	2D vertical MCM (WAG ratio=1)	80×1×20	5.3	0.0055	50	200	80	0.095	0.025	—	0.1	1408
4d	2D vertical MCM	80×1×20	5.3	0.0055, 0.049	50	200	0.8	0.63	1.0	—	0.1	1408, 156
4e	2D vertical MCM	80×1×20	5.3	0.0055	5	200	80	0.095, 0.63, 0.80	0.025, 0.5, 1.0	—	0.1	1408
4f	2D vertical MCM	80×1×10	5.3	0.0055, 0.016, 0.049	50	200	80	0.095, 0.26, 0.8, 0.63, 0.95	0.025, 0.1, 0.5, 1.0, 3.0	—	0.1	1408, 469, 156
					Anisotropic							
5	2D vertical MCM	80×1×10	5.3	0.0055, 0.016, 0.049	1.6	200	80	0.095, 0.26, 0.63, 0.95	0.025, 0.1, 1.0, 3.0	—	0.1	1408, 469, 156
6a	3D MCM	80×4×10	5.3	0.0055	50	200	80	0.63	1.0	1.0	0.1	1408
6b	3D MCM	80×4×10	5.3	0.0055, 0.049	50	200	80	0.63	1.0	1.0	0.1	1408, 156

*K_v = 0.05 for normally distributed permeability field.

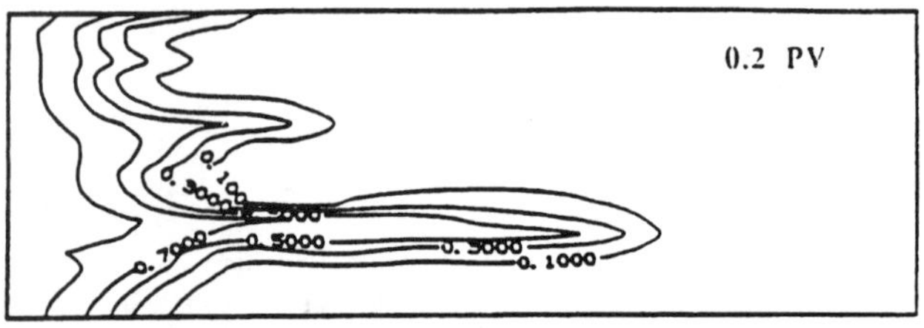

Fig. 1—Overall mole fraction contours of the displacing fluid for a matched-density, FCM displacement.

Peclet numbers of less than ≈10. However, recent enhancements by Liu *et al.*[13] have removed this limitation by adding a flux limiter that makes the method total variation diminishing. Liu *et al.* have also changed the time integration from a first- to a second-order correct method that increases both the accuracy and stability of the solution (larger Courant numbers may now be used).

The hydrocarbon fluid properties are modeled with the Peng-Robinson EOS or a modified version of the Soave-Redlich-Kwong EOS. A rigorous Gibbs stability test is done before all flash calculations to determine the number of phases present, and a phase identification test is done after all flashes to label each phase consistently for subsequent property calculations.[14] Several alternative four-phase relative permeability models are available as options. The full physical dispersion tensor, capillary pressure, and gravity are included. Chang[11] gives the details of all physical property models.

Results and Discussion

Table 1 summarizes the six cases studied and the corresponding dimensionless numbers. Because conditions vary over many orders of magnitude, only describing simulated conditions with dimensionless numbers has a reasonable chance of being useful. The MCM cases in Table 1 were designed to encompass conditions typical of CO_2 field floods. Thus, the mobility ratios are unfavorable but much lower than mobility ratios of many miscible displacements simulated by others. A range of N_g was included because N_g can vary with injection rates, permeability, and density differences. The effective length/thickness ratio, R_L, is equally important in determining the tendency for gravity override and other crossflow phenomena because it controls the approach of the displacement to vertical equilibrium.[15] Both the aspect ratio, L/h, and the vertical/horizontal permeability ratio affect R_L. Two Peclet numbers, a dimensionless ratio of viscous to capillary forces, N_{RL},[11] and three heterogeneity parameters complete the set of scaling groups in Table 1.

To trigger viscous fingers, a small variation, K_v, in permeability was used ($K_v = 0.05$). Krueger[16] and Chang[11] confirmed that a small perturbation was enough to trigger viscous fingers for adverse mobility ratio displacements. Their studies were also used to determine the number of gridblocks required with our third-order numerical method to simulate unstable flow adequately. Both mesh refinement studies and comparisons with experimental data were used for this purpose. Krueger showed that a 20×20 grid gave almost the same recovery curve as a finer 100×72 grid and that both agreed with Blackwell *et al.*'s[17] matched-density FCM experiment at a viscosity ratio of 86. Chang showed that a 75×20 grid gave good agreement with corefloods that used CO_2 to displace oil under unstable conditions (gravity override); these were eight-component MCM displacements.

Fig. 1 shows Chang's[11] FCM, matched-density simulation with the same numerical method and a 30×20 grid. The figure shows the overall mole fraction contours of the displacing fluid at 0.2 PV injected (PVI); the displacing fluid was injected at the left and the initial fluid was produced at the right. The viscosity ratio was 86, longi-

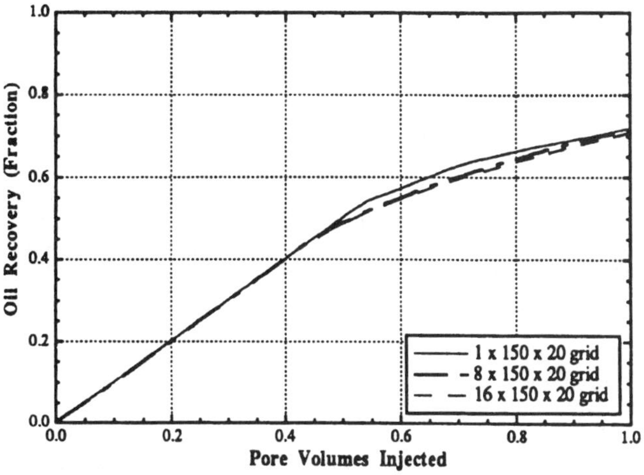

Fig. 2—Comparison of hydrocarbon recovery for 2D and 3D simulations of van der Poel experiment at $N_g = 0.021$.

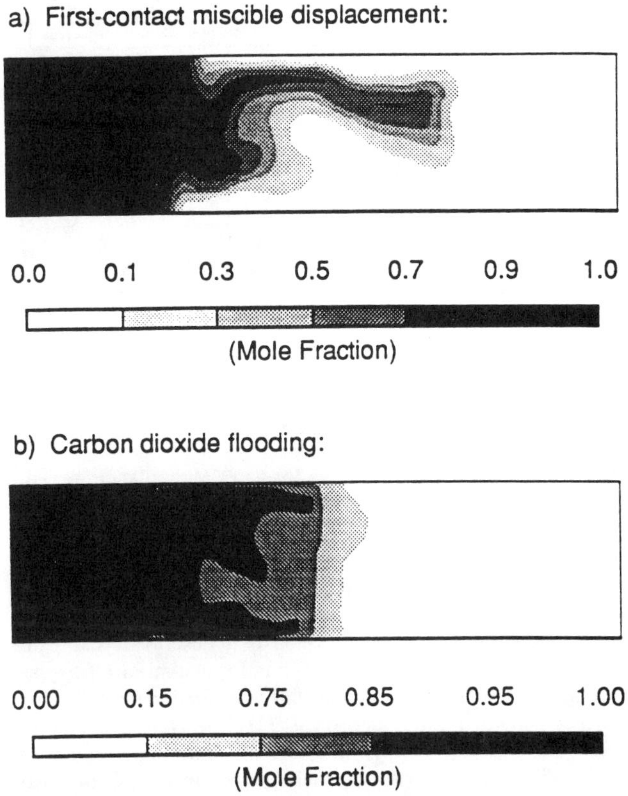

Fig. 3—Comparison of overall mole fraction contours of displacing fluids at 0.3 PVI for an FCM displacement and CO_2 flooding.

tudinal dispersivity was 0.1 ft ($N_{Pe,\ell} = 60$), transverse dispersivity was zero, and $R_L = 3$.

Chang also simulated van der Poel's[18] FCM beadpack experiments and found agreement within 5% when he used a 150×20 grid and very small dispersivities that are typical of glass beadpacks. Fayers and Newley[8] previously simulated these same experiments to investigate the transition from gravity-override- to fingering-dominated displacements. They speculated that there might be some 3D effects on the displacement. To investigate this possibility, we repeated these simulations in three dimensions with $8 \times 150 \times 20$ and $16 \times 150 \times 20$ grids. **Fig. 2** shows that these results are very close to the 2D results. Small differences can be seen in contour plots of overall mole fraction, but these differences are actually smaller than those that result from small perturbations in the description of the beadpack [e.g., the assumed permeability field variation coefficient (assumed to be random) or the longitudinal or transverse dispersivities]. Chang[11] reports these and other results, and Christie *et al.*'s[6] recent results also show almost no difference between FCM displacements in two and three dimensions.

Phase-Behavior Effects. A direct comparison was made of an FCM displacement and an MCM CO_2 flood under the same conditions (Table 1, Case 2a and 3a, respectively). Gravity effects were not included in these simulations. For both runs, an 80×20 grid was used. The reservoir length in the direction of bulk flow was 1,000 ft and the width was 20 ft. The longitudinal dispersivity was 5 ft ($N_{Pe,\ell} = 200$), and the transverse dispersivity was zero. The viscosity ratio for the FCM flow was 10. Chang's[11] three-component characterization of a Maljamar separator oil was used for the MCM CO_2 flood. CO_2 flood pressure and temperature were 1,200 psi and 90°F, respectively. The calculated oil/CO_2 viscosity ratio was 10.8. Dria *et al.*'s[19] relative permeability parameters corresponding to a modified Corey's model were used. The endpoint mobility ratio was 5.3.

Fig. 3 shows a comparison of overall mole fraction contours of the displacing fluid at 0.3 PVI for the FCM displacement with those of the MCM CO_2 flood. The FCM displacement showed much stronger fingering than did the CO_2 flood, which was almost completely stabilized by phase-behavior effects. **Fig. 4** shows a comparison of the mobilities. The FCM flow mobility was computed as $1/\mu$, while relative mobilities for the CO_2 flood were computed as k_{rj}/μ_j. Because of viscous crossflow and physical dispersion, the FCM flow mobility at the transition zone was reduced to about one-half that of the pure solvent but was still about five times that of the initial fluid. Relative mobilities for the CO_2 flood were lower than those for the FCM displacement because two-phase relative permeabilities lower phase mobilities during the MCM CO_2 flood but do not affect the mobility of the single-phase FCM displacement. Also, exchange of components between the oil and CO_2-rich phases causes the viscosity contrast between them to decrease. Both effects reduce growth of viscous fingers. This example illustrates the need to consider phase-behavior effects in CO_2 flooding; all further simulation examples in this paper include phase-behavior effects.

CO_2 Flooding in Heterogeneous Reservoirs. In the two cases discussed above, the effects of gravity and phase behavior on viscous fingering were investigated separately. Now, the effects of phase behavior, gravity, physical dispersion, capillary pressure, and heterogeneity were combined and simulated to study viscous fingering under more realistic conditions for field-scale CO_2 flooding. In the first study, the permeability fields were generated statistically as functions of Dykstra-Parsons coefficient and dimensionless range. These permeability fields included low to high heterogeneities and short to long dimensionless ranges. Isotropic ($k_x = k_z$) and anisotropic ($k_x = 1{,}000\ k_z$) cases were studied. To investigate gravity effects, simulations were made by use of a vertical cross-sectional geometry with different N_g. All these examples were for secondary floods above the MMP. Chang[11] also studied CO_2 floods below the MMP and found similar results.

The 2D heterogeneous permeability fields were generated statistically by use of the turning-band method[20] with a spherical variogram model. All permeability fields were log-normally distributed.

We chose 16 permeability fields comprising four levels of V_{DP} and λ_{Dx} for these studies. The V_{DP} values were 0.095, 0.26, 0.63, and 0.80 unless otherwise indicated. These values represent reservoirs from very low to very high heterogeneity. λ_{Dx} values were chosen to represent porous media from little to extremely strong correlation. λ_{Dx} values were 0.025, 0.1, 1, and 3, with some exceptions as noted. The dimensionless spatial correlation of permeability values in the direction transverse to bulk flow, λ_{Dz}, was held constant at 0.1. **Figs. 5 and 6** show contour plots for these 16 permeability fields.

Mean permeability was 24 md, porosity was 0.202, longitudinal dispersivity was 5 ft ($N_{Pe,\ell} = 200$), and transverse dispersivity was 0.005 ft ($N_{Pe,t} = 80$). The phase-behavior and relative-permeability data were the same as those for the previous MCM example. Chang[11] gives input data details.

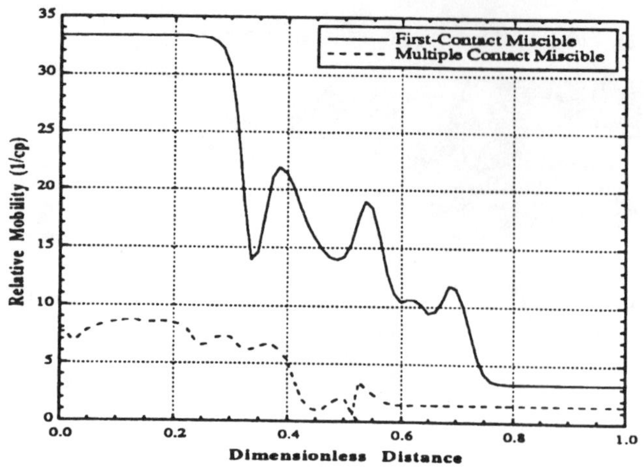

Fig. 4—Comparison of mobilities for an FCM displacement and an MCM CO_2 flood at y_D=0.25 and 0.3 PVI.

At least four displacement patterns can be found for MCM displacements in the 16 permeability fields: channeling, fingering, gravity override, and dispersive mixing. Although fingering and channeling are often used synonymously to describe unstable displacements, we use the term channeling for nonuniform flow patterns that are dominated by permeability distribution for the reasons discussed previously. Comparison of FCM flow patterns at unit mobility ratio with those of MCM flow at nonunit mobility ratio was used to distinguish between channeling and fingering. Comparison of flow patterns of MCM displacements with and without gravity was used to distinguish between channeling and gravity override.

Forty-eight simulations were made for the MCM CO_2 flood in isotropic reservoirs (Case 4) for N_g = 0.0055, 0.016, and 0.049 and each of the 16 permeability fields. The effective R_L for these isotropic reservoirs was 50, a value high enough to expect vertical equilibrium.[15] An 80×20 grid was used for these simulations.

Additional simulations were made at unit mobility ratio and at nonunit mobility ratio without gravity (with 80×20 areal grids instead of 80×20 vertical grids) to help to identify the dominant flow pattern. **Figs. 7 and 8** show example comparisons for one of the permeability fields. Fig. 7c shows that the displacement was fingering because comparison with the FCM unit mobility ratio displacement (Fig. 7a) shows almost uniform flood fronts whereas the MCM nonunit mobility displacements with (Fig. 7c) and without gravity (Fig. 7b) show large instabilities in the flood fronts. This was one of only four fingering results in our entire study and occurred under conditions of very small heterogeneity, which are unlikely to be realized in the field. Fig. 7f shows that the displacement for this permeability field is channeling because, at both unit mobility ratio (Fig. 7d) and at nonunit mobility ratio with (Fig. 7f) and without gravity (Fig. 7e), the flood front correlates strongly with the permeability distribution. By chance, this particular realization gave high permeability at the top of the reservoir (Fig. 6b). As a check on our interpretation, additional realizations were done whenever such conditions as these occurred. Although the quantitative results vary, the qualitative conclusion regarding identification of flow regime change did not change in any of the cases.

Fig. 8c shows a dispersive flow example. Comparison with either Fig. 8a or 8b leads to the conclusion that the flow is not channeling, fingering, or gravity unstable; instead, a mixing zone develops as a result of the physical dispersion plus the mixing caused by the heterogeneities of small λ_D. Fig. 8f shows an example of gravity override. Comparison with Fig. 8d shows that this is not channeling, and Fig. 8e shows that, without gravity, a finger develops in the center of the reservoir rather than a gravity tongue at the top of the reservoir.

Fig. 9 is a schematic diagram of the flow regimes at N_g = 0.0055. Waggoner *et al.*[7] introduced this method of representing flow patterns to summarize their FCM displacement studies. Both Waggoner *et al.* and we found fingering only at very low heterogeneity. Because phase-behavior effects further reduce the tendency for fingering, our fingering region is even smaller than theirs. The low V_{DP} and λ_{Dx} correspond more to laboratory than to field scale. Displacements were channeling for the permeability fields at $\lambda_{Dx} \geqq 0.5$ and dispersive at other simulated conditions. **Fig. 10** shows the effect of mesh size on oil recovery for two of the permeability fields. The oil recovery changed $<2\%$ when the grid was changed from 80×20 to either 80×10 or 80×40. Therefore, additional 2D simulations were made with an 80×10 grid.

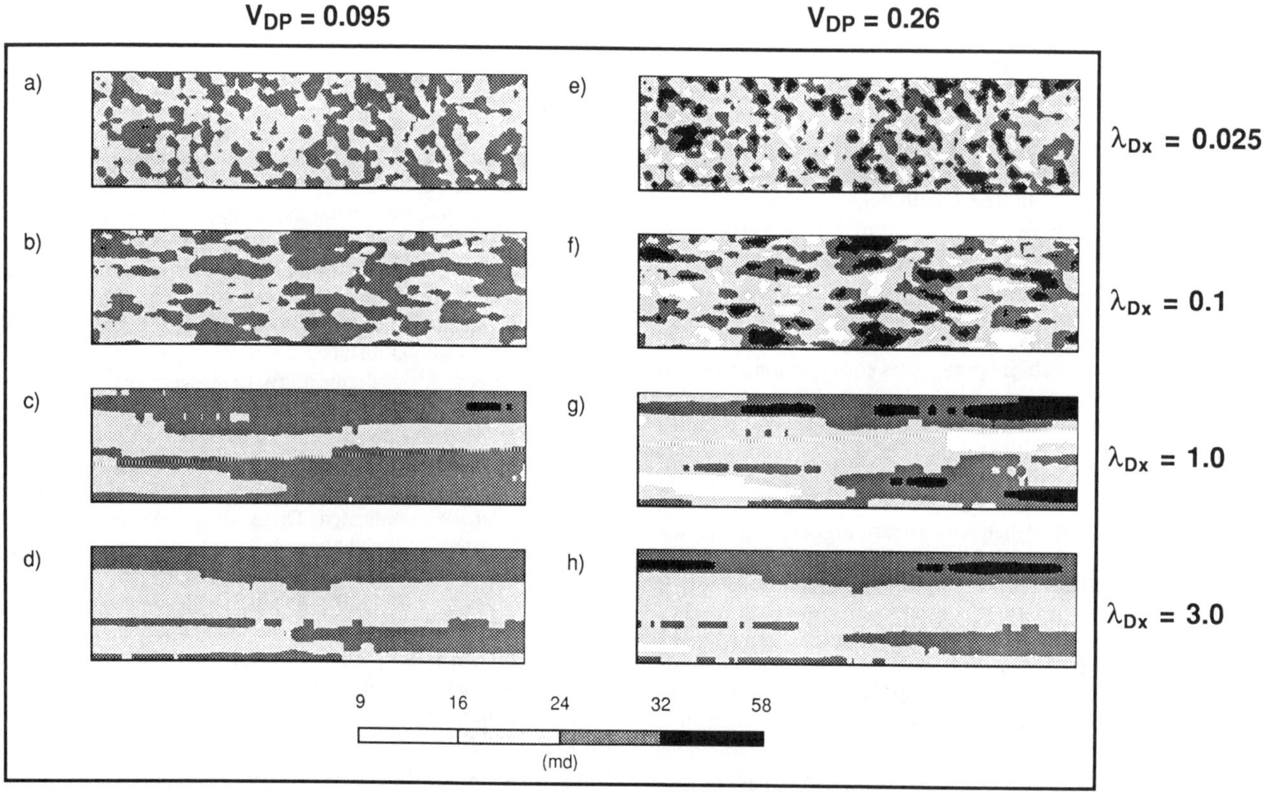

Fig. 5—Permeability fields for Dykstra-Parsons coefficients of 0.095 and 0.26.

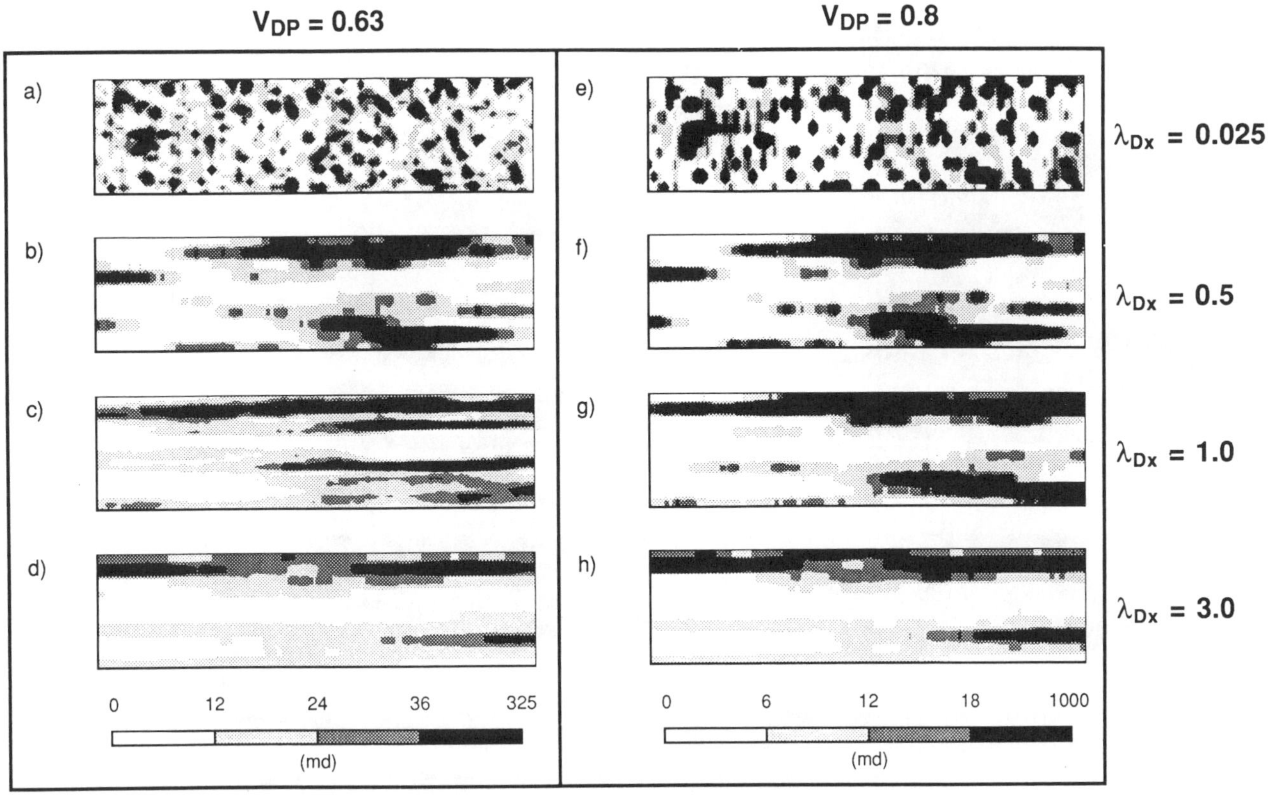

Fig. 6—Permeability fields for Dykstra-Parsons coefficients of 0.63 and 0.8.

Fig. 11 is a schematic diagram of the flow patterns at $N_g = 0.016$. The MCM displacements were channeling for the runs with $\lambda_{Dx} = 1$ and 3. Displacements showed gravity override with $V_{DP} = 0.095$ and 0.26 with $\lambda_{Dx} = 0.1$. The dispersive pattern was found for displacements with low λ_{Dx}. No fingering was found for this set of simulation runs. For displacements at $N_g = 0.049$ (**Fig. 12**), the flow pattern was gravity override at low V_{DP}, dispersive when V_{DP} was increased at low λ_{Dx}, and channeling at high λ_{Dx}.

For the anisotropic reservoirs (Case 5), the displacements were all channeling at all N_g studied. R_L for these anisotropic reservoirs was reduced from 50 to 1.6 by reducing vertical permeability 1,000-fold. This low k_z value tends to suppress crossflow and reduce the tendency for both gravity override and fingering.

Several perturbations were made for isotropic simulations at $Ng = 0.0055$ (Case 4) to evaluate sensitivity to aspect ratio, WAG ratio, 3D flow, and transverse dispersion. **Fig. 13** shows that oil recov-

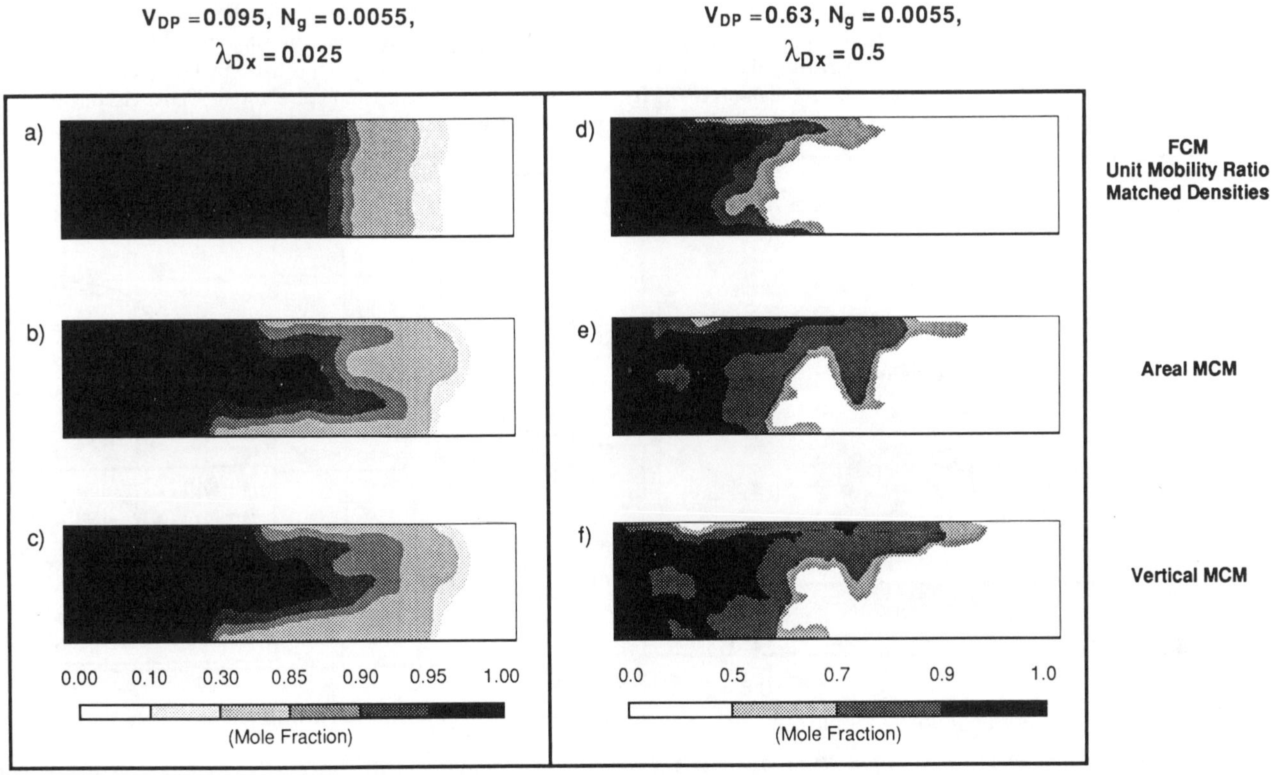

Fig. 7—CO_2 concentration profiles for unit mobility; areal and vertical displacements illustrating fingering and channeling flow patterns.

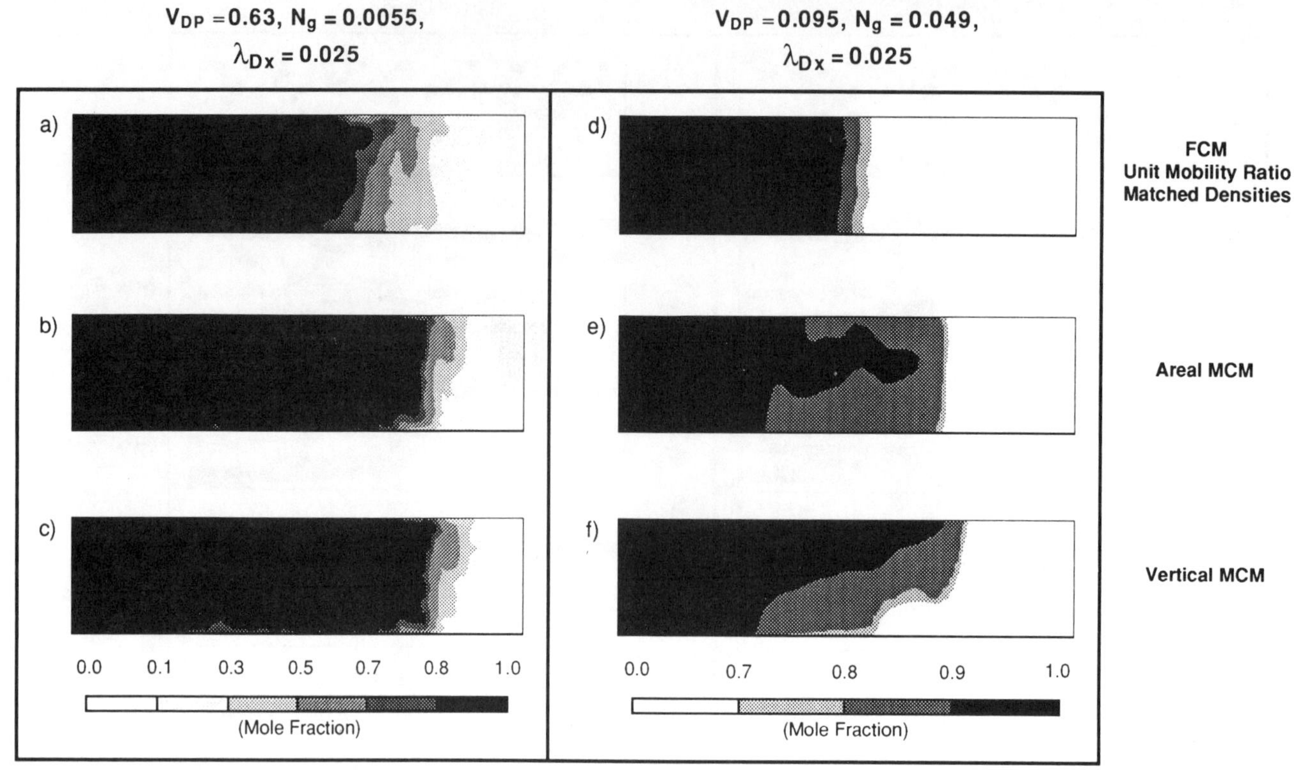

Fig. 8—CO_2 concentration profiles for unit mobility; areal and vertical displacements illustrating dispersive and gravity-override flow patterns.

ery was reduced significantly as R_L was decreased from 50 to 5 (these values are also the aspect ratios for isotropic reservoirs), but that qualitative flow patterns did not change. This was the most significant parameter studied. Khataniar[21] discusses additional studies at low aspect ratio and at a higher mobility ratio than in our study. **Fig. 14** shows that oil recovery was increased when the WAG ratio was increased from 0 to 1. In this case, the flow field did change because it went from weakly fingering at a WAG ratio of zero to gravity override at a WAG ratio of one.

Fig. 15 shows that oil recovery decreased only slightly when the third dimension was included by use of four gridblocks rather than one in the y direction. **Fig. 16** shows that increasing the transverse dispersivity from 0.005 ft used in all previous simulations to 0.5 ft increased the 3D simulated oil recovery. **Fig. 17** shows little difference exists between 2D and 3D oil recoveries at $N_g = 0.0055$ and 0.049 and $\alpha_t = 0.5$ ft. As even finer grids become affordable, their use is recommended. CPU time on a Cray Y-MP for these 3D, 3,200 gridblock simulations was only 0.00035 sec/gridblock-timestep, but the run times are still on the order of 2 hours. Thus, making large numbers of runs and refining the grid as we would like is still costly.

Conclusions

Field-scale simulation study results showed that secondary CO_2 flooding was dominated by viscous fingering only when the Ng was small and the permeability fields had low Dykstra-Parsons coefficients and very short spatial correlations. Thus, viscous fingering was not found to be a dominant flow pattern for field-scale CO_2

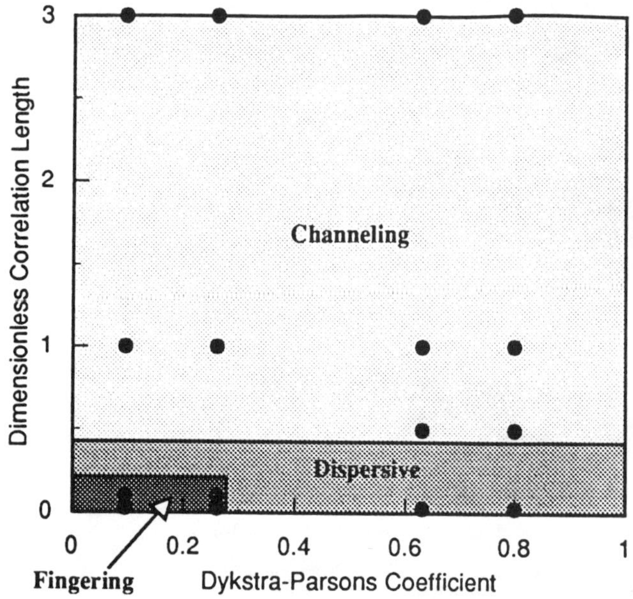

Fig. 9—Flow regime map for isotropic MCM CO_2 flood at $N_g = 0.0055$.

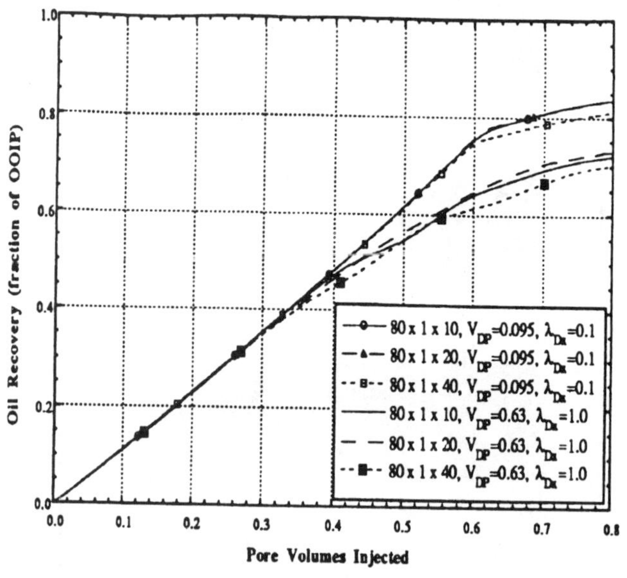

Fig. 10—Comparison of oil recoveries for three mesh sizes with two permeability fields.

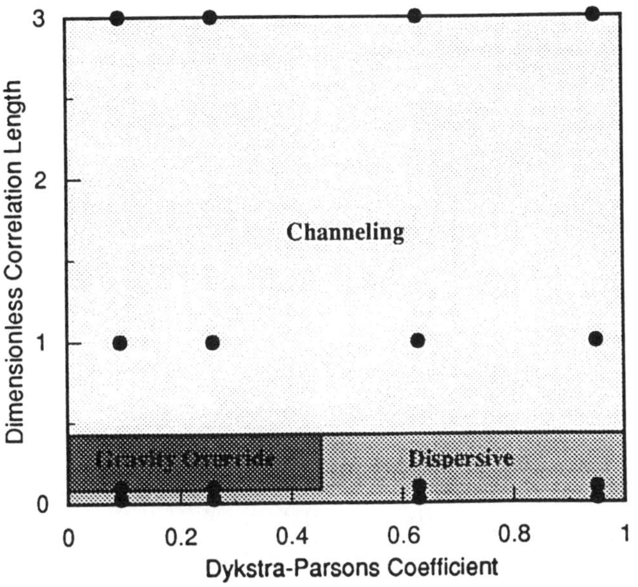

Fig. 11—Flow regime map for isotropic MCM CO_2 flood at $N_g=0.016$.

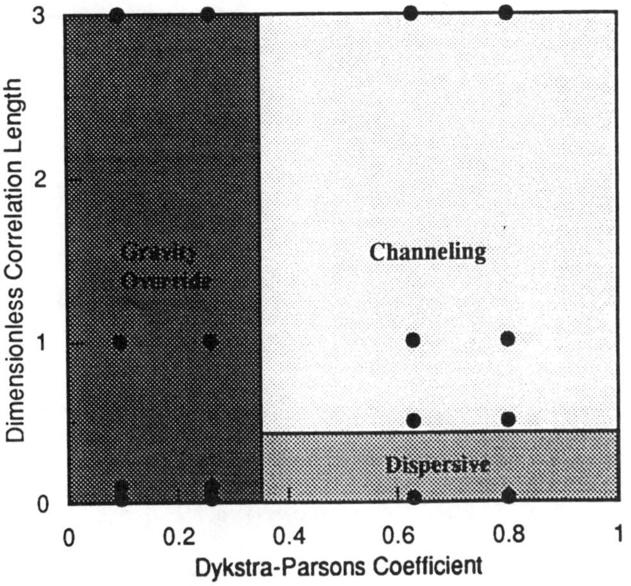

Fig. 12—Flow regime map for isotropic MCM CO_2 flood at $N_g=0.049$.

flooding under secondary conditions even without use of WAG injection, which would tend to reduce fingering even more. An important contributing factor to this result was inclusion of phase-behavior effects for the first time; these reduce the tendency for fingering. In addition, the presence of capillary forces will also reduce the tendency to finger because capillary forces act to smear fingering. Capillary numbers used are typical of field scale.

Under all other conditions, the dominant flow patterns were channeling, gravity override, and dispersive. The flow pattern was channeling when the permeability field had a large Dykstra-Parsons coefficient and large range. For large vertical permeability, the flow pattern was gravity override at high gravity number and channeling at low N_g for permeability fields with small Dykstra-Parsons coefficients and large range. For permeability fields with high heterogeneities and small range, the displacements were dispersive. For anisotropic runs with $k_x = 1{,}000\ k_z$, the flow pattern was channeling at all Dykstra-Parsons coefficients and ranges. This shows that both R_L and N_g must be high for gravity to dominate.

These conclusions are based on simulations that are typical of field-scale CO_2 floods for two classes of reservoirs, one with high vertical permeability (many sandstones) and the other with low vertical permeability (many carbonates). These results clearly point to the benefit of identifying flow regimes for particular reservoir conditions of interest and then emphasizing the dominant flow regime in subsequent studies. Simply assuming that fingering will be important because the injected gas has a low viscosity and thus a mobility ratio greater than one or that gravity will be important because the injected gas has a low density and thus high N_g is clearly wrong. Under laboratory conditions of low R_L and low permeability variation, fingering is indeed common and must be taken into account, but fingering will be much less likely to be important under reservoir conditions of high R_L. Sorbie *et al.*[22] made this same observation in their recent FCM displacement study. They found that as the displacement moves away from the vertical equilibrium conditions (i.e., as R_L decreased from 60 to 1), the fingering and channeling regions became larger and the dispersive region smaller.

Although our numerical method reduces numerical dispersion and grid orientation effects to very low levels, we recommend that additional simulations with finer grids be done when feasible, especially in three dimensions where we present only limited results. More recently, Tchelepi and Orr[23] have shown that unstable FCM displacements in 3D homogeneous and heterogeneous reservoirs can be very different from displacements in 2D reservoirs, especially when the effects of gravity are considered and when the viscous and gravity forces are of comparable magnitude. Simulations with more-complex phase behavior, different relative permeability models and reservoir heterogeneity types, and tertiary flooding conditions with and without WAG injection would be desirable. Roper *et*

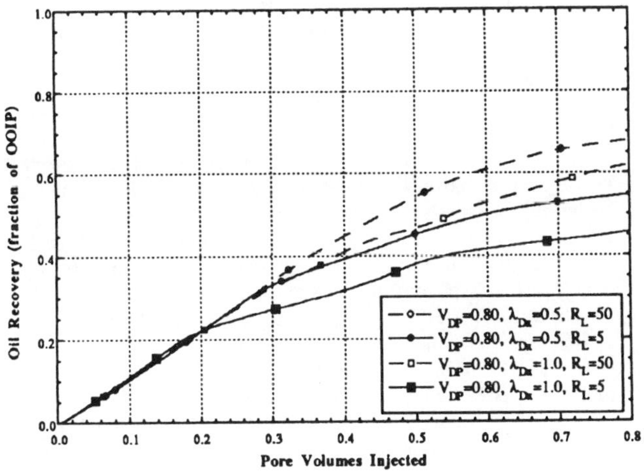

Fig. 13—Oil recovery as function of λ_{Dx} and R_L (OOIP = original oil in place).

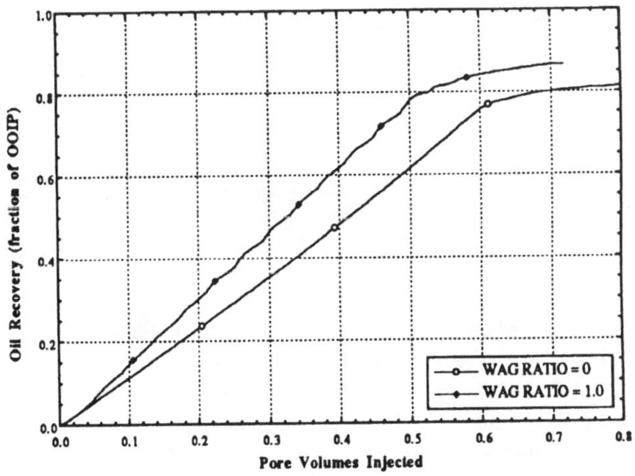

Fig. 14—Oil recovery as a function of WAG ratio.

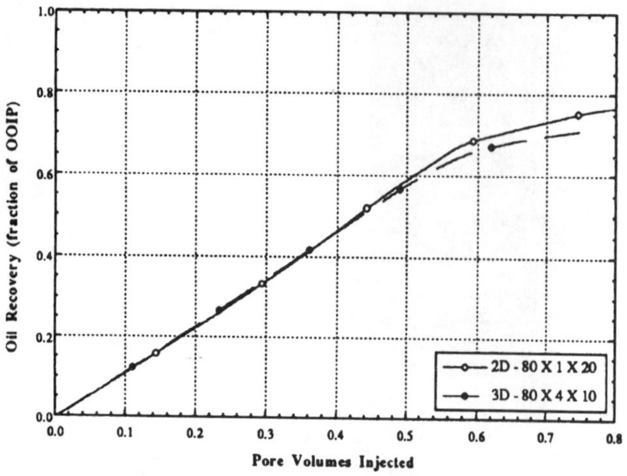

Fig. 15—2D and 3D oil recoveries at $N_g = 0.0055$.

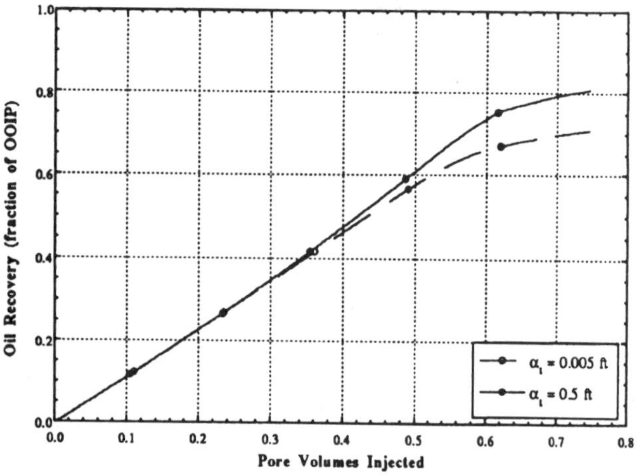

Fig. 16—3D oil recovery as a function of transverse dispersivity.

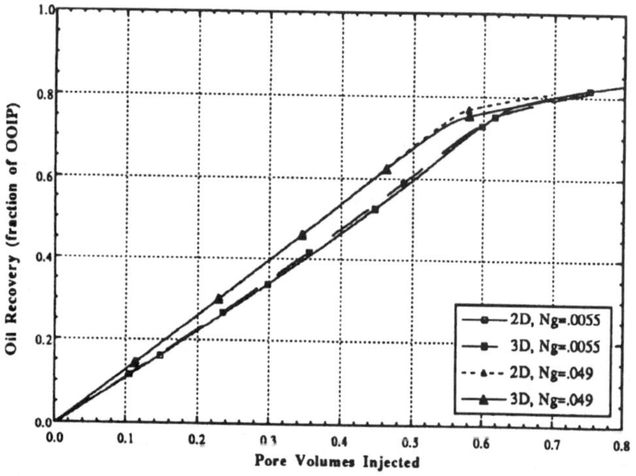

Fig. 17—Comparison of 2D and 3D oil recoveries at N_g=0.0055 and $N_g = 0.049$.

al.[24] have used our compositional simulator to study the effects of mixing and relative permeability on the injectivity of tertiary CO_2 floods. They also found that phase behavior is very important and must be accounted for. With phase behavior and mixing accounted for, CO_2 injectivity is seen to be much higher for reservoirs with crossflow than those without. Simplified analytical models and simulation with mixing parameter models do not predict these results mechanistically, whereas compositional simulation can.

Nomenclature

g = gravitational acceleration constant
h = reservoir thickness, L, ft
k^o_{rj} = endpoint relative permeability of Phase j
k_x, k_z = absolute permeability in the x and z directions, L^2, md
K_v = coefficient of variation, dimensionless
L = length of reservoir, L, ft
M^o = endpoint mobility ratio $= \dfrac{k^o_{r2}\ \mu_1}{\mu_2\ k^o_{r1}}$
N_g = gravity number $= \dfrac{k_x \rho g k^o_{r2}}{u\ \mu_2}\ \dfrac{h}{L}$
$N_{Pe,\ell}$ = longitudinal Peclet number $= L/\alpha_\ell$
$N_{Pe,t}$ = transverse Peclet number $= h^2/\alpha_{tj} L$
N_{RL} = modified Rapoport and Leas number $= \dfrac{158 u \mu_3 L}{C_{pc} \sigma_{23} k^o_{r3} \sqrt{\phi k_x}}$
R_L = effective length/thickness ratio $= \dfrac{L}{h}\sqrt{\dfrac{k_z}{k_x}}$
u = Darcy velocity, L/t, ft/D
V_{DP} = Dykstra-Parsons coefficient
$\alpha_{\ell j}$ = longitudinal dispersivity of Phase j, L, ft
α_t = transverse dispersivity, L, ft
α_{tj} = transverse dispersivity of Phase j, L, ft
$\lambda_{Dx}, \lambda_{Dy}, \lambda_{Dz}$ = dimensionless range in the x, y, and z directions $= \lambda_x/L, \lambda_y/L, \lambda_z/L$
λ_{rj} = relative mobility of Phase j, Lt/m, cp^{-1}
$\lambda_x, \lambda_y, \lambda_z$ = range in the x, y, and z directions, L, ft
μ_j = viscosity of Phase j, m/Lt, cp
ρ = mass density, m/L^3, lbm/ft^3
$\Delta\rho$ = density difference between displaced and displacing fluid, m/L^3, lbm/ft^3

Acknowledgments

We thank John Trangenstein for useful discussions and suggestions during the doctoral studies of Yih-Bor Chang. We also thank Joanna Castillo for graphics help and S. Khataniar for some of the computer runs. We thank the industrial sponsors of the Enhanced Oil & Gas Recovery Research Program of the Center for Petroleum & Geosystems Engineering at the U. of Texas at Austin and Cray Research Inc. for financial support of this research. Computing resources for this work were provided by the U. of Texas Center for High Performance Computing and the Dept. of Petroleum Engineering.

References

1. Peaceman, D.W. and Rachford, H.H. Jr.: "Numerical Calculation of Multidimensional Miscible Displacement," *SPEJ* (Dec. 1962) 327; *Trans.*, AIME, **225.**
2. Christie, M.A. and Bond, D.J.: "Detailed Simulation of Unstable Processes in Miscible Flooding," *SPERE* (Nov. 1987) 514; *Trans.*, AIME, **283.**
3. Araktingi, U.G. and Orr, F.M.: "Viscous Fingering in Heterogeneous Porous Media," paper SPE 18095 presented at the 1988 SPE Annual Technical Conference and Exhibition, Houston, Oct. 2–5.
4. Ewing, R.E., Russell, T.F., and Young, L.C.: "An Anisotropic Coarse-Grid Dispersion Model of Heterogeneity and Viscous Fingering in Five-Spot Miscible Displacement That Matches Experiments and Fine-Grid Simulation," paper SPE 18441 presented at the 1989 SPE Symposium on Reservoir Simulation, Houston, Feb. 6–8.
5. Christie, M.A.: "High-Resolution Simulation of Unstable Flows in Porous Media," *SPERE* (Aug. 1989) 297; *Trans.*, AIME, **287.**
6. Christie, M.A., Muggeridge, A.H., and Barley, J.J.: "3D Simulation of Viscous Fingering and WAG Schemes," *SPERE* (Feb. 1993) 19; *Trans.*, AIME, **295.**
7. Waggoner, J.R., Castillo, J.L., and Lake, L.W.: "Simulation of EOR Processes in Stochastically Generated Permeable Media," *SPEFE* (June 1992) 173; *Trans.*, AIME, **173.**

8. Fayers, F.J. and Newley, T.M.J.: "Detailed Validation of an Empirical Model for Viscous Fingering With Gravity Effects," *SPERE* (May 1988) 542.
9. Christie, M.A., Jones, A.D.W., and Muggeridge, A.H.: "Comparison Between Laboratory Experiments and Detailed Simulations of Unstable Miscible Displacement Influenced by Gravity," paper presented at the 1989 North Sea Oil and Gas Reservoirs Conference, Trondheim, May 8–10.
10. Fayers, F.J., Barker, J.W., and Newley, T.M.J.: "Effects of Heterogeneities on Phase Behavior in Enhanced Oil Recovery," paper presented at the 1989 IEA Collaborative Project on Enhanced Oil Recovery, 10th Intl. Workshop & Symposium, Palo Alto, CA, Oct. 4–6.
11. Chang, Y.: "Development and Application of an Equation of State Compositional Simulator," PhD dissertation, U. of Texas, Austin (1990).
12. Chang, Y., Pope, G.A., and Sepehrnoori, K.: "A Higher-Order Finite Difference Compositional Simulator," *J. Pet. Sci. Eng.* (Nov. 1990) 35.
13. Liu, J., Pope, G.A., and Sepehrnoori, K.: "Improved Finite-Difference Scheme for Reservoir Simulation," 1991 Eighth Annual Report of Enhanced Oil Recovery Research Program of the Center for Petroleum & Geosystems Engineering, Austin (1991).
14. Perschke, D.R.: "Equation-of-State Phase Behavior Modeling for Compositional Simulation," PhD dissertation, U. of Texas, Austin (1988).
15. Zapata, V.J. and Lake, L.W.: "A Theoretical Analysis of Viscous Crossflow," paper SPE 10111 presented at the 1981 SPE Annual Technical Conference and Exhibition, San Antonio, Oct. 5–7.
16. Krueger, C.: "Numerical Simulation and Modeling of Viscous Fingering," MS thesis, U. of Texas, Austin (1989).
17. Blackwell, R.J., Rayne, J.R., and Terry, W.M.: "Factors Influencing the Efficiency of Miscible Displacement," *Trans.,* AIME (1959) **216**, 1.
18. van der Poel, C.: "Effect of Lateral Diffusivity on Miscible Displacement in Horizontal Reservoirs," *SPEJ* (Dec. 1962) 317; *Trans.,* AIME, **225**.
19. Dria, D.E., Pope, G.A., and Sepehrnoori, K.: "Three-Phase Gas/Oil/Brine Relative Permeabilities Measured Under CO_2 Flooding Conditions," *SPERE* (May 1993) 143; *Trans.,* AIME, **295.**
20. Yang, A.-P. and Lake, L.W.: "The Accuracy of Autocorrelation Estimates," *In Situ* (1988–89) **12**, No. 4, 227.
21. Khataniar, S.: "A Numerical Study of the Performance of Unstable Displacements in Heterogeneous Media," PhD dissertation, U. of Texas, Austin (1991).
22. Sorbie, K.S. *et al.*: "Flow Regimes in Miscible Displacements in Heterogeneous Correlated Random Fields," paper SPE 24140 presented at the 1992 SPE/DOE Symposium on Enhanced Oil Recovery, Tulsa, April 22–24.
23. Tchelepi, H.A. and Orr, F.M. Jr.: "The Interaction of Viscous Fingering, Permeability Heterogeneity, and Gravity Segregation in 3D," paper SPE 25235 presented at the 1993 SPE Symposium on Reservoir Simulation, New Orleans, Feb. 28–March 3.
24. Roper, M., Pope, G.A., and Sepehrnoori, K.: "Effects of Mixing and Relative Permeability on Tertiary CO_2 Injectivity," 1991 Annual Report of EOR Research Program, Center for Petroleum & Geosystems Engineering, Austin (1991).

SI Metric Conversion Factors

cp	× 1.0*	E − 03 = Pa·s
ft	× 3.048*	E − 01 = m
°F	(°F–32)/1.8	= °C
lbm	× 4.535 924	E − 01 = kg
md	× 9.869 233	E − 04 = μm^2
psi	× 6.894 757	E + 00 = kPa

*Conversion factor is exact.

Yih-Bor Chang works in reservoir simulation and phase-behavior modeling at Western Atlas Software. His research interests include reservoir engineering, modeling, and simulation and EOR. Chang holds a BS degree in chemistry from Tunghai U., Republic of China, and MS and PhD degrees in petroleum engineering from the U. of Texas (UT) at Austin. **Min T. Lim** is a research associate in petroleum engineering at UT. Previously, he worked for Petronas in Malaysia. His research interests include reservoir engineering, miscible gas EOR, and reservoir simulation and characterization. He holds BS, MS, and PhD degrees in petroleum engineering from UT. Lim was a 1985–87 Kuala Lumpur Section director and 1987–89 membership chairman. **Gary A. Pope** is director of the Center for Petroleum & Geosystems Engineering at UT. He holds the Getty Oil Co. Centennial Chair in petroleum engineering and was formerly chairman of the Dept. of Petroleum Engineering, where he has taught since 1977. His research interests are in EOR, reservoir engineering and simulation, use of tracers for reservoir characterization, and phase behavior. Previously, Pope worked in production research at Shell Development Co. He holds a BS degree from Oklahoma State U. and a PhD degree from Rice U., both in chemical engineering. Pope, a 1991 Reservoir Engineering Award Winner, has served on several SPE committees. **Kamy Sepehrnoori** is director of the computing and database research program of the Center for Petroleum & Geosystems Engineering at UT. He holds the Frank W. Jessen Professorship in petroleum engineering and has served on the C. of Engineering faculty since 1981. He is a specialist in computational methods. His research interests include reservoir simulation, numerical solution of differential equations, applied mathematics, EOR, database development, and vector and parallel computations. He holds a BS degree in mechanical engineering, an MS degree in engineering, and a PhD degree in petroleum engineering, all from UT.

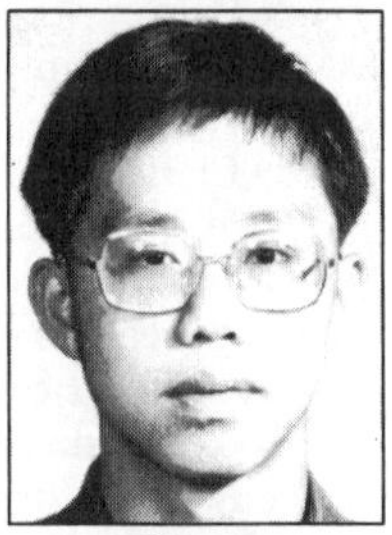
Chang

Lim

Pope

Sepehrnoori

Comparison of Laboratory- and Field-Observed CO_2 Tertiary Injectivity

Peter G. Christman, SPE, Shell Western E&P Inc., and
Sheldon B. Gorell, SPE, Shell Development Co.

Summary. Although CO_2 injectivity should be significantly greater than brine injectivity because CO_2 has a much lower viscosity than brine, this behavior is not always seen, as shown in a Denver Unit field test. This paper examines features that cause differences in CO_2 injectivity with a model that uses simple nondispersive flow with a series of constant-composition slugs to approximate the analytical solution (normally a sequence of shocks and tails) in a sequence of noncommunicating layers. Because of its simplicity, this model identifies the primary features that result in the different observed CO_2 injectivities more clearly than the finite-difference model.

This paper shows that the qualitative differences between Cedar Creek anticline corefloods and field behavior result solely from differences in geometry. That is, a single set of centrifuge-measured, quasinative-state, secondary-drainage relative permeabilities can be used to predict both laboratory and field behavior. Primary factors that contribute to the differences between the two field tests are fluid/rock properties, effective wellbore radius (or skin), and heterogeneity in the layering.

Introduction

In the past decade, Shell Western E&P Inc. has conducted two tertiary CO_2 pilots in carbonate reservoirs. The first was in the Denver Unit (Wasson field). It experienced CO_2 injectivity that was surprisingly lower than the preflood brine injectivity. CO_2 injectivity in the second pilot at the South Pine Unit (Cedar Creek anticline) was significantly higher than brine injectivity. Ironically, this behavior was also surprising because of both the Denver Unit experience and the laboratory experiments conducted on South Pine core material before CO_2 injection.

The Denver Unit pilot was history-matched with a simulator featuring permeability reduction factors similar to models used by Claridge,[1] Todd *et al.*,[2] and Chase and Todd.[3] Without the permeability-reduction factors, this model predicted CO_2 injectivity greater than brine injectivity. The majority of the Denver Unit tertiary CO_2 corefloods also displayed CO_2 injectivities greater than the brine injectivity. **Fig. 1** shows the normalized injectivity of 20 tertiary CO_2 corefloods conducted in Denver Unit core. (Details about these experiments are found in Ref. 4.) Only two of these corefloods displayed low CO_2 injectivities consistent with the pilot behavior. The simulator curve plotted in Fig. 1 represents a linear finite-difference simulation with pilot-tuned reduction factors of a coreflood. Patel *et al.*[4] showed that mixed wettability, and the resulting relative-permeability hysteresis, was the critical property in the only two corefloods that displayed low CO_2 injectivity.

Before CO_2 injection at the South Pine pilot, tertiary corefloods were conducted to study CO_2 injectivity. Frozen core material, taken from the pilot injection well, was mounted in coreholders, and every effort was made to preserve the native-wettability state. As **Fig. 2** shows, these experiments displayed lower CO_2 injectivities than brine, in qualitative agreement with the mixed-wettability Denver Unit counterparts. When the South Pine pilot CO_2 injectivity was much higher than the brine preflood injectivity, it seemed to be completely at odds with coreflood results and the Denver Unit pilot (assumed to be a good analog).

To understand fully the differences in the CO_2 injectivity observed here, it is necessary to examine both the fluid/rock properties and the flow behavior in the geometries involved in the different tests. In this paper, we compare the mixed-wettability Wasson relative-permeability curves with a set of quasinative-state, secondary-drainage relative permeabilities that were measured in Cedar Creek core material. Then we discuss a model that considers the flow of constant-composition slugs in a noncommunicating layered system to show that the rather surprising CO_2-injectivity behavior is caused primarily by differences in flow geometry and tertiary oil-bank properties.

Throughout this paper, properties that are representative of the field (such as API gravity and rock type) are referred to by their field designation (Wasson or Cedar Creek). Results specific to the pilot or injectivity test are identified by the appropriate operating unit name (Denver Unit or South Pine Unit).

Fluid Mobilities

Clearly, any study of injectivity must begin with relative permeability because it directly affects the fluid-bank mobilities. Patel *et al.*[4] showed that mixed wettability and the resulting relative-permeability hysteresis explained the low CO_2 injectivity observed at the Denver Unit (Wasson field) pilot. The imbibition (waterflood) and secondary-drainage (CO_2 flood) curves for San Andres rock in the Wasson field are presented in **Fig. 3.** Similar curves were measured by Schneider and Owens[5] on native-state San Andres rock by a steady-state method.

The major characteristic of the hysteresis in Fig. 3 is the increase in the immobile water saturation from 15% on imbibition to 25% on secondary drainage. This increase occurs because the water is displaced by secondary drainage from oil-wet pores in a mixed-wettability medium; i.e., part of the water, the nonwetting phase, is trapped by the wetting oleic phase. We have directly observed this behavior in 2D mixed-wettability glass models on secondary drainage. In his extensive literature survey,

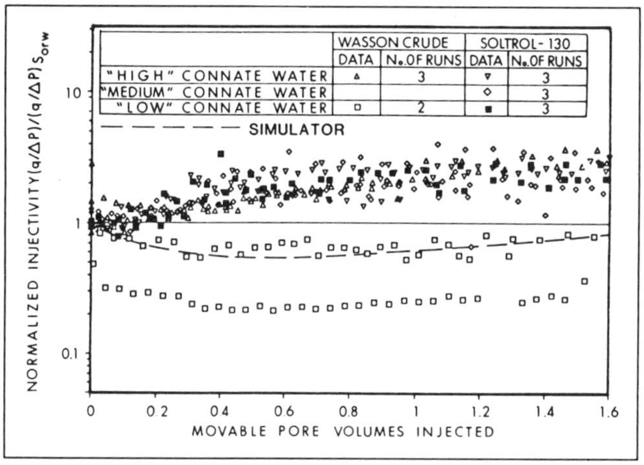

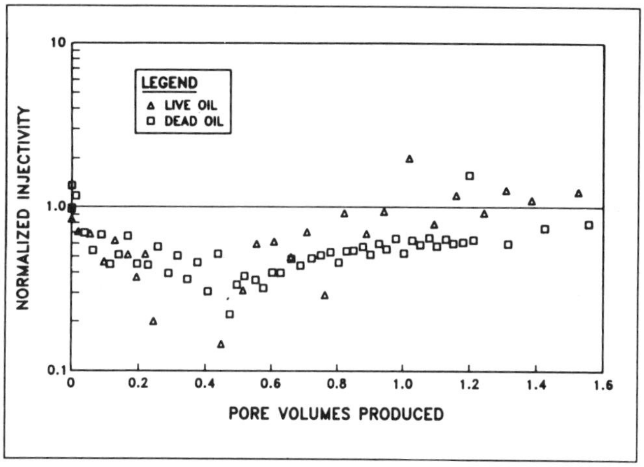

Fig. 1—Injectivity during tertiary solvent corefloods in 9-in.-long Denver Unit, Wasson field, core assemblies (after Patel *et al.*[10]).

Fig. 2—Injectivity during quasinative-state corefloods in 9-in.-long South Pine, Cedar Creek anticline, core assemblies.

Anderson[6] presents several examples of water trapping in mixed-wettability media, inferred from capillary pressure measurements. The trapped water reduces the water relative permeability on secondary drainage and, more importantly, reduces the oil endpoint relative permeability.

In a purely tertiary-solvent displacement, normally conducted in laboratory corefloods and aspired to in field pilots, the only required relative-permeability curves are those measured on secondary drainage. With frozen core material from the South Pine (Cedar Creek) pilot injector, quasinative-state centrifuge experiments were performed. Details about the experimental procedure and analysis are presented in the Appendix. **Fig. 4** shows the results of these experiments and those of the Wasson secondary-drainage curves. The primary differences between the two sets of curves are that (1) the endpoint ratio (k_{rw}/k_{ro}) is reversed between the two sets, indicating that the Cedar Creek rock is more water-wet; and (2) the residual oil saturation is lower at Cedar Creek than at Wasson (35 vs. 40%, respectively).

The first difference results in a much lower oil mobility with respect to water mobility at S_{or} for Wasson. The second difference results in a larger volume of oil available to build an oil bank in Wasson. Both of these result in a greater influence of the oil bank on the tertiary CO_2 injectivity at Wasson.

The other important factor that determines fluid mobility is, of course, fluid viscosity. **Table 1** gives the viscosities of the three primary fluids at the two fields of interest. The differences in the CO_2 and water viscosities result primarily from the differences in the temperatures of the two fields.

Model Description

To understand how differences in process variables affect injectivity and flow behavior in tertiary miscible displacements, a computer program was developed to model the process. This program applies simplified fluid mechanics to multilayered systems. It approximates the fluid flow of a tertiary miscible displacement with a series of constant-composition slugs separated by shocks moving with different velocities. The fluid flow in individual layers is completely decoupled by assuming that there is incompressible flow with constant pressure drop between the injector and the producer (or at some fixed distance) and that there is no vertical communication between the layers. This allows the flow in each layer to be analyzed independently. The effects of multiple layers can then be accounted for by appropriately adding the single-layer contributions. For ease of notation in the following formulation, however, only a single layer will be considered. Results are presented for both linear and radial geometry—the two ideal geometries most appropriate for laboratory and pilot studies.

Our use of this simple, analytical-based model is not meant to imply that other models, such as a finite-difference simulator, cannot be used to model CO_2-injectivity behavior. Given the proper input parameters, a finite-difference model will give essentially the same results. Parametric studies, however, are often difficult and convoluted, and such artifacts as numerical dispersion[7] can obscure useful information. Finite-difference simulators tend to be large, which limits the number of gridblocks and layers that can be used in a reasonable calculation. In contrast, because of the simple way in which fluid flow within a layer is treated in our model, more than 50 discrete layers can be incorporated easily into a calculation. The simplicity of this model makes it much easier to identify the primary features that result in the different observed CO_2 injectivities.

Fluid Mechanics. Viewed simply, the fluid mechanics follow from a generalized version of the Buckley-Leverett equation.[8] A point

"The simplicity of this model makes it much easier to identify the primary features that result in the different observed CO_2 injectivities."

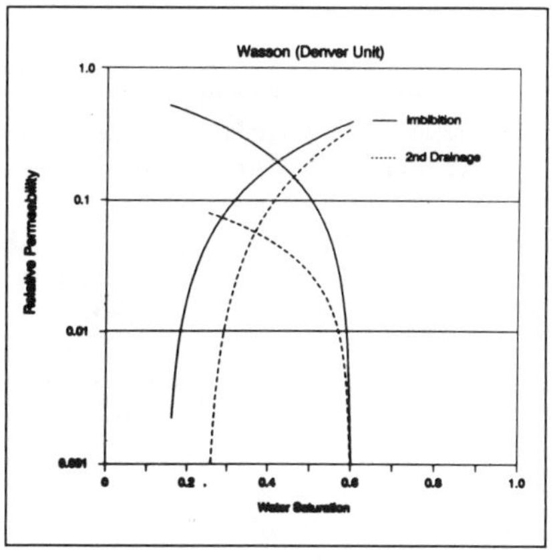

Fig. 3—Imbibition and secondary-drainage relative permeabilities for Wasson.

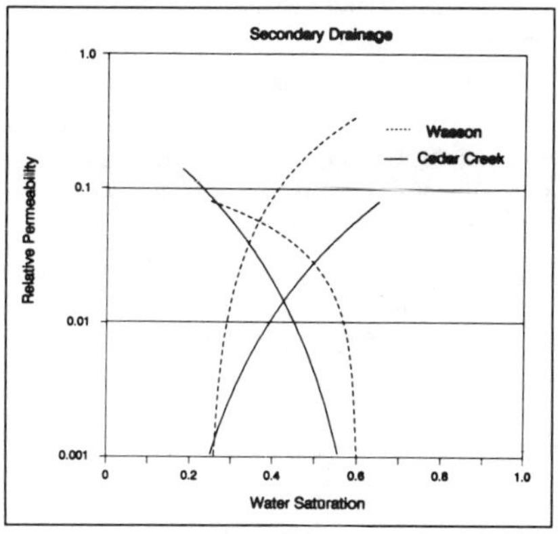

Fig. 4—Comparison of secondary-drainage relative permeabilities for Cedar Creek and Wasson.

TABLE 1—FLUID VISCOSITIES (cp) AT WASSON AND CEDAR CREEK

Field	Oil	Water	CO_2
Wasson	1.30	0.73	0.063
Cedar Creek	1.72	0.40	0.049

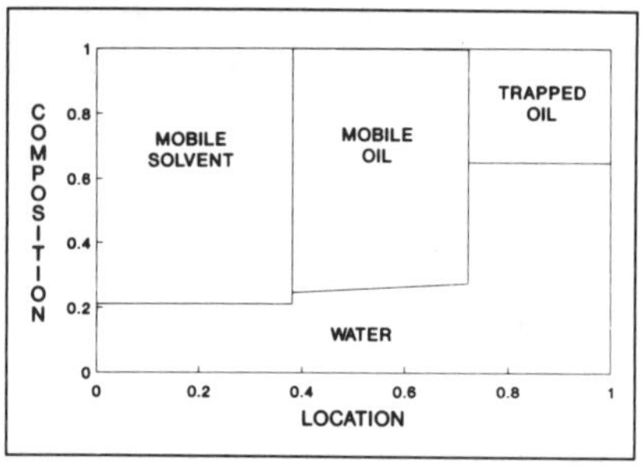

Fig. 5—Composition profile for tertiary CO_2 injection of Cedar Creek relative permeabilities and fluid properties.

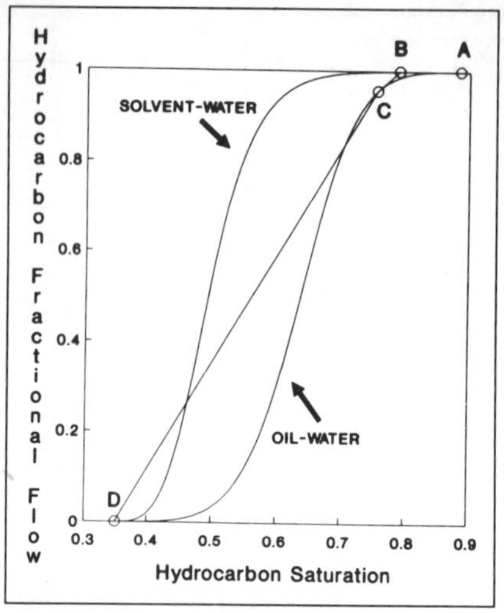

Fig. 6—Fractional-flow construction of Cedar Creek relative permeabilities and fluid properties.

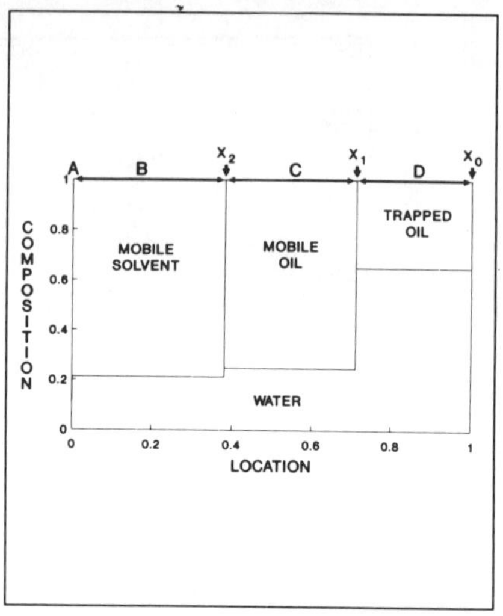

Fig. 7—Constant-bank-composition profile of Cedar Creek relative permeabilities and fluid properties.

"A tertiary displacement of residual oil by CO_2 can be represented by the travel of three shock fronts, with different velocities separating four slugs: waterflood residual, mobile oil bank, CO_2-rich phase, and follow-up brine."

of fixed composition j has velocity

$$dx/dt_D|_{(dS_i=0)} = \partial f_i/\partial S_i = v_{Dj}. \quad \ldots (1)$$

Flow within each layer is based on the analytical method of characteristics and graphical fractional-flow solutions for Eq. 1 applied to tertiary miscible displacements.[9] **Fig. 5** shows a sample composition profile for tertiary CO_2 injection with South Pine relative permeabilities and fluid properties. An interesting feature of these and other dispersion-free results is that the analytical solution can be reasonably approximated as the flow of constant-composition banks. The discontinuities between banks are treated as shock fronts, and their velocities are obtained by requiring the conservation of mass. Although this approximation cannot be justified in general, we have found with many different relative-permeability and fluid-viscosity values that it is reasonable for the parameters of interest in CO_2 flooding.

If we assume that all changes are discontinuous, the result is a set of constant-composition banks described by the integral form of Eq. 1. **Fig. 6** plots the fractional-flow construction for continuous CO_2 injection into a system at waterflood residual, and **Fig. 7** shows the corresponding constant-composition-bank profile. The construction in Fig. 6 involves complete shocks along the solvent/water fractional-flow curve (Point A to Point B, where Point A is the boundary condition), from the solvent/water curve to the oil/water curve (Point B to Point C), and along the oil/water curve (Point C to Point D, where Point D is the initial condition). Line BC is found by requiring conservation of mass (oil, CO_2, and water) at the back of the oil bank. An extension of Line BC (not shown) intersects the hydrocarbon saturation axis at S_t (B), which is the amount of oil trapped in the solvent bank. The front velocities are calculated directly from the integral form of Eq. 1 as

$$v_{Df} = (f_i^u - f_i^d)/(S_i^u - S_i^d), \quad \ldots\ldots\ldots (2)$$

where subscript i=oil, water, or solvent, and superscripts u and d refer to the properties directly upstream and downstream, respectively, of the shock. Eq. 2 is valid for all three components. Graphically, the front velocities are the slopes of the lines connecting the labeled points in Fig. 6. The formulation conserves mass but, by assuming complete shocks, it introduces small errors into the velocities and saturations. The analytic solution (Fig. 5) has a small tail within the oil bank, and the velocity of the front of the oil bank is the slope of the line tangent to the oil/water fractional-flow curve through the initial condition. There is little difference, however, between Line CD and the tangent line in Fig. 6. Even with this error, the slug approximation in Fig. 7 is a reasonable representation of the more exact solution shown in Fig. 5.

If we consider tertiary miscible displacements where all the fronts are initiated at time zero and the coherence condition holds (i.e., faster fronts are always ahead of slower fronts), then the size and location of the slugs (fluid banks) in each layer are determined solely by the individual-layer PV's injected. The result is that the fluid-flow problem can be solved once and applied individually to each layer. It is very useful and often necessary, however, to model follow-up brine injection. When this is considered, the problem is complicated because a fast water front can be behind and overtake slower water fronts. When a slower front is overtaken, this creates an immobile front at the point of intersection. Furthermore, the state of each layer is different because the individual injected PV's (dimensionless time) vary from layer to layer, even though the dimensional time when the follow-up slug is injected is the same for all layers.

When follow-up brine is initiated, the boundary condition becomes 100%-water fractional flow. Because each slug is assumed to have a constant composition, follow-up brine is modeled as a complete shock moving through the system. There is

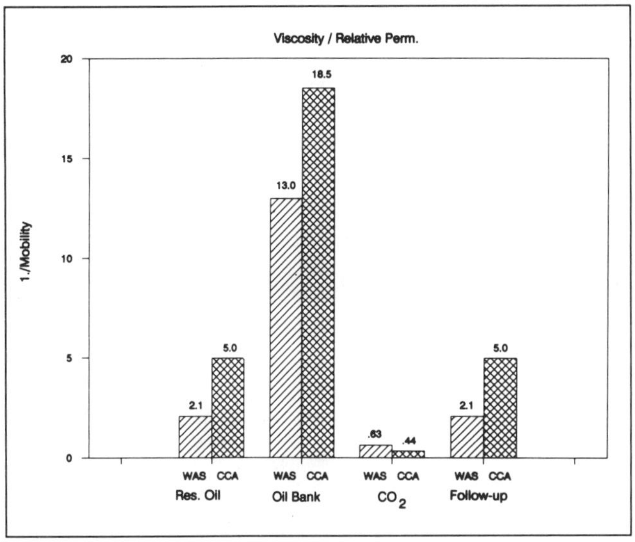

Fig. 8—Calculated slug mobilities for Wasson (WAS) and Cedar Creek (CCA) tertiary displacements.

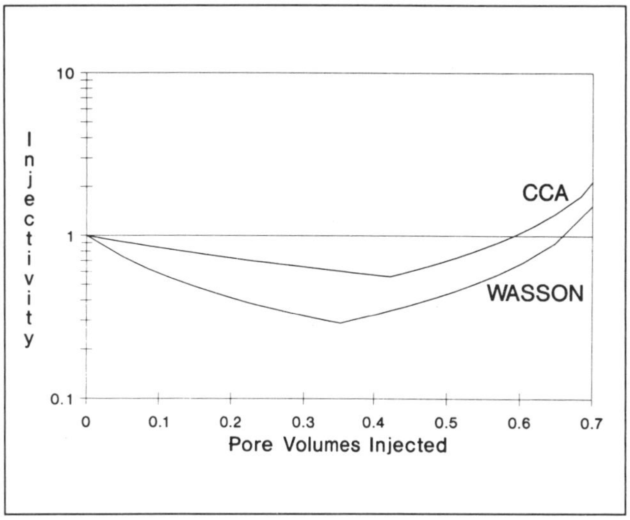

Fig. 9—Comparison of linear (coreflood) behavior of miscible tertiary CO_2 displacements using Wasson and Cedar Creek fluid properties.

a series of slugs in the system when this is initiated, and the brine bank will see (and possibly overtake) each one sequentially, starting with the slowest. The brine bank is assumed to trap whatever mobile fluid was present in the slug being overtaken. If f_w=fractional flow of water within the slug being overtaken and S_w=water saturation of the slug, then the velocity of the follow-up brine front is

$$v_{Dbr}=(1-f_w)/(1-S_{or}-S_w). \quad \ldots\ldots(3)$$

To illustrate this, consider the case of follow-up brine after 100% CO_2 injection. Before brine injection, the construction shown in Fig. 6 is applicable. When brine injection begins, it "sees" a region with the composition of Point A, and the velocity of the brine front is Slope AD. After overtaking the small Slug A, it next sees Slug B. The brine-front velocity becomes Slope BD. Finally, the brine front overtakes the oil bank, Slug C, and moves with velocity CD. Note that this construction is approximate and intended only to account for some of the qualitative features of follow-up brine injection. These constructions are all derived by assuming complete shocks between regions of constant composition. Subject to this simplifying assumption, the velocity at the back of the oil bank, where oil is being retrapped, is the same as the velocity of the front of the bank, where oil is being mobilized.

Linear and Radial Geometry. With this simplified view of fluid flow within a layer, tractable analytical expressions that describe the observable time-dependent flow behavior can be obtained. This is fairly straightforward in a linear system. Because each slug has a constant composition and mobility, the pressure gradient is constant across any individual slug and makes discontinuous changes at front locations. For a system comprising N slugs, Darcy's law can be written as

$$\left(\frac{\Delta p}{L}\right)kA=q\sum_{j=0}^{N}\frac{\Delta x_j}{\lambda_j}. \quad \ldots\ldots\ldots(4)$$

In radial flow, the fronts move with a velocity that varies with $1/r$. The linear solution of the fluid mechanics can be applied directly to a radial system by a simple coordinate transformation.

$$\mathrm{d}r_D^2/\mathrm{d}t_D|_{\mathrm{d}S_i=0}=\partial f_i/\partial S_i=v_{Dj}. \quad \ldots.(5)$$

This is the radial version of the Buckley-Leverett equation (Eq. 1), where r_D^2 has been substituted for x. The left side of Eq. 5 represents the constant pseudovelocity, which is used to track the front locations in the radial model.

The radial version of Darcy's law, which is analogous to Eq. 4, is

$$\Delta pk2\pi h=$$

$$q\left[\frac{\ln\left(\frac{1}{r_{DM}}\right)}{\lambda_{M-1}}+\sum_{j=M}^{P-1}\frac{\ln\left(\frac{r_{Dj}}{r_{Dj+1}}\right)}{\lambda_j}+\frac{\ln\left(\frac{r_{DP}}{r_{wD}}\right)}{\lambda_P}\right]^{-1}, \quad \ldots\ldots\ldots\ldots(6)$$

where r_{wD} is the dimensionless effective wellbore radius. It is implicitly assumed in Eq. 5 that the volume of the wellbore is negligibly small. In Eq. 6, however, the wellbore radius is an important term that cannot be eliminated by assuming it to be small. This apparent discrepancy does not cause any problems. It is assumed that the pressure drop within the effective wellbore, including any skin effects, is zero. The parameter P is used to track the number of slugs outside the effective wellbore, as they are the only slugs that contribute any resistance to flow.

The character of Eq. 6 is quite different from Eq. 4, its linear counterpart. In radial geometry, the interior slugs (i.e., those bordered by two moving fronts) have a constant contribution to flow resistance. The relative importance of these interior slugs is deter-

"Eq. 8 shows that r_{wD} affects the initial brine or base injectivity and determines the relative importance of the slugs created by CO_2 injection."

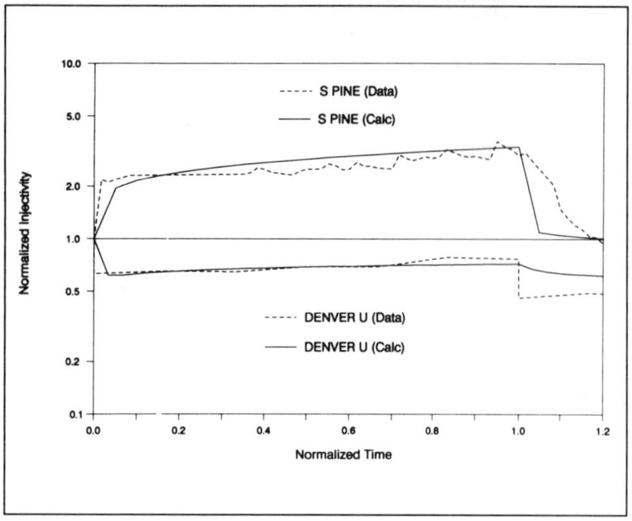

Fig. 10—Comparison of calculated and observed injectivities at the Denver Unit (Wasson) and South Pine (Cedar Creek) pilots.

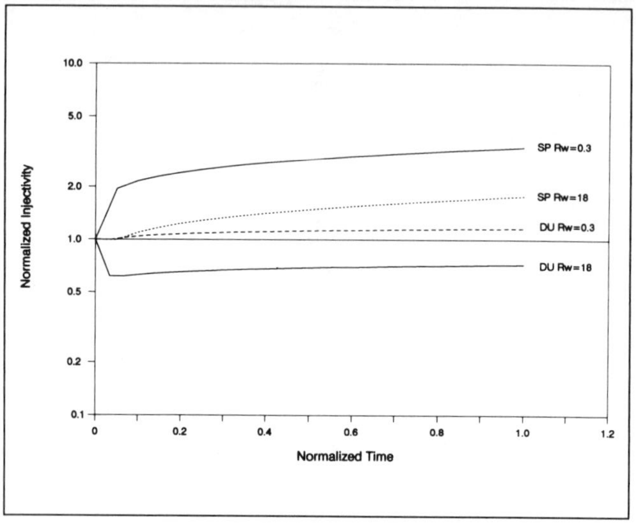

Fig. 11—Sensitivity of CO_2 injectivity to the effective wellbore radius.

mined solely by the mobility of each slug and the ratio of the pseudovelocities of the front and back of the slug. Only the two slugs bordered by the fixed boundaries of the system (the effective wellbore and the outer radius) have contributions that vary with time. In the linear case, all the slugs are growing except the slug being displaced out of the system. This results in very different injectivity behavior between a displacement that occurs in a radial vs. a linear system.

Field, Laboratory, and Model Comparisons

A tertiary displacement of residual oil by CO_2 can be represented by the travel of three shock fronts, with different velocities separating four slugs: waterflood residual, mobile oil bank, CO_2-rich phase, and follow-up brine. **Fig. 8** shows the mobilities of the four slugs calculated for conditions at the two fields of interest. For both sets of conditions, the CO_2-rich slug and the oil bank are the highest- and lowest-mobility slugs, respectively. The ratio of oil-bank mobility to waterflood-residual mobility is much smaller for Wasson conditions than Cedar Creek conditions (0.16 vs. 0.27, respectively), indicating that the oil bank will have a more dominant effect under Wasson conditions. In the following, the predictions from this simple model are compared to the observed field and laboratory behavior.

Corefloods (Linear Geometry). The normalized injectivity for linear geometry is determined by expanding Eq. 4 to account for moving fronts. Eq. 7 represents the condition of a core when all three of the fronts resulting from CO_2 injection are in the system.

$$I=(q/\Delta p)/(q/\Delta p)_{t=0}$$
$$=(1/\lambda_0)\div\{[(1-t_D v_{D1})/\lambda_0]$$
$$+[t_D(v_{D1}-v_{D2})/\lambda_1]+[(t_D v_{D2})/\lambda_2]\}. \quad \ldots\ldots\ldots\ldots (7)$$

The term in the numerator represents the resistance of the system at waterflood residual and the term in the denominator represents the dynamically changing resistance under tertiary miscible conditions. From Eq. 7, we see that normalized injectivity is most strongly affected by the lowest-mobility slug, which in both systems is the oil bank. Furthermore, because the fronts must move in a coherent manner, the relative importance of each slug grows until it reaches the physical end of the system.

The strong effect that the low-mobility oil bank has on injectivity is seen in **Fig. 9,** where the calculated normalized injectivities for Wasson and Cedar Creek conditions are presented. In both cases, the injectivity decreases as the size and influence of the oil bank grows. This is in qualitative agreement with behavior observed in laboratory corefloods in mixed-wettability Wasson and Cedar Creek cores (Figs. 1 and 2). At oil breakthrough, which occurs earlier for Wasson conditions because of the higher residual oil saturation, production of the oil bank results in an increasing injectivity as the CO_2 slug grows. Even though the injectivity under Cedar Creek conditions decreases in this geometry, it never falls to as low a value as under Wasson conditions. The influence of the Cedar Creek oil bank is smaller because the residual oil is smaller (0.35 vs. 0.40), resulting in a smaller oil bank, and the oil-bank mobility is 27% of the waterflood residual mobility as compared to 16% under Wasson conditions.

Pilots (Radial Geometry). Eq. 8 is the radial analog of Eq. 7 for the normalized injectivity of a tertiary miscible displacement in a single layer.

$$I=(q/\Delta p)/(q/\Delta p)_{t=0}$$
$$=[\ln(1/r_{wD}^2)/\lambda_0]\div\{[\ln(1/t_D v_{D1})/\lambda_0]$$
$$+[\ln(t_D v_{D1}/t_D v_{D2})/\lambda_1]$$
$$+[\ln(t_D v_{D2}/r_{wD}^2)/\lambda_2]\}. \quad \ldots\ldots\ldots (8)$$

As in Eq. 7, the numerator represents the resistance of the system at waterflood residual. The denominator is made up of three terms that represent the resistance of the slugs. Two of the slugs are bordered by the fixed boundaries of the system, and the oil bank is bordered on both sides by moving fronts. The term in the denominator representing the oil bank remains unchanged because the time-dependent factors multiplying the pseudovelocities cancel each other. The resistance of the oil bank is felt immediately after it leaves the effective wellbore and does not change, even though it continues to grow in volume until it moves outside the outer radius. As the oil is displaced outward, the effects of the growing oil bank are exactly compensated for by the slower velocity of the radial displacement.

Eq. 8 shows that r_{wD} affects the initial brine or base injectivity and determines the relative importance of the slugs created by CO_2 injection. As r_{wD} decreases, the term in the numerator increases dramatically—i.e., the dimensional water injectivity decreases. This has the effect of decreasing the base injectivity used for the normalization. Furthermore, as r_{wD} decreases, the relative importance of the slug nearest the injector increases because of the increasing value of $\ln[t_D(v_{D2}/r_{wD}^2)]$. This results in the mobility of the CO_2-rich slug having a proportionally greater influence on the observed injectivity in a radial displacement than in a linear one. Although a small effective wellbore radius results in less absolute injectivity for both brine and CO_2 injection, it has the effect of increasing CO_2 injectivity relative to brine injectivity.

Fig. 10 compares the calculated normalized-injectivity behavior with the observed normalized injectivity at the two pilots. The time scale is normalized so that CO_2 injection begins at zero and ends at one. The actual periods of CO_2 injection were 6 months at the Denver Unit and 2 months at South Pine. The match of both of these pilots is excellent over the CO_2-injection period, and there is qualitative agree-

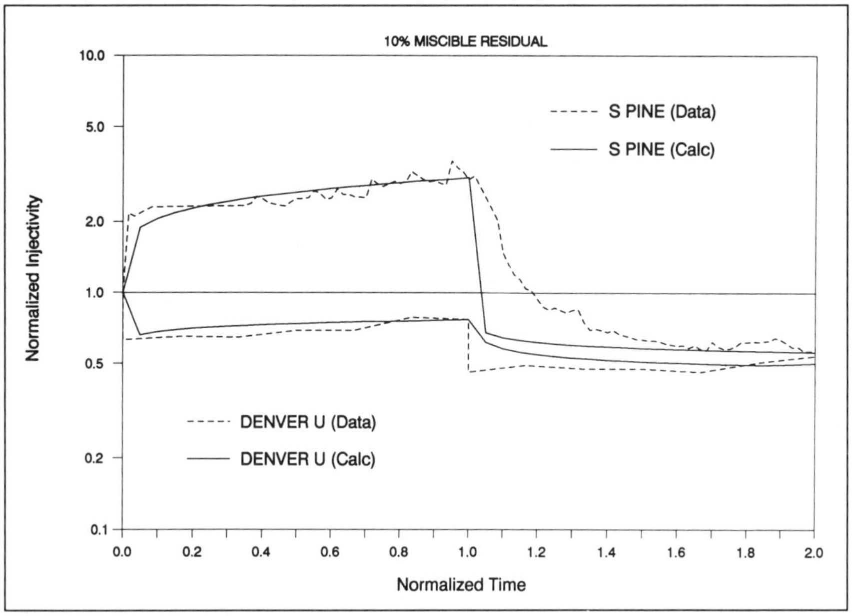

Fig. 12—Comparison of calculated and observed injectivities with a 10%-miscible residual at the Denver Unit (Wasson) and South Pine (Cedar Creek) pilots.

"The results show that differences between linear and radial flow can result in qualitatively similar (Wasson) or different (Cedar Creek) laboratory and field CO_2 injectivities."

ment after post-CO_2 brine injection is initiated. Injectivity in the Denver Unit decreases sharply as the influence of the low-mobility oil bank is felt and rises gradually as the relative importance of the higher-mobility CO_2-rich slug behind the oil bank increases, at the expense of the waterflood residual slug in front of the oil bank. In contrast, injectivity at South Pine initially rises sharply as the near-wellbore replacement of the less mobile water by the more mobile CO_2 dominates the effect of the oil-bank creation, followed by a slow rise in injectivity similar to the Denver Unit.

The pilots at both the Denver Unit[4] and South Pine[10] were operated with a single injector surrounded by nearby sampling and logging wells. Shut-in producers 600 to 700 ft [183 to 213 m] away, which were used as passive pressure observers, served as the outer radius of the modeled system. In both cases, fluid was sampled intermittently to maintain radial symmetry around the injector. To model the Denver Unit pilot, 21 layers were used to describe 176 ft [53.6 m] of net pay with a Dykstra-Parsons coefficient of 0.54. For the South Pine pilot, 10 layers were used to model a 12-ft [3.66-m] interval with a Dykstra-Parsons coefficient of 0.69. The layering description in both cases was determined from vertical characterization of the pilot areas. The injector in the Denver Unit pilot had a negative skin ($s=-4.1$, $r_w=18$ ft [5.5 m]), while the injector in the South Pine pilot had no skin at all ($r_w=0.29$ ft [0.09 m]). From the modeling details, it is apparent that the rather dramatic differences in the behavior of these two pilots result from (1) relative differences in the fluid-bank mobilities, caused by relative permeability and fluid viscosities; (2) differences in the effective wellbores of the injectors; and (3) differences in the heterogeneity of the systems. The smaller contrast between the oil-bank and waterflood-residual mobilities was also seen in the corefloods. However, the small effective wellbore radius and the greater degree of heterogeneity at South Pine further reduced the influence of the oil bank, resulting in higher injectivity. In the Denver Unit, the influence of the oil bank dominates the normalized injectivity. The normalized injectivity is plotted in **Fig. 11** for the two pilots with varying effective wellbore radius as a parameter. Note that if the wellbores are exchanged, the differences between the CO_2 injectivities for the two fields are reduced.

Finally, note that the follow-up brine injectivity in Fig. 10 is less than the initial water injectivity for Wasson conditions. This is because the low-mobility oil bank remains in the system after CO_2 injection stops and its influence is still felt. Although there are other reasons why the follow-up brine injectivity is lower than the initial injectivity (e.g., phase behavior and three-phase trapping[5]), the creation of the oil bank is one of them. **Fig. 12** shows a relatively good match of the follow-up brine injectivity at both pilots. This match was obtained by modifying the completely miscible model to include an *ad hoc* residual (10%) to the solvent displacement. This "miscible residual" was added to the trapped saturation in an attempt to mimic three-phase trapping. This reduced the follow-up injectivity significantly by reducing the mobility of the follow-up brine slug, while having only a small effect on the injectivity during CO_2 displacement.

The degree of agreement between theory and observed injectivity in these pilots is really quite remarkable. The model is tractable because it is assumed that the system is radial, miscible, incompressible, and dispersion-free, with no fingering or vertical communication. Results from logging-observation wells confirmed the validity of the radial-flow assumption and indicated that there was little vertical crossflow in the pilot areas. The arrival times of the oil bank in different observation wells indicated the existence of a stable front. The CO_2/oil interface showed much more dispersion than modeled here, while maintaining radial symmetry. Compressibility effects were small because CO_2 was injected immediately after a brine preflood, maintaining the same bottomhole injection pressure with both fluids. Ignoring phase-behavior effects has little effect on the injectivity during the CO_2-injection phase but clearly must be accounted for if follow-up brine injectivity is to be understood. Quantitative agreement is seen here precisely because the model assumptions conform well to these specific field applications. There are many other field applications for which this might not be true.

Conclusions

We show that some surprising injectivity behavior seen in tertiary CO_2 displacements is a result of the interaction of fluid properties and displacement geometries. Permeability-reduction factors or sudden changes in wettability are not required. Native-state relative permeabilities combined with detailed reservoir description should yield the correct behavior.

These results are quite encouraging because a simple model correctly accounts for the primary differences among these two pilots and their corresponding laboratory experiments. The results show that differences between linear and radial flow can result in qualitatively similar (Wasson) or different (Cedar Creek) laboratory and field CO_2 injectivities. The interplay of the various slugs created by a tertiary miscible displacement can be quite different between these two basic geometries. We see a wide range of behavior for these fluid properties, primarily because the oil bank has the lowest mobility. The influence of the oil bank grows before breakthrough for a linear displacement, but it is felt instantly for a radial displacement and remains unchanged. These two geometries represent two extremes that should bracket the behavior of a more conventional pattern displacement.

The success of this simple model, however, should not be used to argue that tertiary CO_2 displacements are easily understood as purely miscible displacements. Correct modeling of follow-up brine injectivity requires a more complicated model that quantitatively accounts for the existence of three phases interacting in a porous medium.

Nomenclature

a = empirical-model parameter
A = cross-sectional area
f_i = fractional flow of Component i
h = thickness of radial layer
I = normalized injectivity
J = Leverett J function
k = permeability
k_{ri} = relative permeability of Component i
L = dimensional distance between inlet and outlet
p = pressure
Δp = pressure change
q = volumetric flow rate
r_D = dimensionless radial distance
r_{Dj} = dimensionless radial distance traveled by jth front
r_w = effective wellbore radius
r_{wD} = effective dimensionless wellbore radius
s = skin
S_i = saturation of Component i
S_{or} = residual trapped-phase saturation
S_t = trapped hydrocarbon saturation
S_w = water saturation
t_D = dimensionless time (PV's injected)
v_D = dimensionless velocity
v_{Db} = dimensionless velocity of shock fluid front
v_{Dbr} = dimensionless velocity of follow-up brine front
v_{Dj} = dimensionless velocity or pseudovelocity of fixed composition j
x = dimensionless linear distance in x direction
Δx_j = fractional thickness of Slug j
λ_j = mobility of jth slug, $\left(\sum_i \frac{k_{ri}}{\mu_i}\right)$
μ_i = viscosity of pure Component i

Subscripts

i = oil, water, or solvent
j = indices M, N, and P
M = index of lowest-numbered front that has not exited through outlet
N = index of largest-numbered front in system
o = oil
P = index of largest-numbered front that exited wellbore
w = water

Superscripts

d = downstream
u = upstream

Acknowledgments

We thank P.D. Patel for his encouragement in the initiation of this work, especially in the early model development. A.A. Cherubini, J.J. Evans, and J.A. Rohan performed the experiments on equipment designed by C.F. van Egmond and W.O. Lease. We thank W.C. Miller and P.A. Good for supplying us with the field data used here. Discussions with M.B. Moranville and P.T. Baxley improved the clarity of the presentation. Finally, we thank Shell Development Co. for permission to publish this paper.

References

1. Claridge, E.L.: "CO_2 Flooding Strategy in a Communicating Layered Reservoir," *JPT* (Dec. 1982) 2746–56.
2. Todd, M.R., Cobb, W.M., and McCarter, E.D.: "CO_2 Flood Performance Evaluation for the Cornell Unit, Wasson San Andres Field," *JPT* (Oct. 1982) 2271–82.
3. Chase, C.A. and Todd, M.R.: "Numerical Simulation of CO_2 Flood Performance," *SPEJ* (Dec. 1984) 597–605; *Trans.*, AIME, **277.**
4. Patel, P.D., Christman, P.G., and Gardner, J.W.: "Investigation of Unexpectedly Low Field-Observed Fluid Mobilities During Some CO_2 Tertiary Floods," *SPERE* (Nov. 1987) 507–13.
5. Schneider, F.N. and Owens, W.W.: "Relative Permeability Studies of Gas-Water Flow Following Solvent Injection in Carbonate Rocks," *SPEJ* (Feb. 1976) 23–30; *Trans.*, AIME, **261.**
6. Anderson, W.G.: "Wettability Literature Survey—Part 4: Effects of Wettability on Capillary Pressure," *JPT* (Oct. 1987) 1283–1300.
7. Lantz, R.B.: "Quantitative Evaluation of Numerical Diffusion (Truncation Error)," *SPEJ* (Sept. 1971) 315–20; *Trans.*, AIME, **251.**
8. Buckley, S.E. and Leverett, M.C.: "Mechanism of Fluid Displacement in Sands," *Trans.*, AIME (1942) **146,** 107–16.
9. Gorell, S.B.: "Modeling the Effects of Trapping and Water-Alternating-Gas Injection on Tertiary Miscible Displacements," paper SPE 17340 presented at the 1988 SPE/DOE Symposium on Enhanced Oil Recovery, Tulsa, April 17–20.
10. Good, P.A. and Downer, D.G.: "Cedar Creek Anticline CO_2 Injectivity Test: Design, Implementation, and Analysis," paper SPE 17326 presented at the 1988 SPE/DOE Symposium on Enhanced Oil Recovery, Tulsa, April 17–20.
11. O'Meara, D.J. Jr. and Lease, W.O.: "Multiphase Relative Permeability Measurements With an Automated Centrifuge," paper SPE 12128 presented at the 1983 SPE Annual Technical Conference and Exhibition, San Francisco, Sept. 5–8.
12. Hassler, G.L. and Brunner, E.: "Measurement of Capillary Pressures in Small Core Samples," *Trans.*, AIME (1945) **160,** 114–23.
13. O'Meara, D.J. Jr. and Crump, J.G.: "Measuring Capillary Pressure and Relative Permeability in a Single Centrifuge Experiment," paper SPE 14419 presented at the 1985 SPE Annual Technical Conference and Exhibition, Las Vegas, Sept. 22–25.
14. Leverett, M.C.: "Capillary Behavior in Porous Solids," *Trans.*, AIME (1941) **142,** 152–68.

Appendix—Centrifuge Relative Permeability Experimental Procedure and Analysis

Core material obtained from the South Pine Unit (Cedar Creek anticline) pilot was frozen at the wellsite and transported to the laboratory. Every effort was made to maintain the wettability of the samples as received, but the wettabilities could have changed in the coring process or during handling. One-inch [2.54-cm] -diameter frozen cores were then mounted in Hassler core holders, and the resident fluids were replaced with synthetic brine and a refined oil (Soltrol^{TM} 130) with an interfacial tension of 30 dynes/cm [30 mN/m]. Cores were then placed in an automated centrifuge, as described by O'Meara and Lease,[11] and

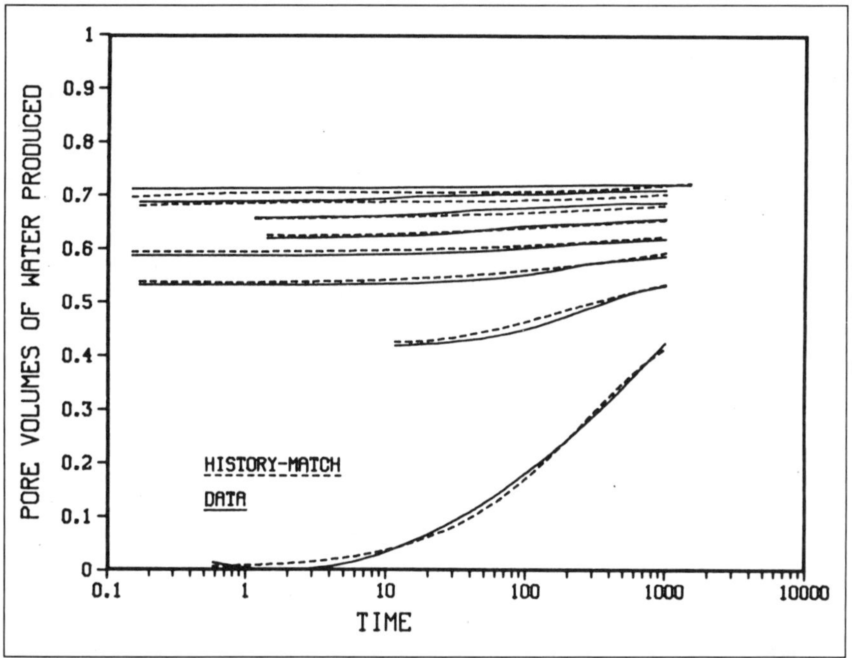

Fig. A-1—Comparison of history-match (dashed) and smoothed experimental production data (solid).

TABLE A-1—EMPIRICAL MODELS OF THE LEVERETT *J* FUNCTION AND RELATIVE PERMEABILITIES FOR SECONDARY-DRAINAGE CEDAR CREEK CORE

Parameter	J	k_{rw}	k_{ro}
a_0	0.80122	0.394	1.008
a_1	0.300	3.157	7.607
a_2	0.115	0.115	0.043

$$J = a_0[S_w/(1-a_2)]^{a_1}$$
$$k_{rw} = a_0[(S_w-a_2)/(1-a_2)]^{a_1}$$
$$k_{ro} = a_0[(S_o-a_2)/(1-a_2)]^{a_1}$$

were centrifuged for 1,000 minutes at eight different speeds. Finally, the endpoint production was analyzed by a conventional Hassler and Brunner[12] technique to obtain the capillary pressure.

The transient production data were history matched with a modified version of the program developed by O'Meara and Crump.[13] This work differs significantly from their work in that here (1) both the displaced-phase and invading-phase relative permeabilities are determined because both fluid viscosities are comparable; and (2) capillary pressure is held fixed at the value determined from the conventional analysis—i.e., it is not corrected for relative-permeability effects that delay the ultimate production at each speed. When all three curves were allowed to vary, uniqueness was a significant problem. Thus it was very important that permeability of the sample was relatively high and that the time at each speed was long enough to make the conventional capillary-pressure analysis valid.

In **Fig. A-1,** the smoothed transient production data from a 144-md, 24%-porosity core is plotted with the history match used to estimate the relative permeabilities (the clock is reset at the beginning of each speed). The quality of the match between experimental and simulated data is very good. Note that this is a semilog plot, which greatly emphasizes the production at long times. Nonetheless, the production leveled out at each of the speeds after 1,000 minutes, with the exception of the first speed. It is important that the production be small at the end of each speed so that the assumptions required for conventional capillary-pressure analysis hold.

Table A-1 presents the empirical-model parameters with numerical values for Cedar Creek anticline relative permeability and capillary pressure. The Leverett *J* function,[14] which results from the conventional analysis of these endpoint data, has a very low and flat plateau, corresponding to a capillary pressure of 0.5 psi [3.45 kPa] for all $S_w > 25\%$. This low value is a strong indicator of mixed wettability. The relative permeability curves shown in Fig. 4 also show mixed-wettability characteristics.

SI Metric Conversion Factors

cp × 1.0*	E−03	= Pa·s
in. × 2.54*	E+00	= cm

*Conversion factor is exact.

Provenance

Original SPE manuscript, **Comparison of Laboratory and Field Observed CO_2 Tertiary Injectivity,** received for review April 17, 1988. Paper accepted for publication April 15, 1989. Revised manuscript received Aug. 31, 1989. Paper (SPE 17335) first presented at the 1988 SPE/DOE Enhanced Oil Recovery Symposium held in Tulsa, April 17–20.

Authors

Christman

Gorell

Peter G. Christman, a senior research engineer at Shell Western E&P Inc. in Houston, works on CO_2 displacement processes and reservoir simulation. He is the reservoir engineer for the South Wasson Clearfork Unit. Christman holds a BS degree in chemistry and an MS degree in chemical engineering from the U. of Wisconsin and a PhD degree in chemical engineering from the U. of Texas. A senior research engineer in the Petroleum Engineering Research Dept. at Shell Development Co. in Houston, Sheldon B. Gorell has interests in reservoir simulation and characterization, and miscible EOR processes. Gorell holds a BS degree from the U. of Rochester and a PhD degree from Stanford U., both in chemical engineering.

SPE

Society of Petroleum Engineers

SPE 22653

A Laboratory and Field Injectivity Study: CO_2 WAG in the San Andres Formation of West Texas

J. Prieditis, C.R. Wolle, and P.K. Notz, Texaco Inc.
SPE Members

This paper was prepared for presentation at the 66th Annual Technical Conference and Exhibition of the Society of Petroleum Engineers held in Dallas, TX, October 6–9, 1991.

ABSTRACT

Laboratory measurements of CO_2 and brine injectivities in restored San Andres carbonate cores at reservoir conditions are reported. Included are results both with unsaturated CO_2 and brine as well as with brine equilibrated with CO_2 and CO_2 equilibrated with brine. A waterflood, oilflood, and miscible flood followed by many cycles of CO_2 and brine injection were done. Important results are the following: endpoint CO_2 relative permeabilities were substantially smaller than both water and oil endpoint relative permeabilities making CO_2 less mobile than would be anticipated just from viscosity ratios, the irreducible water saturation increased after an oilflood and persisted during subsequent CO_2 floods, residual CO_2 saturations to saturated brine were about the same magnitude as the residual oil saturations to waterflood, and the injectivity of unsaturated brine was much higher than saturated brine because of absorption of the CO_2 phase.

The laboratory work was done in support of an injectivity field trial. The results of an injection of two cycles of CO_2 and of brine into a well in the Mabee Field in the San Andres formation are reported. Both the CO_2 and the brine field injectivities were significantly greater than the pre-CO_2 waterflood injectivity on a volumetric basis. Field injectivities were somewhat greater than would be anticipated from the laboratory data. A simple analytical injectivity model could approximate some of the features of the field test but was not completely satisfactory.

References and figures at end of paper.

INTRODUCTION

Early in the study of Mabee Field as a candidate for a CO_2 project, the question arose whether CO_2 could be injected at reasonable rates and pressures. A related question was whether or not the water injection rates would be reduced following CO_2 injection. Information from other CO_2 projects in the San Andres formation in the Permian Basin showed no consistent behavior for CO_2 and water injection; both increases and decreases in injectivity relative to pre-CO_2 water injection rates have been reported.

To answer the question of what CO_2 and water injectivities to expect, laboratory tests were performed, and field injectivity tests were conducted in two wells. Two wells were deemed sufficient to give representative results at reasonable cost, and results would be obtained even if one well were to experience problems. Since the injected volumes of CO_2 were small, no production response was anticipated. In fact, problems were experienced with one of the two wells, and the results for only one well are discussed.

Several sets of laboratory tests were conducted. The objective of these tests was to measure relative permeability relationships of relevance for CO_2 flooding. These included relative permeabilities for an initial waterflood followed by an oil flood as well as permeabilities during several subsequent CO_2 and brine WAG cycles. Results for both saturated and unsaturated CO_2 and brine were obtained. The importance of relative permeabilities for understanding CO_2 and brine injectivity behavior as well as the importance of making measurements in properly prepared and restored cores has been pointed out by several authors.[1,2,3,4]

Another important objective was to determine how well the laboratory data agreed with the field data and whether the two types of data could be related using simple analytical models from the literature[2].

LABORATORY INJECTIVITY TESTS

Fluid and Core Data

A total of four core plugs were tested in the laboratory. The designations of the plugs as well as the depths from which they were taken and their permeabilities are presented in Table 1. Each core plug was 1.5 inches (3.81 cm) in diameter and 3.0 inches (7.62 cm) in length. Internal taps for pressure measurement had been installed to help reduce capillary end effects, but pressure drops across the entire core plug were used in computing the permeabilities since they were found to give essentially the same results as pressure measurements between the two internal taps.

Since the cores had not been preserved, the plugs were cleaned, resaturated with fluids, and aged for 25 days or longer. The objective of the cleaning and restoration process was to restore wettabilities and initial saturations which would be representative of the reservoir.

Two basic strategies were used in the restoration process. The first was to saturate the cores with brine immediately after cleaning and then drive out the brine with oil to establish the initial irreducible brine saturation. This was done with core plugs 2-3 and 2-6. However, there was a concern that exposing rock surfaces first to brine would not permit them to become as oil wet as they might actually be in the reservoir. Consequently, in accordance with a previously suggested procedure, the alternate strategy was to first saturate cores with oil[5,7]. This was done with cores 2-1 and 2-2. A small slug of brine was then injected and displaced with additional oil. By restoring the core plugs in these two different ways it was hoped to bracket the performance that could be expected in the reservoir. The viscosities of the fluids used in the flow tests are presented in Table 2.

CO_2 Solubility in Reservoir Brine. The solubility of CO_2 in synthetic Mabee reservoir brine (37,500 ppm TDS) at 110°F was determined to be

150 scf/res bbl (27 std m^3/m^3) at 1800 psia (12,410 kPa) and

144 scf/res bbl (26 std m^3/m^3) at 1500 psia (10,340 kPa)

Overall Summary of Flooding Steps

The basic goal of the flooding and injection sequence was to represent what would occur in the reservoir as closely as possible. The flooding sequence consisted of the following steps after each core plug was restored and aged:

1. A waterflood in order to simulate the actual waterflooding of the reservoir.

2. An oilflood to simulate the creation and movement of an oil bank in the reservoir since oil is mobilized into an oil bank by the CO_2.

3. Injection of a mixture consisting of 75% by volume CO_2 and 25% by volume butane (abbreviated C_4). The volumes were determined at 2000 psia (13,800 KPa). This mixture was designed to be first contact miscible with the oil since multi-contact miscibility can not be attained in a short core. This simulates CO_2 displacing oil at the back of the oil bank.

4. Injection of several cycles of CO_2 and brine to simulate the WAG process. Both saturated and unsaturated CO_2 and brine were used.

Water and Oil Floods

After cleaning and restoring each core plug, a **waterflood** followed by an oilflood was performed. An overall summary is presented in Table 3. The water and oil floods were conducted using synthetic brine and stock tank oil with no back pressure at 110°F (43°C).

The standard unsteady-state JBN (Johnson, Bossler, and Naumann[6]) technique was used to generate relative permeabilities from the production and pressure histories. All relative permeabilities discussed here are relative to the absolute permeability unless otherwise stated.

The oil and water relative permeability curves generated during the waterflood crossed at less than 50 percent water saturation for all the samples. Also, the water curve endpoint relative permeabilities were 0.20 or greater. Both these characteristics indicate a tendency toward oil wet rather than water wet behavior according to the commonly used criteria provided by Craig.[5] Fig. 1 shows typical waterflood unsteady state relative permeability curves.

Core plugs 2-1 and 2-2, which were saturated with oil rather than brine immediately after cleaning, exhibited on average a larger water permeability endpoint relative to oil. This indicates that core plugs 2-1 and 2-2 were probably more oil wet on average than the other two, an indication which is supported by their subsequent oil flood behavior. Otherwise, the curves for the four core plugs are very similar.

An **oilflood** was done next. In the tertiary CO_2 flooding process, an oil bank is built up ahead of the injected CO_2. In an oil bank, the water saturation decreases while the oil saturation increases. The direction of these saturation changes is simulated by an oilflood in the laboratory. However, complete curves could not be generated. No additional water was produced after oil breakthrough for cores 2-1 and 2-2. Water continued to be produced slowly after oil breakthrough during the oilflood for core plugs 2-3 and 2-6, but the JBN method did not provide very good curves. Therefore, only endpoint values are reported.

Two important differences between the oilflood and the waterflood characteristics were the following:

1. An oilflood following a waterflood was unable to drive the water saturation down to the original low levels. The **irreducible water saturation (S_{wir}) after an oilflood increased substantially.**

2. The maximum oil relative permeability k^o_{ro} **attained at the end of the oilflood was less than the value at the initial irreducible water saturation.** The amount of reduction varied from core to core, but the new maximum oil relative permeability was 0.33 to 0.67 times the value at the initial irreducible water saturation.

The hysteresis or shift in curves for an oilflood following a waterflood has been observed by other investigators[1,2,3]. The main **effect of the new increased irreducible water saturation (S_{wir}) is an earlier CO_2 breakthrough as well as an earlier production of oil** than would be predicted using the initial smaller connate water saturation.

Experimental CO_2 - WAG Cycles

After the oilflood, a mixture of 75 volume percent CO_2 and 25 volume percent butane was injected. This mixture was designed to be first contact miscible with the oil. The residual oil saturation after the miscible solvent flood appeared to be less than 5%. Following injection of the first contact miscible mixture, injection of pure CO_2 was alternated with injection of brine in a simulation of the WAG process.

Results for both saturated and unsaturated CO_2 and brine were obtained. Saturated CO_2 refers to CO_2 equilibrated with brine while saturated brine refers to brine equilibrated with CO_2. Unless otherwise noted, all tests were done at 110°F (43°C) and a back pressure of 1800 psia (12,410 kPa).

General Observations. Some general observations about the CO_2 WAG cycles are given first. A detailed discussion follows later. It should be noted that these injectivities are for linear flow and would be altered for radial flow.

1. CO_2 injectivity was greater than brine injectivity in all the core tests. However, even though the viscosity of CO_2 is about 11 times less than that of brine, in no case was CO_2 injectivity 11 times greater than brine injectivity. This is because the endpoint CO_2 relative permeability was lower than that of brine.

2. Representative CO_2 endpoint permeabilities were about one-tenth of the oil relative permeabilities at the initial irreducible water saturation.

3. CO_2 injectivity was less in cores 2-1 and 2-2 which were saturated with oil first in the restoration process. The water endpoint permeabilities relative to the oil endpoints were higher on average for these two cores. Based on the results of the core tests, a way to anticipate CO_2 injectivity is to look at the ratio of the water endpoint relative permeability over that of the oil for a waterflood. The larger is this ratio (i.e., the larger is the relative permeability of water), the poorer the CO_2 injectivity is expected to be.

4. CO_2 injectivities did not consistently decrease or increase during the cycles.

5. Injectivity results for saturated and unsaturated CO_2 injection were essentially the same. The small amount of water absorbed by the CO_2 did not significantly change the injectivity behavior.

6. Unsaturated brine reduced the CO_2 saturation to low levels. This reduction substantially increased unsaturated brine injectivity.

7. The final saturated brine injectivity in the WAG cycles was about the same as the previous waterflood injectivity. However, it took more pore volumes of saturated brine to reach the endpoint when CO_2 was being displaced than when oil was being displaced.

Experimental CO_2 and Brine Injectivities as a Function of Pore Volumes Injected. Figs. 2 through 8 show histories of how CO_2 and brine injectivities change as a function of injected pore volumes for each of the cores. Both saturated and unsaturated fluid injections are shown. What is plotted in these figures is the normalized injectivity,

$$\text{Normalized Injectivity} = \frac{\left(\dfrac{Q}{\Delta p}\right)_{CO_2 \text{ or brine}}}{\left(\dfrac{Q}{\Delta p}\right)_{\text{max during waterflood}}} \quad (1)$$

where Q is the injected flow rate

and Δp is the pressure drop across the core.

Injectivities increased both for CO_2 and brine as additional pore volumes were injected. There was some leveling off of injectivity. There was a large variation in final CO_2 injectivity from core to core, with a range of about 1.6 to 5.6 times the maximum waterflood injectivity.

The largest variation of CO_2 injectivity was between cores which were saturated first with oil immediately after cleaning (i.e., cores 2-1 and 2-2) and those which were saturated first with brine (i.e., cores 2-3 and 2-6). The CO_2 injectivity was less on average in the cores saturated with oil first. These cores appeared to be more oil wet as indicated by their higher endpoint water relative permeabilities and also by the plug flow like character of the oilfloods. Although the data is limited, a tentative conclusion is that the magnitude of the CO_2 injectivity increase over the previous waterflood injectivity will become smaller as the rock becomes more oil wet.

There was also variation in CO_2 injectivity from cycle to cycle for some cases; however, there was no pattern of either the CO_2 or brine injectivity decreasing or increasing from cycle to cycle. Although CO_2 viscosity was about 11 times less than brine viscosity, in no case was CO_2 injectivity 11 times greater than brine injectivity. Unsaturated and saturated CO_2 injectivities were about the same. The relative injectivity of CO_2 was greatest in core 2-3 and smallest in core 2-2. Representative values of final CO_2 injectivities are presented in Table 4. The injectivities are normalized with respect to the maximum waterflood injectivity as shown by Eqn. (1).

Endpoint saturated brine injectivities were about the same as maximum waterflood injectivities. The behavior reported by Schneider and Owens[1] in which endpoint brine injectivities following a solvent flood are lower than those for the initial waterflood was not observed here. Surprisingly, though, it took longer to establish the endpoint injectivities when brine was used to displace CO_2 rather than oil. Unsaturated brine was very effective in reducing the CO_2 saturation and increasing the brine injectivity if enough pore volumes of unsaturated brine are injected.

Fig. 9 shows a comparison of the three types of brine displacement test. The much greater unsaturated brine injectivity is especially evident. Fig. 10 shows the effect of unsaturated brine on CO_2 saturation. Unsaturated brine reduces the CO_2 saturation by actually dissolving and absorbing the CO_2. The residual CO_2 saturation decreased substantially if enough brine was injected. The values for the residual CO_2 saturation given in Fig. 10 do not take into account the instantaneous amount of CO_2 dissolved in the brine. The CO_2 saturations were determined from the volume of CO_2 expanded to atmospheric pressure from about 1800 psia (12,410 k Pa). Because of these results, a general observation is that **any process or mechanism which reduces the CO_2 saturation also increases brine injectivity substantially**.

Normalized brine injectivities are also presented in Table 5. The tabular data shows that it was difficult for saturated brine to displace CO_2. About 10 pore volumes (Vp) of brine were used for the tests in which oil was displaced. Since the injectivities at 10 Vp of saturated brine injection are typically less than 1.0, CO_2 was typically harder to displace than oil. In contrast, unsaturated brine was very effective in displacing CO_2. The maximum injectivity of unsaturated brine was typically at least two times the maximum injectivity for the initial waterflood and was between about 50 and 75 percent of the maximum injectivity that could be attained based on the absolute permeability.

Relative Permeabilities and Saturations. A summary of endpoint CO_2 relative permeabilities and saturations is presented in Table 6. The CO_2 relative permeabilities are representative values and are from saturated data except for core 2-2. For core 2-2, the value at the end of the $CO_2 + C_4$ injection phase is used. These may not be true endpoints because some change continued to occur.

To emphasize different features, the CO_2 relative permeabilities in Table 6 are presented with three different normalizing factors: the absolute permeability of the core, the oil permeability at the initial irreducible water saturation, and the maximum permeability at the end of the initial waterflood. The endpoint permeabilities of CO_2 relative to brine are provided because these values best represent how CO_2 injectivity compares to brine injectivity; the higher these values are, the higher are the relative injectivities of CO_2.

The permeabilities relative to the initial oil permeability show an important feature of CO_2 injection behavior in these tests. The permeability of CO_2 relative to the oil permeability at the initial irreducible water saturation is fairly constant and equal to about 0.1. Consequently, **endpoint CO_2 relative permeability is substantially less than oil relative permeability**. Whether this is a feature to be expected of all cores or just peculiar to this set of cores is not yet known. The common practice of using oil relative permeability to represent CO_2 relative permeability as is done, for example, by Christman and Gorell[2] is not always valid.

The **irreducible brine saturations** to CO_2 injection are large and close to the irreducible water saturations to oil flooding. The **residual CO_2 saturations** after flooding with many pore volumes of saturated brine are about equal to the residual oil saturations after waterflooding. As shown in Table 8, endpoint relative permeabilities are very similar for saturated brine and the initial waterflood. The relative permeability of the unsaturated brine reaches 50 to 75 percent of the absolute permeability.

Estimates of CO_2 Injectivity from Water-Oil Relative Permeabilities. Although the data is limited, the results of these core tests suggest a way to estimate or anticipate CO_2 injectivity at least qualitatively for cores if only water-oil relative permeability curves for a waterflood are available.

Since the CO_2 endpoint relative permeability appears to involve a fairly constant reduction of the oil endpoint relative permeability, the important feature in the water-oil curves is the ratio of the water relative permeability endpoint to the oil relative permeability endpoint. If this ratio is low (about 0.2 or less), then CO_2 injectivity can be expected to be several times the water injectivity. In contrast, if ratio is high (greater than 0.5), then CO_2 injectivity may not be much greater than brine injectivity. This suggests that the more oil wet the rock, the smaller will be the CO_2 injectivity (because oil wet rock has a high relative water permeability).

Mobility Ratios

A mobility ratio gives an idea of how favorable or unfavorable a displacement is; the larger the mobility ratio is, the more unfavorable the displacement and the more subject it is to instability. The experimental data collected here provide information with which to construct approximate mobility ratios. The mobility ratio is defined here as the endpoint mobility of the displacing fluid divided by the endpoint mobility of the displaced fluid,

$$\frac{k_{ri}^{o} / \mu_i}{k_{rj}^{o} / \mu_j} \qquad (2)$$

where

the subscript i refers to the displacing fluid

the subscript j refers to the displaced fluid

k^o_r is the endpoint relative permeability

μ is the fluid viscosity

Mobility ratios can be compared with inverse viscosity ratios to see how relative permeabilities affect the overall ratio. The inverse viscosity ratio is given by

$$\frac{1 / \mu_i}{1 / \mu_j} \qquad (3)$$

The live oil viscosity and **endpoint relative permeabilities** are used in the ratios presented in Table 8. Viscosities at 110°F (43°C) from Table 2 and permeabilities for core 2-3 from Tables 3 and 6 are used. Core 2-3 had the highest CO_2 injectivities and, thus, has the most unfavorable mobility ratios. In all cases except for oil displacing brine, the displacements are more favorable when relative permeabilities are taken into account. The effect of relative permeability is to make displacements more stable.

The information in Table 8 for core 2-3 can be phrased and viewed in the following manner.

1. Although CO_2 is 41.8 times less viscous than the oil, it is on average only 11.5 times more mobile than the oil.

2. Although CO_2 is 11.5 times less viscous than the brine, it is on average only 5.6 times more mobile than the brine.

3. Although the brine is 3.6 times less viscous than the oil, it is on average only 0.8 times as mobile.

As mobility ratios increase, injectivity ratios also increase. Thus, while the low relative permeability of CO_2 reduces CO_2 injectivity, it has the beneficial effect of making the mobility ratio more favorable for CO_2 as a driving fluid.

FIELD INJECTIVITY TEST

This test was not intended to be a pilot project and no production increase was expected or anticipated; the purpose was only to establish injectivity behavior.

Procedure

The basic procedure for conducting the injectivity test was as follows:

1. Run a step-rate test to determine fracture pressure
2. Prior to switching to CO_2 injection, run electric line and surface readout equipment. Record bottom-hole pressure and temperature, and continue monitoring until bottom-hole pressure stabilizes during CO_2 injection.
3. Commence CO_2 injection at rates up to 1 MMSCF/D (28,300 m^3/day)
4. After 30 days switch to water injection for 30 days
5. Switch back to CO_2 for 30 days, at rates up to 1.5 MMSCF/D (42,400 m^3/day)
6. Resume water injection
7. Strive to inject CO_2 and water below fracture pressure
8. Immediately prior to switching injectants in any cycle, run an injection profile

Results

Tests were conducted in two wells, but only the test in one of the wells is reported in detail. In this well, the base water injection rate prior to CO_2 was 200-250 BWPD (32-40 m^3/day) at 750-780 psi (5170-5380 kPa). Fracture pressure was approximately 800 psi (5,520 kPa) surface and 2900 psi (20,000 kPa) bottom-hole. CO_2 injection was initiated on October 22, 1990 at rates initially in the 600-800 MCFD (17,000-22,700 m^3/day) range. A measured bottom-hole pressure of 2880 psi (19,860 kPa) equated to a surface pressure of 1170 psi (8070 kPa), and bottom-hole temperature appeared to stabilize at 97°F (36°C). Injection of CO_2 continued, at rates ranging up to 1.4 MMCFD (39,600 m^3/day), until November 21. During this period approximately 32 MMCF (906,000 m^3) of CO_2 were injected.

The well was switched to water injection status but did not take any water; wellhead pressure was too high to be overcome by injection plant pressure. On November 27, a pump truck was brought to the well and after 8-10 barrels (1.3-1.6 m^3) of water were pumped, the well began taking water at initial rates of 900 BWPD (143 m^3/day) at 640 psi (4,410 kPa). Injection dropped gradually, stabilizing at 350-400 BWPD (57-64 m^3/day) at 750 psi (5,170 kPa). On December 26, CO_2 injection in the second cycle began, at a rate slightly exceeding 1 MMCFD (28,300 m^3/day). The rate was gradually increased until it reached approximately 2.5 MMCFD (70,800 m^3/day), with wellhead pressures not exceeding 1170 psi (8,070 kPa). Injection was terminated on January 18, 1991 following cycle injection of approximately 42.5 MMCF (1,204,000 m^3). A pump truck was required to switch the well to water injection, but it was not immediately successful. On January 24, water injection was established, and injection again stabilized in the 350-400 BWPD (57-64 m^3/day) range.

Injection profiles run during each cycle (water and CO_2) showed very little change in where the injected fluid was going. Both water and CO_2 were contained within the perforated interval, with the exception of a minor loss of CO_2 above the perforated interval noted during the second CO_2 cycle. CO_2 and brine injectivities normalized as shown by Eqn. 1 (with the pressure drops understood as those in the field) are plotted against total pore volume injected in Fig. 11. A factor of 0.40 RB/MSCF (7.1 E-5 m^3/std m^3), corresponding to 105°F (41°C) and 2500 psia (17,240 kPa), was used to represent near wellbore conditions. CO_2 and water injectivities were both substantially greater than pre-CO_2 water injectivity.

Test results in the other well are not reported here because poor wellbore conditions caused excessive losses of injected CO_2 outside the productive interval, and excessive bottom-hole pressures unrelated to CO_2 injection resulted in much lower injection rates than expected.

COMPARISON OF LABORATORY AND FIELD DATA

An important objective was to determine how well the laboratory data agreed with the field data and whether simple models could be used to relate the two types of data. Although the cores were not taken from one of the wells in which the injectivity test was done, the cores were representative.

The high brine injectivity following the CO_2 cycles was not expected, and a complete explanation is not available. Possible explanations are rock dissolution, change in rock wettability to a more oil wet condition, and reduction of the CO_2 saturation around the well bore because of the solubility of CO_2 in the brine.

The only explanation supported by laboratory data is one based on the decrease of the CO_2 saturation in the near wellbore region. There was no apparent rock dissolution in the laboratory tests even after more than 100 pore volumes of injected CO_2 and water; nor was there any apparent change in wettability as would be indicated by an increase of saturated brine injectivity. The endpoint injectivity of saturated brine displacing CO_2 was about the same as the endpoint injectivity of water displacing oil. However, in the unsaturated CO_2 and water cycles, the water injectivity did increase because of a reduction in the CO_2 saturation. When the CO_2 saturation decreased from 0.25 to about 0.1, the brine injectivities increased by 2 to 3 times. If brine absorbed the equilibrium amount of CO_2 reported above (150 scf/RB brine), then the CO_2 saturation could decrease from a saturation of 0.25 to 0.1 out to a distance of 35 feet (10.7 m) from the well bore. Using a simple model such as proposed by Martin[8], the reduction in CO_2 saturation out to 35 feet (10.7 m) associated with an effective brine permeability increase by a factor of 2 to 3 could account for an increase in brine injectivity by a factor of 1.4 to 1.7 assuming a wellbore radius of 0.33 feet (0.10 m) and an outer radius of 660 feet (201 m).

If this mechanism occurred here, then there is a question why such a mechanism would not always occur in a CO_2 flood. Brine injectivities have not often increased in CO_2 floods. One factor may be that the amount of potential improvement by reduction of the CO_2 saturation depends on how small the water endpoint relative permeability is. If the endpoint is already large (indicative of oil wet conditions), then the potential for increased injectivity is small. However, if the endpoint is small, then the potential for increased injectivity is much larger. As discussed before, a large water endpoint relative permeability also tends to decrease CO_2 injectivity relative to brine. The actual water relative permeability endpoints may have been even smaller for the case here than indicated by the laboratory results.

The increased brine injectivities complicate the interpretation of the CO_2 injectivities. The CO_2 injectivities should not be compared to the pre-CO_2 brine if something such as rock dissolution has occurred. However, the following conclusions can be made. CO_2 injectivities were greater than pre-CO_2 brine injectivities both in the laboratory and the field. The CO_2 injectivities measured in cores 2-1 and 2-2, which were saturated with oil first in the restoration process, were definitely too small and not representative of the larger CO_2 injectivities observed in the field test. These cores were probably made too oil wet. The highest laboratory injectivities (i.e., those measured in cores 2-3 and 2-6) are more representative of the field, but even these are probably not high enough if the field CO_2 injectivity is compared to pre-CO_2 brine. The following are the maximum injectivities of CO_2 relative to brine for the laboratory and the field data:

Field CO_2 Cycles

1st	2nd	
2.8	4.4	relative to waterflood
1.6	2.6	relative to brine during WAG

Laboratory

Core 2-3	Core 2-6	
5.6	3.5	relative to waterflood

Injectivities in a linear core can not be directly compared with those in a radial system with several banks of fluids. Christman and Gorell[2] have proposed a simple analytical method for injectivity prediction in a radial system which uses relative permeability data. The basis of their model is the assumption that several constant mobility and saturation banks occur in the reservoir. Their method was extended here to the injection of two CO_2 slugs alternated with two brine slugs.

The model could only be used to approximate some of the qualitative features of the tests. In order to approximate the high injectivity in the field, endpoint relative permeabilities were used. Using endpoint data makes the banks have slightly higher mobilities and lower velocities than using fractional flow methods. For CO_2, data from core 2-3 was used because it had the largest CO_2 permeability while an increase in brine endpoint relative permeability by a factor of 2.0 was used to approximate the brine injectivity increase in the field. The analytical method becomes cumbersome when there are multiple WAG cycles because banks overtake each other. It was assumed that whenever a brine front overtook an oil bank, a residual oil saturation to water was left behind[2]. Also, whenever CO_2 overtook a waterflood residual oil saturation, an oil bank was assumed to form.

Both one layer and two layer models were used. Two cases for the two layer model were done. In one case a 10 to 1 permeability contrast was assumed with the high permeability zone comprising 20% of the net pay while in the other a 100 to 1 permeability contrast was assumed with the high permeability zone comprising 9% of the zone. A wellbore radius of 0.33 feet (0.10 m) and outer boundary radius of 660 feet (201 m) were assumed. A porosity-thickness product of 5 feet (1.5 m) was used. The results are shown in Fig. 11.

Even though endpoint relative permeabilities were used, the assumption of a very high permeability thief zone was still required to approximate the high injectivity of the second CO_2 cycle. This assumption over predicts the injectivity of the first CO_2 cycle. The assumption of either a one layer model or of a smaller permeability contrast approximates the first CO_2 cycle more closely but under predicts the injectivity of the second cycle. In general, a model with a high permeability zone (or thief zone) has greater injectivity; consequently, a well with a thin thief zone taking a large proportion of the CO_2 would be expected to have high injectivity. The only indication of a possible thief zone in this case was that CO_2 showed up in the wellbore of a west offset injection well 1320 feet (402 m) away.

These results show some limitations of a simple analytical model. The large increase in CO_2 injectivity in the second cycle over that in the first cycle was greater than could be easily modeled even though the second CO_2 slug was assumed to have the same mobility as the first. This is an optimistic assumption since the second CO_2 slug in the WAG process can not really be expected to drive the water saturation down as easily and uniformly as the oil bank which is built up ahead of the first CO_2 slug. A very high second CO_2 slug injectivity could be explained by a process increasing total permeability such as rock dissolution, but the absence of water injectivity increase from cycle to cycle would argue against such a process.

CONCLUSIONS

Laboratory Tests

1. Final CO_2 linear injectivities were on average about 3.4 times greater than the initial waterflood injectivities. The injectivities of saturated and unsaturated CO_2 were about the same for the pore volumes injected. CO_2 mobility ratios were more favorable than expected from just viscosity ratios.

2. Endpoint CO_2 relative permeabilities were found to be substantially smaller than both the brine and oil endpoint relative permeabilities. This is the reason why CO_2 injectivities are not as great as would be expected just from viscosity ratios. The practice of using oil relative permeability to represent CO_2 relative permeability may not always be valid.

3. As the water endpoint permeability increases in relation to oil for the waterflood, CO_2 injectivity relative to brine decreases. This observation can be used to anticipate CO_2 injectivities from water-oil relative permeability data.

4. A substantial increase, or hysteresis, in the irreducible water saturation after an oilflood was confirmed in these tests. The high irreducible water saturation remained about the same for the subsequent CO_2 floods.

5. The residual CO_2 saturations to saturated brine were about the same as the residual oil saturations to waterflood. Residual CO_2 saturations to unsaturated brine were dependent on pore volumes injected and reached very low levels.

6. The final injectivities of saturated brine were about the same as those of the initial waterflood. In contrast the final unsaturated brine injectivities were 2 to 3 times greater than those of the initial waterfloods.

7. The rock material did not dissolve after many pore volumes of CO_2 WAG injection.

Field Test

A number of conclusions can be drawn from the CO_2 injectivity test:

1. CO_2 injectivity on an equivalent reservoir barrel basis was substantially greater than pre-CO_2 water injectivity.

2. Water injection rates following the CO_2 cycles were greater than the pre-CO_2 rates.

3. Wellbore condition is critical in CO_2 injectors to maintain injection in the pay zone.

4. Re-establishing water injection following CO_2 requires additional well head pressure.

Comparison of Laboratory and Field Results

1. Laboratory and field tests both experienced CO_2 injectivity greater than pre-CO_2 water injectivity, but the field CO_2 injectivity was greater than anticipated from the laboratory tests. The CO_2 injectivity in the cores which were made more oil wet was especially low.

2. CO_2 being absorbed by the brine may explain part of the high post-CO_2 water injectivity.

3. A simple analytical model could only approximate some of the features of the field injectivity test.

NOMENCLATURE

k_r relative permeability

k^o_{ri} endpoint relative permeability of phase i

Q injected flow rate

Δp pressure drop

S_{wir} irreducible water saturation

S_{or} residual oil to waterflood

μ_i viscosity of phase i

ACKNOWLEDGEMENTS

The authors thank Texaco Inc. and Texaco USA for permission to publish this paper.

REFERENCES

1. Schneider, F. N. and Owens, W. W.: "Relative Permeability Studies of Gas-Water Flow Following Solvent Injection in Carbonate Rocks," S.P.E.J. (Feb. 1976) 23-30.

2. Christman, P. G. and Gorell, S. B.: "A Comparison of Laboratory and Field-Observed CO_2 Tertiary Injectivity," paper SPE/DOE 17335, presented at the 1988 SPE/DOE Symposium on Enhanced Oil Recovery, Tulsa, April, 1988.

3. Potter, G. F.: "The effects of CO_2 Flooding on Wettability of West Texas Dolomitic Formations," paper SPE 16716, presented at the 62nd Annual Technical Conference and Exhibition of the SPE, Dallas, September, 1987.

4. Hawkins, J. T. and Bouchard, A. J.: "Reservoir Engineering Implications of Capillary Pressure and Relative Permeability Hysteresis," 1989 SCA Conference Paper Number 8909.

5. Craig, F. F. Jr.: The Reservoir Engineering Aspects of Waterflooding, Monograph Series, SPE, 3

6. Johnson, E. F., Bossler, D. P., and Naumann, V. O.: "Calculation of Relative Permeability from Displacement Experiments," Trans AIME, 216, p.370 (1959).

7. Schneider, F. N. and Owens, W. W.: "Sandstone and Carbonate Two- and Three-Phase Relative Permeability Characteristics," S.P.E.J. (March 1970) 75-84; Trans., AIME, Vol. 249.

8. Martin, F. D.: "Injectivity Improvement in the Grayburg Formation at a Waterflood in Lea County, NM," paper SPE 12599, presented at the 1984 Permian Basin Oil & Gas Recovery Conference, Midland, March, 1984.

TABLE 1
Test Core Plug Depths and Permeabilities

Core plug number	Depth (ft)	Absolute Nitrogen Permeability (md)
2-1	4725	14.8
2-2	4726	8.9
2-3	4728	2.9
2-6	4735	8.5

TABLE 2
Fluid Viscosities

	Viscosities (cp)	
Fluid	73°F	110°F
Brine	1.0	0.63
CO_2	0.08	0.055
Stock Tank Oil	-	4.7
Live Oil	-	2.3

TABLE 3
Oilflood and Waterflood Endpoint Relative Permeabilities and Saturations

Core number	Initial S_{wir}	k^o_{ro} at Initial S_{wir}	k^o_{rw} at end of Water flood	S_{or}	k^o_{ro} at end of Oil flood	S_{wir} after Oil flood
2-1	0.25	0.66	0.24	0.28	0.43	0.54
2-2	0.23	0.74	0.35	0.40	0.26	0.50
2-3	0.19	0.97	0.20	0.25	0.39	0.32
2-6	0.26	0.65	0.24	0.31	0.42	0.60
Core average	0.23	0.76	0.26	0.31	0.38	0.49

where

S_{wir} is the irreducible water saturation

S_{or} is the residual oil to waterflood

k^o_{rw} is the endpoint (maximum) relative permeability of oil at irreducible water saturation

k^o_{ro} is the endpoint (maximum) relative permeability of water at the residual oil saturation

TABLE 4
Representative Values of Normalized CO_2 Injectivity

Core number	Representative CO_2 injectivities relative to initial waterflood		
	End of CO_2 + C_4	Unsaturated CO_2 final	Saturated CO_2 final
2-1	2.6	2.7	2.4
2-2	2.2	1.6	-
2-3	3.8	5.6	5.6
2-6	4.7	3.5	3.2

TABLE 5
Saturated and Unsaturated Brine Injectivities

Core number	Brine injectivities relative to initial waterflood		
	Saturated brine 10 Vp	Saturated brine final	Unsaturated brine final
2-1	0.40	0.86	2.2
2-2	-	-	1.9
2-3	0.64	1.05	2.4
2-6	1.04	1.20	3.0

TABLE 6
CO_2 Endpoint Relative Permeabilities and Saturations

Core number	Endpoint CO_2 relative permeabilities: Relative to absolute	Relative to initial oil	Relative to brine	CO_2 residual to sat'd brine	S_{wir} after CO_2
2-1	0.061	0.092	0.21	0.24	0.48
2-2	0.074	0.10	0.19	-	0.52
2-3	0.107	0.11	0.49	0.37	0.41
2-6	0.071	0.11	0.28	0.25	0.49

TABLE 7
Saturated and Unsaturated Brine Endpoint Relative Permeabilities

Core number	Brine Endpoint Relative Permeabilities: Initial Waterflood	Saturated brine	Unsaturated brine
2-1	0.24	0.21	0.53
2-2	0.35	-	0.67
2-3	0.20	0.21	0.48
2-6	0.24	0.29	0.72

TABLE 8
Approximate Mobility and Viscosity Ratios*

Displacing Fluid / Displaced Fluid	Mobility Ratio	Inverse Viscosity Ratio
CO_2 / oil	11.5	41.8
CO_2 / brine	5.6	11.5
oil / brine	0.5	0.3
brine / oil	0.8	3.6

* data for core 2-3

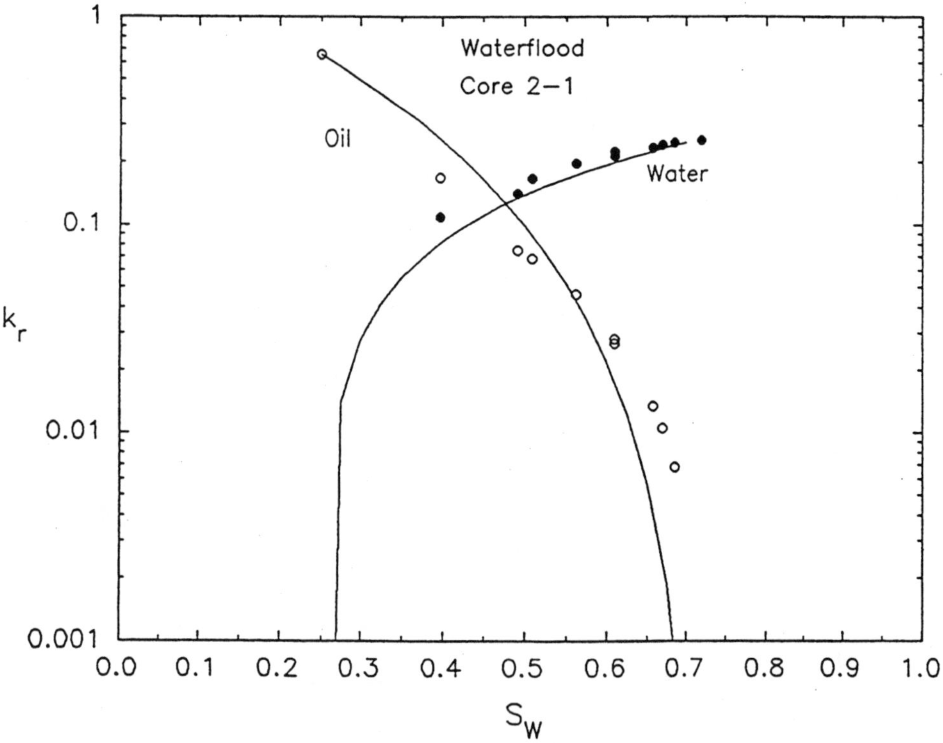

Figure 1. Waterflood Relative Permeability Curves for Core 2−1

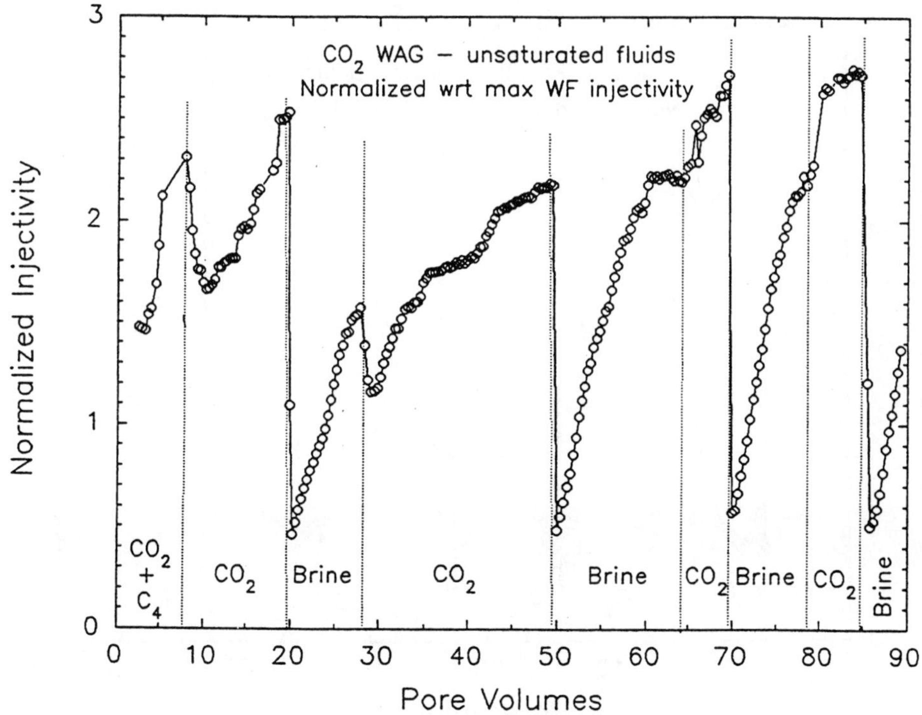

Figure 2. Injectivity of Unsaturated Fluids − Core 2−1

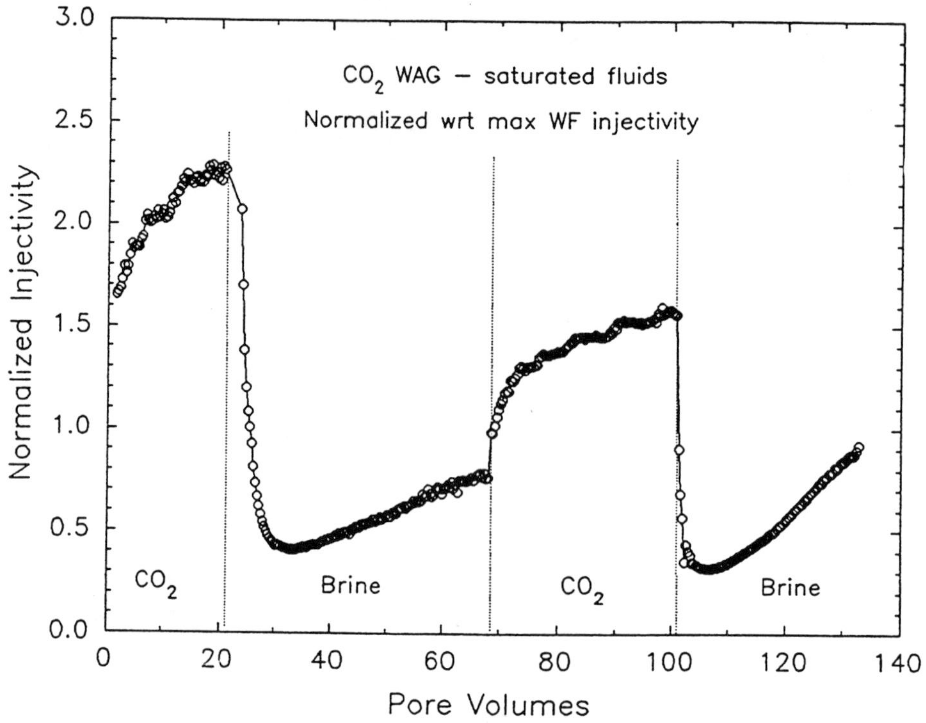

Figure 3. Injectivity of Saturated Fluids – Core 2–1

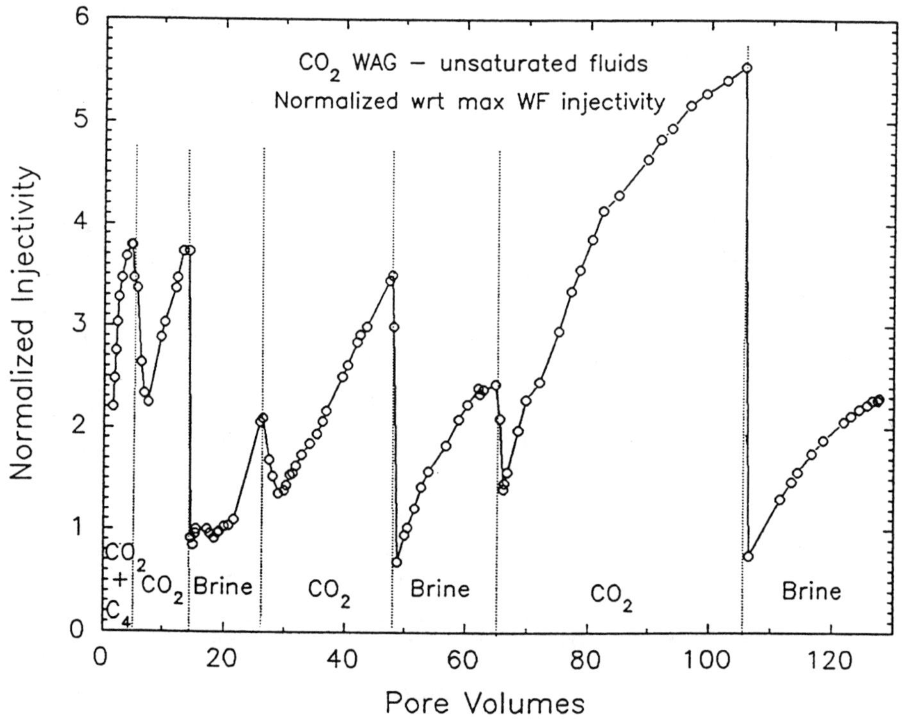

Figure 4. Injectivity of Unsaturated Fluids – Core 2–3

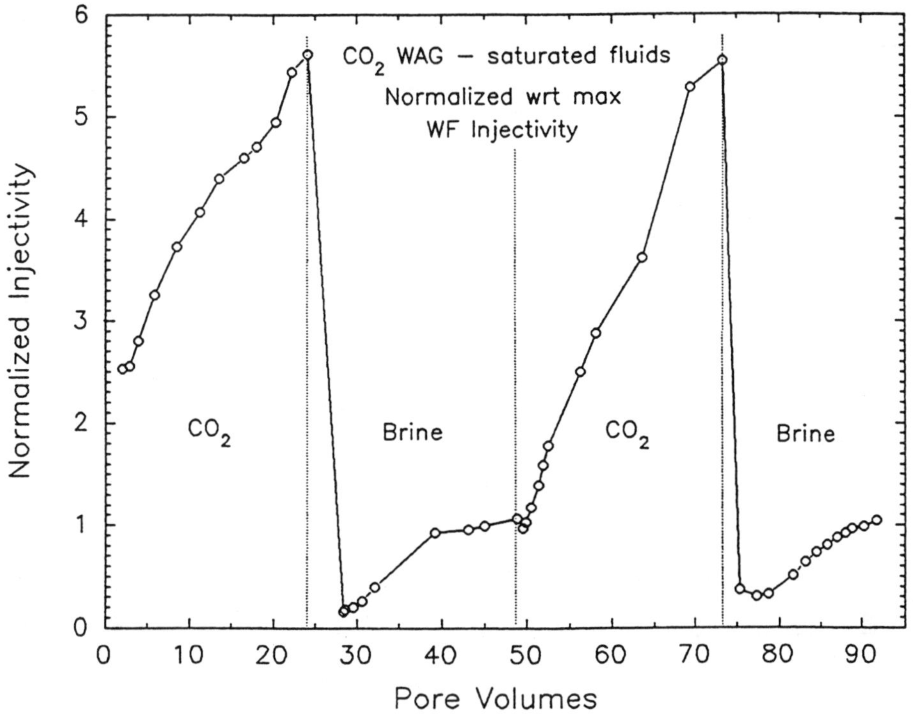

Figure 5. Injectivity of Saturated Fluids – Core 2–3

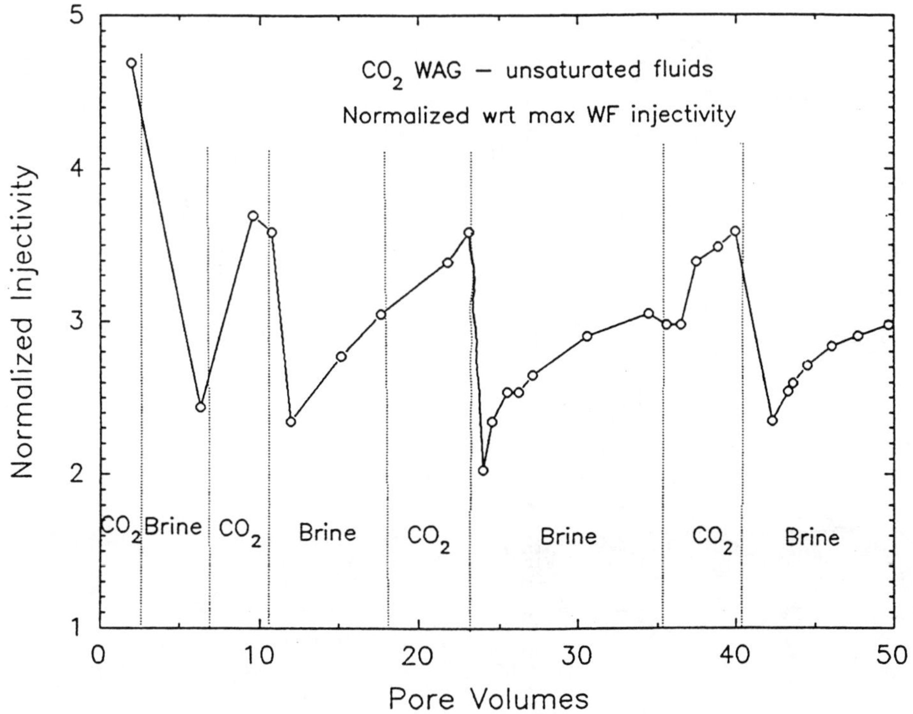

Figure 6. Injectivity of Unsaturated Fluids – Core 2–6

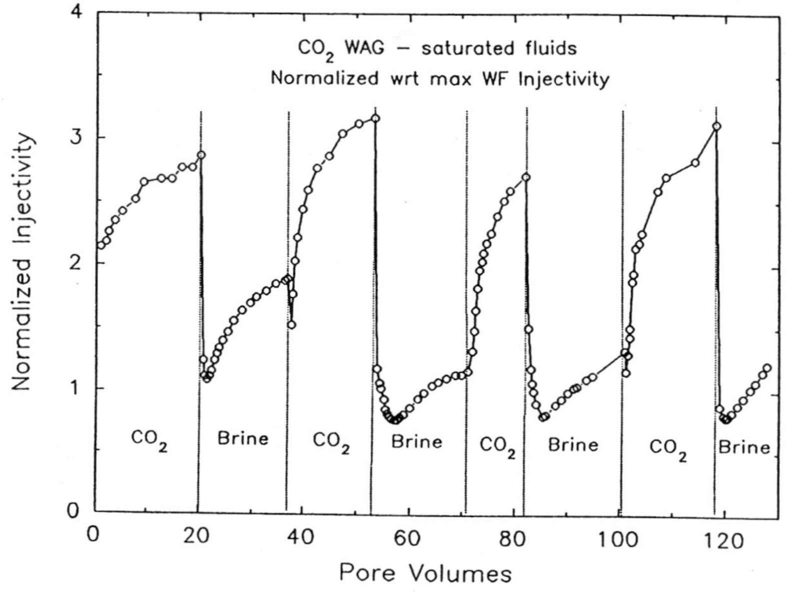

Figure 7. Injectivity of Saturated Fluids – Core 2–6

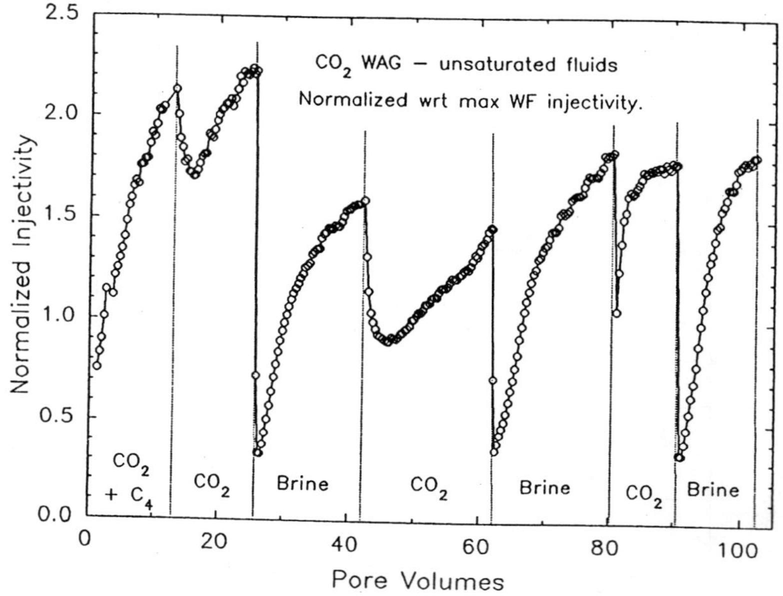

Figure 8. Injectivity of Unsaturated Fluids – Core 2–2

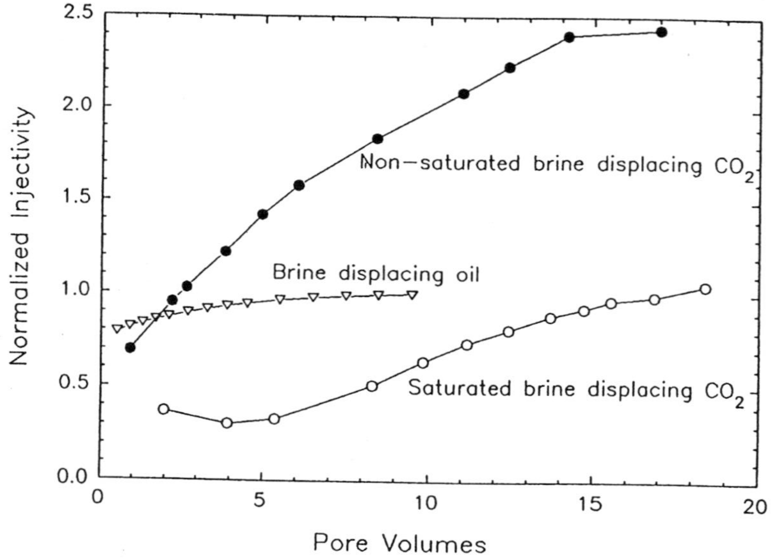

Figure 9. Comparison of saturated and unsaturated brine displacements for core 2–3

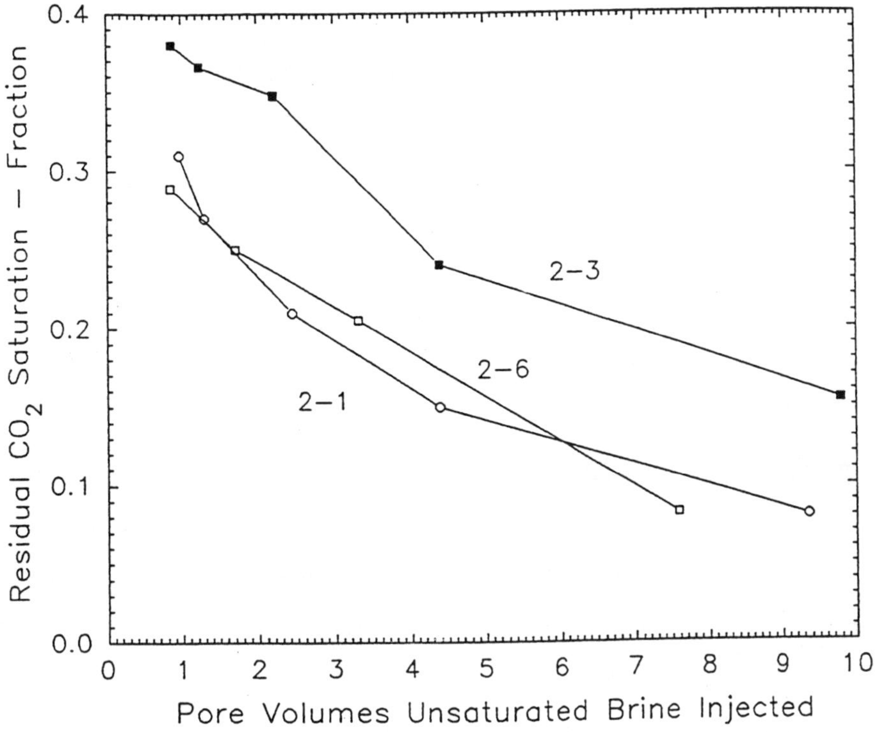

Figure 10. Apparent residual CO_2 saturation as a function of pore volumes of unsaturated brine injected

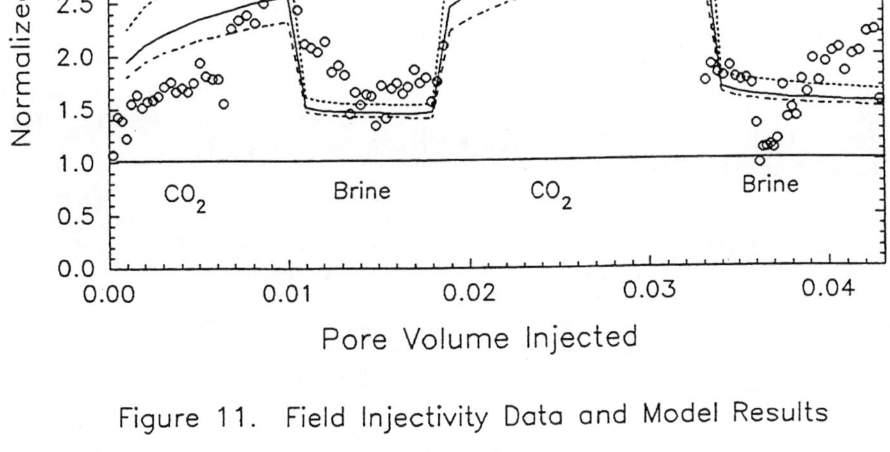

Figure 11. Field Injectivity Data and Model Results

SPE

Society of Petroleum Engineers

SPE 22651

Mechanisms of Miscible Oil Recovery: Effects of Pressure on Miscible and Near-Miscible Displacements of Oil by Carbon Dioxide

J-G.J. Shyeh-Yung, Exxon Production Research Co.
SPE Member

This paper was prepared for presentation at the 66th Annual Technical Conference and Exhibition of the Society of Petroleum Engineers held in Dallas, TX, October 6–9, 1991.

ABSTRACT

This paper presents results obtained in a laboratory study investigating incentives for near-miscible CO_2 processes. Reservoir-condition corefloods were completed in both idealized and West Texas-type fluid/rock systems at pressures above and below the minimum miscibility pressure (MMP). Coreflood oil recoveries and CO_2 mobilities are presented to examine effects of pressure, initial saturation, and volume of CO_2 injection.

Results show that decreases in tertiary oil recovery at pressures below the MMP are not as great as suggested by slim tube tests. The presence of water in the core apparently reduces the pressure sensitivity. In addition, it was found that CO_2 mobility is reduced by the presence of a miscible flood residual oil saturation (S_{orm}), which increases as the pressure decreases. Both observations suggest that lower pressure (near-miscible) CO_2 EOR processes have potential to economically recover oil.

Two oil recovery mechanisms, low interfacial tension (IFT) displacement and extraction of oil components, were found to be important. This is deduced from analyses of produced oil compositions. Lower IFT (higher pressure) and lower initial water saturation enhance displacement and reduce pore level bypassing. Extraction dominates after solvent breakthrough; extraction increases with pressure and decreases with water saturation.

The work suggests that, whenever possible, slim tube tests be replaced with reservoir-condition corefloods which provide more realistic data not only on pressure sensitivity, but also on miscible flood residual oil saturation and solvent mobility.

References and illustrations at end of paper.

INTRODUCTION

Multiple-Contact-Miscible (MCM) CO_2 injection processes are proven, economically viable means to increase crude oil reserves. Conventionally, these processes are operated so as to remain at pressures above the minimum miscibility pressure (MMP), which is determined for each crude and solvent combination. Pressures slightly below the MMP are referred to as near-miscible conditions.

A lower-pressure process is potentially more attractive from both economic and operational standpoints. Both solvent volume requirements in SCF (Standard Cubic Feet) and costs for compression are reduced at lower pressures. Some reservoirs are at a pressure lower than the MMP, while other miscible projects are experiencing pressure decline; in these cases, a lower-pressure operation may be more feasible. Therefore, the effect of pressure on oil recovery, particularly at pressures below the MMP, has been an area of increasing interest.

Several other incentives for lower-pressure CO_2 processes may include: (1) low interfacial tension (IFT), near-miscible displacements can be nearly as effective as a zero IFT miscible displacement; and (2) possible sweep efficiency improvement may occur at lower pressures due to formation of two liquid phases as reported earlier[1].

Further motivation for our study of near-miscible processes comes from the question of whether a field "miscible process" is truly miscible. It is generally believed that miscibility may repeatedly develop and break down in a reservoir due to dispersion arising from viscous fingering and reservoir heterogeneity[2]. Thus, many "miscible" processes are in fact a mixture of miscible and low IFT, near-miscible processes. Understanding displacement mechanisms at near-miscible conditions may help us to better simulate field-scale miscible processes.

A research program, whose results are presented in this paper, was initiated to study lower-pressure near-miscible CO2-injection processes. Objectives of this coreflooding program are two fold: (1) developing techniques to determine accurately effects of pressure on CO_2 flood S_{orm} and CO_2 mobility, and (2) improving our understanding of CO_2 displacement mechanisms as a function of pressure. In this study, we were especially interested in determining whether the MMP is as critical as implied by slim tube displacement tests. Also, we wish to explore the effect of pressure on CO_2 mobility.

To accomplish these objectives, coreflooding experiments were conducted in an idealized fluid, water-wet core system and in a more realistic MCM, mixed-wet system. Displacement efficiencies as measured by S_{orm} and CO_2 mobilities were investigated as a function of pressure, water saturation, pore volumes of CO_2 injection and pre-equilibration. Improved understanding of oil displacement mechanisms was achieved by analyzing produced oil compositions from these corefloods.

The first section of this paper provides background information. The second section describes experimental materials and methods. The third section discusses in detail coreflood results, concentrating on oil recoveries, produced oil compositions and solvent mobilities. Finally, conclusions are presented together with suggestions for future work.

BACKGROUND

This section reviews the definition of MMP and the effect of pressure, or CO_2/oil phase behavior, on miscible displacements. Since the term near-miscible refers to pressures slightly below the MMP, MMP becomes an important reference point. Pressure, the experimental variable of interest in this study, directly affects the phase behavior of the CO2/oil system, which in turn affects miscible displacement efficiencies.

Minimum Miscibility Pressure

Minimum miscibility pressure, or MMP, is defined as the lowest pressure at which a crude oil and a solvent develop miscibility dynamically. MMP depends on both crude and solvent compositions and is typically determined experimentally using slim tube tests[3]. The MMP value, determined by slim-tube tests, may be different from that calculated based on phase behavior data. A slim-tube is filled with an oil, which is displaced by solvent at a velocity of approximately 40 ft/day during the test. Water is absent from the slim tube. For a CO_2 MCM process, oil recovery is a function of the test pressure. Oil recoveries from slim tube tests typically are high at pressures above the MMP and decline steeply as pressures are reduced below the MMP. This bend-over point in the recovery versus pressure laboratory data is the MMP.

The MMP determined by slim tube tests has been accepted by the oil industry as the target reservoir pressure for designing a miscible process. Slim tube tests, however, have several drawbacks: (1) the "porous medium" is a bead pack bearing little resemblance to reservoir rocks; (2) water is not present during the test; and (3) often, the MMP value depends on the number and choices of test pressures and can result in a large difference in interpreted values.

Effects of Pressure on Phase Behavior

Effects of pressure on CO_2/crude oil phase behavior have been studied experimentally[4,5] using PVT and slim tube tests. CO_2/crude oil interactions have often been characterized using an equation of state. Effects of phase behavior on coreflood oil recovery also have been investigated[6], but mostly in the absence of water. To obtain realistic data, solvent flood performance should be studied in a core where water is present.

When CO_2 is used as the solvent, four mechanisms may adversely affect oil recovery as pressure is reduced:

1. The ability of CO_2 to extract components from crude oil decreases as pressure decreases. Less oil will be recovered by extraction and miscibility may not develop dynamically.

2. Interfacial tension between the CO_2-rich phase and crude oil increases as the pressure decreases, making displacement of crude oil less efficient.

3. At lower pressures, less CO_2 will dissolve in oil to increase oil volume, and oil recovery due to swelling will be reduced.

4. Lower CO_2 solubility also results in less oil viscosity reduction at lower pressures.

EXPERIMENTS

In the first part, effects of pressure were examined in a CO_2/Decane/Berea-core system. For this ideal system, CO_2 was first-contact-miscible with the oil (decane) at high pressures and immiscible at lower pressures. The core was strongly water-wet. Experiments were conducted to study effects of pressure or interfacial tension (between CO_2 and decane) on CO2 tertiary flood performance. In the second part, experiments were completed in a CO_2/West Texas crude/carbonate-core system. In this more realistic system, CO_2 and crude oil were multiple-contact-miscible. Corefloods were designed to minimize effects of gravity segregation, and viscous fingering, allowing us to focus on accurately measuring S_{orm} and CO_2 mobility at reservoir conditions. Experiments were aimed at resolving effects of pressure, the presence of a slim tube at the core inlet, initial water saturation, and pore volumes of CO_2 injection.

Experimental Apparatus

Figure 1 is a simplified flow diagram of the coreflooding apparatus. A 16-foot long slim tube ahead of the core was used to pre-equilibrate CO_2 and crude oil. The slim tube was bypassed during waterfloods and could be bypassed in CO_2 floods, if desired. Corefloods were conducted vertically, flowing upward for waterfloods and downward for CO_2 floods. A pulse-free syringe pump delivered high-pressure CO_2, at low flow rate, directly from a CO_2 gas cylinder. Pressure control was achieved by an oscillating back-pressure regulator[7]. Oil and water effluents from the core were collected through a multi-port switching valve into test tubes and gas was collected in an automated gas meter. Material balances performed on oil and brine were used to determine final average oil and water saturations in the system. Overall and mid-core pressure drops, to determine fluid mobilities, were measured by transducers and monitored by a laboratory computer.

Fluids

As mentioned earlier, two fluid systems were used in this study. System 1 used idealized fluids, specifically, CO_2 and decane at 160°F [71°C]; System 2 used more realistic multiple-contact-miscible fluids, in this case, a degassed West Texas separator oil and CO_2 at 100°F [38°C].

Properties of CO_2 and its interaction with oil change with pressure. CO_2 and decane is a well characterized system, becoming miscible at 1850 psia and 160°F [71°C]. Equilibrium phase compositions and IFTs as a function of pressure have been measured by Nagarajan and Robinson[9]. An in-house fluid property simulator was used to calculate equilibrium phase viscosities. The crude oil used in System 2 is a degassed crude from a West Texas Field, having a viscosity of 13 cp, a density of 0.91 g/cc. This crude oil is composed of 31% saturates, 25% aromatics and 43% polar components such as asphaltenes, and nitrogen, sulfur and oxygen containing compounds.

In both systems, 3.5 wt% TDS (total dissolved solids) brine was used. Composition of the brine was 0.8% $CaCl_2$ and 2.7% NaCl. Brine was filtered and degassed before use.

Cores

Properties of cores used in these two systems are tabulated in Table 1. To eliminate effects due to core variability, a single piece of Berea sandstone was used in the first series of experiments, and one piece of San Andres carbonate (an outcrop core from western Guadalupe Mountains) was used throughout the second series of tests. Both cores were epoxy encased and cast inside a high-pressure coreholder with a low melting-point lead-bismuth alloy. The cores were cleaned and reused after each CO_2 flood to allow direct comparison of coreflood results. The carbonate core, containing connate water, was soaked with crude oil for five days each time to obtain a mixed wettability. Recirculation crude-soaking procedures were also used in a few experiments to achieve lower irreducible water saturations for comparison. Great care was taken to return the core to the same initial state by monitoring absolute brine permeability and waterflood residual oil saturation. Cleaning procedures were repeated if permeabilities were not restored. In general, this cleaning procedure, described in the Appendix, was effective and nondestructive.

Procedures

For System 1, residual decane saturation in the slim tube was measured when the tube was refilled with decane for the subsequent run. Residual decane saturation in the core was determined by taking the difference between the total amount of decane remaining in the system and the residual decane in the slim tube. No extraction was performed. For System 2, the residual crude oil was removed from the system after each run because CO_2 flood residual oil did not have the same composition as the original oil. The slim tube (when included) and the core were extracted separately using trichloroethane after each CO_2 flood to obtain final crude oil saturations. Final oil saturations determined by extraction (see Appendix) also provided an independent check on the material balance calculations.

Displacement sequences, pore volumes, and rates for a typical experiment are listed in Table 2. Most experiments were tertiary CO_2 slug floods carried out after a 3 pore volume injection (PVI) waterflood. Some secondary CO_2 floods, where step 3 was skipped, were also conducted using System 2.

Produced crude oil samples were collected into several 10 cc and 5 cc tubes as the CO_2 flood progressed. Oil samples from some experiments were analyzed by a standard, gas chromatographic, whole-oil analysis method, using a capillary column and a flame ionization detector.

RESULTS AND DISCUSSION

In this section, results of 33 coreflood experiments are presented and discussed in four sub-areas. First, oil recovery results for the idealized system are presented and analyzed. Second, oil recovery results from the more realistic system are summarized and effects of four experimental parameters are examined. Third, produced oil compositions for some experiments in system 2 are presented to provide insight to oil recovery mechanisms. Lastly, solvent relative permeabilities or mobilities in system 2 are examined and discussed.

Oil Recovery For System 1, Decane/CO_2/Berea Core

Seven tertiary experiments were carried out, and pertinent results from these floods are summarized in Table 3. Oil production from the core was monitored during the CO_2 flood, and production generally slowed after 1 PVI of CO_2. These experiments allowed us to observe coreflood performance with a single-component oil whose composition does not change when displaced by CO_2. Results from these tests can be interpreted without considering the complicated effect of miscibility development since decane and CO_2 are first-contact-miscible at high pressures.

Effects of Pressure or IFT

We observed that the higher the pressure, the larger the oil bank and the later the gas break-through time in these floods. Tertiary oil recovery, graphed versus pressure in Figure 2, increased with pressure and PVI. Oil recoveries increased almost linearly with pressure showing no dramatic change around the critical pressure of 1835 psig (shown by the dashed line on Figure 2).

Pressure and IFT were related for the decane/CO_2 system at 160°F by Nagarajan and Robinson[9]. The IFT decreased from 7.8 to 0 mN/m as pressure was increased from 1000 to 2100 psig. Decane and CO_2 under equilibrium conditions should be miscible at 1835 psig and 160°F [71°C]. However, we found that 2100 psig was required for a single phase to be observed at the sight glass located between the slim-tube and the core under flowing conditions in our experiments.

In Figure 3, oil recovery data were replotted versus IFT. Improved oil recovery is more apparent at IFT lower than 1 mN/m. Also, these graphs show that oil can be recovered more quickly at higher pressures or at zero IFT.

Comparison with Other Low IFT Waterfloods

The effect of IFT on oil recovery for low IFT surfactant floods has been described by capillary number correlations[8], such as the one shown in Figure 4. Capillary number is defined as the ratio of the product of permeability and pressure drop to interfacial tension. In Figure 4, low IFT CO_2 flood results are compared to low IFT surfactant flood results obtained by others, both in Berea cores. For a CO_2 flood, the effect of IFT or capillary number on oil recovery is different from that of low tension waterfloods in two respects:

1. Some residual oil can be recovered by high IFT displacements with CO_2, while high IFT surfactant floods usually do not achieve any additional oil recovery.

2. Some residual oil will remain unrecovered in zero IFT miscible floods in this water-wet rock. In contrast, low IFT or high capillary number surfactant floods will recover essentially all the oil.

Two important characteristics of CO_2 displacements, evident from this comparison, were observed in this decane/CO_2/water-wet core system. First, oil recovery was improved by the diffusive capability of CO_2 which caused mass transfer even under immiscible and near-miscible conditions. Because of this mass transfer, oil can be recovered by oil swelling at pressures much lower than the miscible pressure. Second, at higher pressure miscible conditions, oil recovery was not complete, hindered by the presence of the additional water phase. This water-shielding effect leads to oil by-passing and is more pronounced in the strongly water-wet Berea core[12] used in system 1.

Oil Recovery In System 2, West Texas Crude/CO_2/ Carbonate Core

Results from the decane/CO_2/water-wet system were encouraging because additional oil can be recovered in low pressure, or high IFT (> 0.1 mN/m), tertiary floods. Twenty six more experiments were then carried out in the more realistic multiple-contact-miscible/mixed-wet system.

Pressure sensitivities for these corefloods were compared to slim tube test results, and differences were explained by a study of effects of water saturation. Effect of water saturation was investigated by comparing tertiary CO_2 floods with secondary CO_2 floods. Effects of using a slim tube to pre-equilibrate CO_2 and crude oil were also examined to develop better experimental procedures. Scale-up of laboratory tests to field-scale was attempted by examining effects of pore volume injection (PVI).

Effects of Pressure

Pressures investigated ranged from 700 to 3000 psig. The MMP for this system is 1200 psig. Table 4 summarizes CO_2 tertiary coreflood results obtained at different pressures. Because incorporation of a slim tube affects experimental results, only experiments without a slim tube in front of the core are compared directly in Table 4. Oil recoveries in Table 4 were calculated from the measured S_{orm} as percent OOIP (original-oil-in-place), and percent S_{orw} (waterflood residual oil saturation). S_{orm} is preferred as a measure of CO_2 flood performance over recovery, because it is measured directly in our experiments, while recovery levels depend on initial oil saturation (OOIP or S_{orw}). Standard deviation in the values of measured final oil saturation, S_{orm}, was about 2% pore volume.

Presented in Figures 5 and 6 are data demonstrating the effect of pressure on S_{orm} and tertiary oil recovery, respectively. S_{orm} decreased (recovery increased) with increasing pressure almost linearly and showed no dramatic change of slope at pressures below the MMP as slim-tube results might suggest. When examining the effect of pressure, one should take into consideration that the critical pressure for CO_2 is 1070 psia. Hence, CO_2 not only interacts differently with this crude oil below the MMP of 1200 psig, but also behaves more like a gas at these pressures.

Figure 6 includes the result of a tertiary nitrogen flood conducted at 500 psig as a base immiscible case, where very little oil swelling or thinning was expected. Only 8 percent S_{orw} was recovered after 6 PVI, which was lower than the 30 percent S_{orw} recovery obtained in a 700 psig CO_2 flood. This result indicates that immiscible CO_2 can mobilize more oil than immiscible N_2. This is not due to a better mobility ratio since gas viscosities in these two cases are equivalent. The higher recovery was probably related to the ability of CO_2 to dissolve in oil, a result of its higher density (about 3 times that of N_2), causing oil swelling and viscosity reduction.

Effect of Slim Tube

In several experiments, CO_2 was flowed through a slim tube full of oil before entering the core to allow CO_2 to pre-equilibrate or develop miscibility dynamically with oil outside the core. The concern was that displacements of oil would be immiscible and S_{orm} would be higher in regions of the core where miscibility was not developed. Other researchers[5,10,11] have used 8- to 16-foot cores such that this miscibility development zone was small compared to the total length of the core to determine representative S_{orm} values. We chose to use a 16-foot slim tube instead. However, a 16-foot slim tube (45cc pore volume) caused the introduction of a large oil bank into the core ahead of the CO_2 front.

Experiments were conducted with and without a 16-foot slim tube at different pressures. One test was carried out with an 8-foot slim tube before the core. These data are summarized and compared in Table 5 and in Figure 7. The effect of slim tube appears to be different at pressures above the MMP and at pressures below the MMP. At pressures higher than the MMP, pre-equilibration in a slim tube did not reduce the average residual oil saturation significantly; on the contrary, S_{orm} seemed slightly higher with pre-equilibration.

At 700 psig, S_{orm} at 6 PVI was significantly higher with a slim tube than without, indicating that some oil displaced from the slim tube was left in the core. In other words, displacement of oil in the slim tube may be more efficient than in the core where water was present. This additional oil from the slim tube made interpretation of coreflood results difficult, especially at low pressures. For example, S_{orm} data at 1 PVI (in Table 5) suggest that the slim tube has no effect at all pressures, while results at 6 PVI and 700 psig lead to a different conclusion. Final S_{orm} was much higher with a slim tube at pressures lower than the MMP.

These results indicate that the miscibility development zone for this particular crude at 2 ft/day was probably negligible, compared to the length of the core. This conclusion is deduced from the result that lower S_{orm} was measured without a slim tube than with a slim tube. Therefore, an 18-inch core should be long enough for obtaining accurate S_{orm} data at pressures above 1200 psig. The use of a slim tube for pre-equilibrating crude with CO_2 at lower pressures is not recommended if representative S_{orm} values need to be determined in the core. A rising bubble chamber[2] that only lets the equilibrated CO_2-rich phase enter the core may be a better choice.

However, data obtained using a slim tube in front of the core may simulate conditions in regions far away from the injection well better than data obtained without a slim tube. In the field, an oil bank will form even at low CO_2 pressures, but, may break down as viscous fingers form because of the unfavorable mobility ratio for CO_2 displacing oil. It will then take more continuous CO_2 injection to recover this oil, as suggested by our data at 700 psig.

Effects of Initial Water Saturation

The effect of pressure on CO_2 tertiary coreflood behavior is clearly different from the effect of pressure on slim tube behavior. As mentioned earlier, a slim tube is a highly-permeable bead pack and contains no water phase. One possible reason for the observed CO_2 flood performance difference between a core and a slim tube is that the presence of water in the core may make oil less accessible to CO_2. Therefore, secondary CO_2 floods in cores containing only irreducible water were conducted. Displacement pressures were varied from 700 to 3000 psig as in tertiary floods, and the slim tube was by-passed. On average, initial water saturation was about 40% for secondary CO_2 floods and a waterflood preceding a tertiary CO_2 flood usually raised the water saturation to about 75%.

Results from these secondary CO_2 floods showed that more oil can be recovered than for tertiary floods at similar pressures and CO_2 throughputs, except at the very low pressure of 700 psig. In fact, residual oil saturation after 1 PVI of CO_2 in a secondary flood was about the same as after 6 PVI of CO_2 in a tertiary flood, except at 700 psig. Secondary coreflood results are tabulated in Table 6 and plotted in Figure 8 to compare with the tertiary flood's S_{orm}. Figure 9 compares our slim tube and secondary flood percent OOIP oil recoveries. Unlike tertiary flood results, secondary flood results are quite similar to slim-tube results. Sensitivity of oil recovery to pressure was different above and below the MMP for secondary floods as compared to tertiary floods.

These results, then, support the idea that presence of a water phase in the core hinders CO_2 flood oil recovery performance. There are several explanations for this behavior:

1. Higher initial oil saturation in a secondary flood results in oil being more connected.

2. Lower water saturation in a secondary flood facilitates mass transport of CO_2 to crude oil, allowing more extraction and displacement of oil.

3. Intermediate and heavy components extracted from crude oil will make their way to the flowing CO_2 phase and be produced more easily in a lower water saturation secondary flood than in a tertiary flood.

In other words, water not only caused oil to be by-passed by CO_2, but also retarded extraction of oil. Consequently, we expect the amount of additional recovery for a secondary flood to decrease with pressure since CO_2 loses its capacity to extract at lower pressures. This was substantiated by the observation of similar recoveries at 700 psig.

Watershielding of oil extraction by CO_2, occurring at initial water saturations in the range of 40% to 75% for this study, had an effect on oil bank development and hence on oil recovery even in a mixed-wet core. This is somewhat different from waterblocking phenomena in WAG floods. Oil extraction may not be an important oil recovery mechanism at extremely high water saturations such as exist in WAG floods. Stern[12] has found that oil extraction occurs to a lesser extent in a mixed-wet core for WAG floods than for CO_2 slug floods.

The impact of lower initial water saturations, attainable by continuous recirculation of crude oil through the core, was also investigated. Using the crude recirculation method, the initial water saturation was driven down from 40% to about 26%. Secondary and tertiary CO_2 floods were carried out at this lower initial water saturation. S_{orm} values measured in these floods were similar to those obtained at 40% initial water (see Tables 4 and 6). At initial water saturations lower than 40%, CO_2 flood behavior was not affected because water probably existed only in smaller, water-wet pores.

Better oil recovery at lower initial water saturation may not be a general result. This West Texas crude oil contains more polar compounds than other crude oils we have experience with. Small amounts of oil-water emulsion was observed in effluents from waterfloods and CO_2 floods in this study. The polar nature of this crude may have caused the crude oil to be highly dispersed by the waterflood and adversely affected tertiary oil recovery efficiencies.

Effect of Solvent Throughput

CO_2 throughput was also varied to obtain data more representative of conditions at different regions in a reservoir. Regions closer to an injection well or in direct stream lines will have a much higher solvent throughput than regions closer to a production well or in stagnant zones.

As expected, oil recovery increases with volume of CO_2 injected. Figure 6, discussed earlier, also shows the effect of pressure on tertiary oil recovery after 1 PVI and after 6 PVI of CO_2 without a slim tube preceding the core. This result suggests that approximately 80% of the tertiary recoverable oil for 6 PVI of CO_2 had been recovered after 1 PVI. Effects of IFT dominate at less than 1 PVI of CO_2 and extraction starts to take over as the dominant mechanism of oil recovery after solvent breakthrough. Extraction continues to recover oil at a much lower rate at the expense of large amounts of injected CO_2, by vaporizing components into the moving CO_2 stream. This conclusion is supported by produced oil compositional data presented in the next section.

Produced Oil Compositions

Oil produced from selected CO_2 floods was analyzed by gas chromatography (GC). Produced oil from tertiary floods at 700, 2000, and 3000 psig and secondary floods at 2000 psig were selected. Because of the limitations of the chromatography column and the method used, only crude oil components from the C_{10} to C_{28} range can be reliably quantified. Produced oil compositions were examined by comparing them to the original oil composition. A normalization method, dividing the weight percent of a component in the produced oil sample by the weight percent of the same component in the original crude, to accentuate changes in composition, was adopted. When a component is preferentially extracted, its normalized composition will increase above 1. If no extraction has taken place, or the oil is miscibly displaced, the normalized concentration should remain about 1.

Figure 10 is a plot of normalized produced oil compositions as a function of PVI for the 3000 psig tertiary flood. Extraction started with lower molecular weight components, gradually advanced to heavier components. Extraction was evident immediately after gas breakthrough (about 0.5 PVI); a mixture of extraction and displacement was evident from gas breakthrough to 1 PVI, and oil extraction dominated after 1 PVI of CO_2. As more solvent was injected or more oil was recovered, heavier components were extracted and produced. Presenting oil compositions as a function of PVI gave an erroneous impression that a large quantity of oil was produced by extraction. In fact, average oil saturation in the core was reduced very slowly at the expense of a lot of CO_2. This is apparent by comparing Figure 10 with Figure 11, which plots normalized composition as a function of S_{or} (the average oil saturation remaining in the core during a CO_2 flood).

Effect of pressure on produced oil compositions are presented as a function of S_{or} in Figures 11, 12, and 13 for 3000, 2000, and 700 psig, respectively. Hardly any extraction occurred at 700 psig while evidence of extraction was abundant at 2000 psig and 3000 psig. Extraction increased with increasing pressure. Extraction of the heavier components (C_{20} to C_{30}) did not start until S_{orm} reached 12%.

Produced oil samples from a 2000 psig secondary flood have also been analyzed and compositional data are presented in Figure 14, for comparison with a tertiary flood at the same pressure (Figure 12). More oil was recovered by displacement in the secondary flood than in the tertiary flood. In addition, more oil was recovered by extraction in the secondary CO_2 flood than in the tertiary CO_2 flood as demonstrated by comparing Figure 14 with Figure 12.

In addition, a few samples were also analyzed by GC in the gasoline range and by HPLC (High Performance Liquid Chromatography). The C_4 to C_7 gasoline range of an unprocessed oil sample was only 4.6 wt%. At 3000 psig, these light components were absent in the produced oil after 0.5 PVI of CO_2. Hence, it was decided that C_4-to-C_7 GC analyses would not provide much information and were discontinued.

Samples collected from the 3000 psig flood were also analyzed by HPLC to separate the saturates from aromatics, NSO (Nitrogen, Sulfur and Oxygen compounds) and asphaltenes. Results, shown in Figure 15, indicate that both saturates and aromatics were extracted leaving the NSO and asphaltenes behind until the very end of the flood. A slight preference to extraction of saturates over aromatics is displayed in this compositional analysis.

CO_2 End-Point Mobilities In Crude/Carbonate System

In the crude/mixed-wet carbonate core system, as mentioned earlier, all CO_2 floods were carried out in a single piece of core. Pressure drops were monitored by overall and internal pressure taps to provide data on fluid mobilities. Experimental measurements of CO_2 relative permeabilities in the presence of S_{orm} are needed in CO_2 miscible process simulation, because CO_2 mobilities determine process sweep efficiency and WAG injectivity. To determine the CO_2 mobilities accurately at different S_{orm} levels is difficult by either steady-state or unsteady-state methods. Since pressure had an effect on S_{orm}, these CO_2 floods, which left 24.8% to 0.1% S_{orm} in the core, provided us an opportunity to determine CO_2 end-point permeabilities at various liquid and gas saturations.

CO_2 mobilities calculated in (ft/day)/(psi/ft), based on a CO_2 viscosity of 0.069 cp, are listed along with oil recovery data in Tables 4, 5 and 6. These values are plotted in Figures 16, 17 and 18. Data points for 6 PV injection of CO_2 are plotted using solid symbols in the figures, and they are most representative of end-point mobilities.

kr_{CO_2} and S_{orm}

An attempt to correlate CO_2 permeability to S_{orm} is shown in Figures 16. In general, the CO_2 end-point mobility decreases with increasing residual oil saturations. There is considerable amount of scatter due to variation in water saturation. This correlation has a coefficient, r^2, of 0.38 for a Corey-type relative permeability expression. The dependence of CO_2 permeability on S_{orm} is stronger at low S_{orm} (<15%), but the correlation breaks down at high S_{orm}.

kr_{CO_2} and S_{CO_2}

End-point CO_2 relative permeabilities appeared to correlate better with CO_2 saturation (or residual oil saturation plus brine saturation). Plots of CO_2 relative permeability as a function of CO_2 saturation for this mixed-wet carbonate core are shown in Figure

17. The correlation coefficient, r^2, is 0.87 for Corey type of correlation function where the Corey exponent is 4.46, connate water saturation is 0.38, and critical CO_2 saturation is 0.08. This correlation has been extended by Stern[11] to lower CO_2 saturations. These results show that CO_2 relative permeability is a function of CO_2 saturation only, which is consistent with a three-phase relative permeability formulation such as the Stone model[13].

This dependence of CO_2 permeability on CO_2 concentration suggests that CO_2 mobility may be lower in an immiscible or near-miscible process, which can result in a better overall sweep efficiency. It also helped explain the observation that secondary CO_2 floods, which reached higher solvent saturations, had greater end-point solvent mobilities than tertiary floods carried out at the same pressure.

kr_{CO_2} and Sw

End-point CO_2 permeabilities are not a monotonic function of water saturation. However, a family of curves can be derived from our data for various levels of S_{orm} as shown in Figure 18. Here data points are grouped by three ranges of S_{orm} values. Figure 18 demonstrates that CO_2 permeability is lowered in the presence of S_{orm}. This result suggests that effects of miscible residual oil on solvent relative permeability must be accounted for to accurately predict mobilities and injectivities for CO_2 miscible processes in solvent-swept regions.

ADDITIONAL DISCUSSION AND CONCLUSIONS

Results obtained from this laboratory coreflood study have several important implications regarding CO2 gas injection process design and performance prediction. First, CO_2 near-miscible processes have been shown in this study to recover substantial quantities of oil in linear corefloods. Although oil recovery declined as pressure was lowered, the loss in oil recovery was not great and might be offset economically by the reduction in mass of CO_2 required and better CO_2 sweep efficiency. These benefits may be even more pronounced when the MMP is farther above the critical pressure of CO_2 than what was for the system 2 of this study. Field scale simulations should be conducted to further define sweep efficiency of near-miscible processes.

Concerning laboratory screening tests, results obtained in this study suggest replacing slim tube tests with reservoir-condition coreflood displacement tests. It has been demonstrated that the presence of water phase in CO_2 floods is very important. Slim tube tests, where water is absent, cannot accurately predict oil recovery or S_{orm} for tertiary CO_2 floods. Corefloods may take twice as long to conduct than slim-tube tests, but corefloods provide much more useful data for field process design.

Also, this work strongly supports replacing the mixing model currently used in most miscible simulators with a more realistic three-phase relative permeability formulation. CO_2 end-point mobility in the multiple-contact crude/carbonate-core system was found to be a function of its own saturation. Coreflood simulations to match these experimental data can be carried out to improve relative permeability formulation in miscible simulators.

Finally, both displacement and extraction of oil by CO_2 were observed to be important oil recovery mechanisms that increased with increasing CO_2 pressure. Oil that is bypassed, at the pore level, due to low IFT displacement or watershielding can be recovered by extraction. At pressures lower than the critical pressure of CO_2, oil recovery attributed to these mechanisms are reduced greatly. However, oil swelling and viscosity reduction are still evident in immiscible CO_2 corefloods. More experiments using other crude oils, solvents and cores should be conducted to verify these mechanisms and provide data for developing a simulator more representative of miscible displacement physics.

Conclusions reached from linear corefloods in this study may be summarized as follows:

1. Tertiary CO_2 flood S_{orm} increases linearly (oil recovery decreases) as pressure decreases. No dramatic loss of recovery is observed below the MMP as suggested by slim-tube test results.

2. CO_2 mobility decreases with decreasing pressure, offering possible mobility control benefits for near-miscible CO_2 processes.

3. CO_2 mobilities in the multiple-contact-miscible crude/carbonate system are solely a function of CO_2 saturation. CO_2 mobilities in a MCM system can be reduced one to two orders of magnitude by the presence of residual oil.

4. Reservoir condition corefloods should, whenever possible, replace slim-tube tests as laboratory screening tests for determining effects of pressure on CO_2 gas-injection processes. These same tests also provide more realistic S_{orm} values and CO_2 mobilities that can be used in miscible or near-miscible process design and simulation.

5. Oil recoveries for near-miscible CO_2 floods are not only the result of low IFT displacement but are also influenced by mass transfer and water-shielding effects. This is true for both first-contact-miscible and multiple-contact-miscible systems.

6. Both displacement and extraction of oil by CO_2 increase with higher pressure. Displacement seems to be the dominant mechanism of oil recovery before one pore volume injection of CO_2, after which oil extraction becomes important.

7. Secondary CO_2 floods can recover more oil than tertiary CO_2 floods for crude oils similar to the one used in this study. Higher oil recovery in secondary CO_2 floods probably results from less water-shielding and/or more efficient mass transfer between CO_2 and crude oil.

8. For tertiary floods, using extra long cores or a slim tube for miscibility development is not required at pressures higher than the MMP. For the crude oil used in this study, 18-inch long cores are sufficient to measure S_{orm} accurately.

ACKNOWLEDGMENTS

The author wishes to thank S. M. Klem for his efforts in performing the coreflood experiments; N. R. Daniel for analyzing oil samples by GC; T. R. Reppert and D. Stern for their very helpful discussions. The author also appreciates Exxon Production Research Co. management for their permission to publish this paper.

REFERENCES

1. Henry, R. L. and Metcalfe, R. S.: "Multiple-Phase Generation During Carbon Dioxide Flooding", SPEJ (Aug. 1983), p.595-601.

2. Watkins, R. W.: "A Technique for the Laboratory Measurement of Carbon Dioxide Unit Displacement Efficiency in Reservoir Rock," Paper SPE 7474 presented at the 1978 SPE Annual Technical Conference and Exhibition, Houston, Oct.1-4.

3. Yellig, W. F. and Metcalfe, R. S.: "Determination and Prediction of CO_2 Minimum Miscibility Pressures," J. Pet. Tech. (Jan. 1980) pp. 160-168.

4. Gardner, J. W., Orr, F. M. and Patel, P. D.: "The Effect of Phase Behavior on CO_2 Flood Displacement Efficiency," paper SPE 8367 presented at the 54th Annual Fall Technical Conference and Exhibition, Las Vegas, Sept.23-26, 1979.

5. Shelton, J. L., Yarborough, L.: " Multiple Phase Behavior in Porous Media During CO_2 or Rich-Gas Flooding," JPT (Sept. 1977) pp. 1171-1178.

6. Nghiem, L. X., Li, Y. K.: "Effect of Phase Behavior on CO_2 Displacement Efficiency at Low Temperatures: Model Studies with and EOS," Paper SPE 13116 presented at the 59th Annual Technical Conference and Exhibition, Houston, Sept.16-19, 1984.

7. Lease, W. O., Egmond, C. F.: "An Oscillating Backpressure Regulator for Low Flow Rate, High-Pressure Coreflooding Experiments," SPEJ (Dec. 1982), 899-901.

8. Larson, R. G., Davis, H. T., and Scriven, L. E.: "Displacement of Residual Nonwetting Fluid from Porous Media," Chemical Engineering Science, vol. 36 (1981), pp. 75-85.

9. Nagarajan, N., Robinson, R. L.: "Equilibrium Phase Compositions, Phase Densities and Interfacial Tensions for CO_2 + Hydrocarbon Systems," J. Chem. Eng. Data, 31, p. 168 (1986).

10. Negahban, S., Shiralkar, G. S., Gupta, S. P.: "Simulation of the Effects of Mixing in Gasdrive Core Tests of Reservoir Fluids," SPE Reservoir Engineering (Aug. 1990), 402-408.

11. Yellig, W. F.: "Carbon Dioxide Displacement of a West Texas Reservoir Oil," SPEJ (Dec. 1982), 805-815.

12. Stern, D.: "Mechanisms of Miscible Oil Recovery: Effects of Pore-Level Fluid Distributions," Paper SPE 22652, presented at the 1991 Annual Technology Conference and Exhibition, Dallas, Oct.6-9.

13. Stone, H. L.: "Estimation of Three-Phase Relative Permeability and Residual Oil Data," J. Canadian Pet. Tech. (Oct.-Dec. 1973) 53-61.

SI METRIC CONVERSION FACTORS

psi x 6.895	E + 00	= kPa
cp x 1.000	E - 03	= $N \cdot sec/m^2$
ft/day x 3.528	E - 06	= m/sec
md x 9.869	E - 04	= μm^2

APPENDIX

CORE CLEANING PROCEDURES FOR NEAR-MISCIBLE COREFLOODS

These procedures are carried out in the oven at coreflood temperature:

1. With core inlet valve closed, lower the core pressure to one atmosphere gradually through an OBPR and a gas-meter. Record volumes of gas, water and oil collected during blow-down.

2. Fill the core with Trichloroethane from the core outlet valve using a Beckman pump at 3 cc/min until pump pressure reaches about 2000 psig. Allow the core to soak in solvent for at least one hour.

3. Flush the core with Trichloroethane backward at approximately 3 cc/min flow rate until about 100 ml of effluents are collected. Surges of flow are created by shutting off the core inlet valve periodically to build-up pump pressure and then opening the core inlet valve. A 100 ml centrifuge tube is used for effluent collection because it has easy-to-read graduations at volumes lower than 10 ml and a small amount of water may be produced and need to be read accurately. We normally measure less than 10 ml of crude oil after the solvent is evaporated.

4. Reverse the flow (return to the same direction as the flood) at five-times the flow rate used in the previous step for 3-4 pore volumes until the effluent solvent is clear. The second wash effluents, usually about 250 ml, are collected into another flask. Soaking and pulsing may be necessary to assure good cleaning.

5. Nitrogen purging of the core is conducted next at 500 psig supply pressure from one-to-two hours until no solvent is produced into the flask.

6. Evacuate the core overnight after venting the nitrogen.

7. Effluents from the first and second wash are then evaporated separately in a hood aided by a constant stream of air purge. Evaporation is considered complete when sample volumes ceased to change. Volume and weight of oil left in the centrifuge tube and flask are then determined.

TABLE 1. CORE PROPERTIES

	System 1	System 2
Type	Berea Sandstone	San Andres Carbonate
Wettability	Water-wet	Mixed-wet
Length (inches)	17.75	17.75
Cross-section	1.367" Diameter	1"x1"
Area (cm^2)	9.469	6.824
Pore Volume (cc)	85.2	63.4
Porosity	20%	20.6%
Permeability (md)	70	160
Typical Waterflood Initial Water Saturation, S_{wi}	40%	40%
Typical Waterflood Residual Oil Saturation, S_{orw}	38%	25%

TABLE 2. COREFLOOD SEQUENCES

Displacements	PVI	Flow Rates
1. Brine Into Evacuated Clean Core	1.5	~ 300 cc/hr
2. Oil Flood	2	~ 100 cc/hr
3. Waterflood	3	8 cc/hr (~ 4 ft/day)
4. CO_2 Flood	1	4 cc/hr for the first pore volume
	6	8 cc/hr after 1 PVI
5. Blow Down		
6. Core Cleaning		

TABLE 3. TERTIARY CO_2 FLOOD RESULTS IN DECANE/BEREA, SYSTEM 1

T = 160°F

P	IFT		S_{orm}**		Recovery, $1-(S_{orm}/S_{orw})$	
(psia)	(mN/m)	S_{orw}*	3 PVI	7.5 PVI	3 PVI	7.5 PVI
1015	7.8	0.4	0.36	0.25	0.125	0.375
1660	1.0	0.36	0.21	0.13	0.417	0.639
1821	0.1	0.41	0.19		0.537	
1821	0.1	0.37	0.14	0.07	0.621	0.811
1915+	0	0.39	0.15		0.615	
1970+	0	0.35	0.08	0.07	0.771	0.8
2115+	0	0.4	0.07		0.825	

* S_{orw} = Waterflood Residual Oil Saturation

** S_{orm} = CO_2 Flood Residual Oil Saturation

+ Above the miscible pressure of 1850 psia for decane/CO_2 at 160°F.

TABLE 4. TERTIARY CO_2 FLOOD RESULTS IN SYSTEM 2 (WITHOUT THE SLIM TUBE)

T = 100°F

Oil: Degassed West Texas Crude
Core: Crude Soaked San Andres Carbonate

P (psig)	OOIP*	S_{orw}	CO_2 PVI	S_{orm}	S_{wf}**	Oil Recovery % OOIP	Oil Recovery % S_{orw}	CO_2 Mobility $\left(\frac{ft/day}{psi/ft}\right)$
3000	0.62	0.28	1 6	0.09 0.06	0.45 0.39	86 91	68 80	5.4 9.6
2300	0.62	0.26	1	0.12	0 42	81	56	1.1
2000	0.60	0.29	1 6	0.14 0.11	0.40 0.34	77 81	52 61	--- 2.3
2000	0.74+	0.31	1 6	0.14 0.10	0.43 0.39	81 87	56 69	1.2 2.3
1400	0.63	0.25	1	0.15	0.36	76	39	1.4
700++	0.59	0.25	1 6	0.23 0.17	0.45 0.42	61 70	9 30	0.8 0.9

* OOIP = Original Oil In Place.

** S_{wf} = CO_2 Flood Final Water Saturation.

+ Achieved by crude recirculation during soaking.

++ Below the MMP of 1200 psig for West Texas crude/CO_2 at 100°F.

TABLE 5. EFFECTS OF SLIM TUBE ON TERTIARY CO_2 FLOOD RESULTS IN SYSTEM 2

P (psig)	CO_2 PVI	Length of Slim Tube (ft)	OOIP	S_{orw}	S_{orm}	S_{wf}	Oil Recovery % OOIP	Oil Recovery % S_{orw}	CO_2 Mobility $\left(\frac{ft/day}{psi/ft}\right)$
3000	6	16	0.65	0.27	0.08	0.35	87	70	7.2
		0	0.62	0.28	0.06	0.39	91	80	9.6
2000	6	16	0.60	0.23	0.10	0.40	84	57	5.7
		0	0.60	0.29	0.11	0.34	81	61	2.3
700*	6	16	0.60	0.25	0.25	0.34	58	1	0.8
		0	0.59	0.25	0.17	0.42	70	30	0.9
3000	1	16	0.62	0.24	0.09	0.37	85	62	3.1
		0	0.62	0.28	0.09	0.45	86	68	5.4
2300	1	16	0.61	0.25	0.13	0.37	79	49	3.1
		8	0.62	0.26	0.14	0.39	78	47	2.0
		0	0.62	0.26	0.12	0.42	81	56	1.1
1400	1	16	0.63	0.25	0.18	0.39	72	29	---
		0	0.64	0.25	0.15	0.36	76	39	1.4
700*	1	16	0.61	0.25	0.25	0.39	59	0	0.8
		0	0.59	0.25	0.23	0.45	61	9	0.8

* Below the MMP of 1200 psig for this West Texas crude/CO_2 system at 100°F.

TABLE 6. SECONDARY CO_2 RESULTS IN SYSTEM 2

P (psig)	OOIP	CO_2 PVI	S_{orm}	S_{wf}	Oil Recovery (% OOIP)	CO_2 Mobility $\left(\frac{ft/day}{psi/ft}\right)$
3000	0.60	1	0.06	0.40	89	6.6
		6	0	0.39	100	10.0
2300	0.61	1	0.12	0.39	81	3.2
		6	0.07	0.38	89	6.0
2000	0.74*	1	0.13	0.26	82	6.0
		6	0.06	0.26	91	11.8
1400	0.62	1	0.13	0.37	79	2.3
		6	0.08	0.37	86	6.3
1100**	0.61	1	0.15	0.39	75	0.9
		6	0.09	0.39	84	2.8
700**	0.62	1	0.25	0.36	60	0.6
		6	0.18	0.36	71	1.3

* Achieved by crude recirculation during soaking.

** Below the MMP of 1200 psig for this West Texas crude/CO_2 system at 100°F.

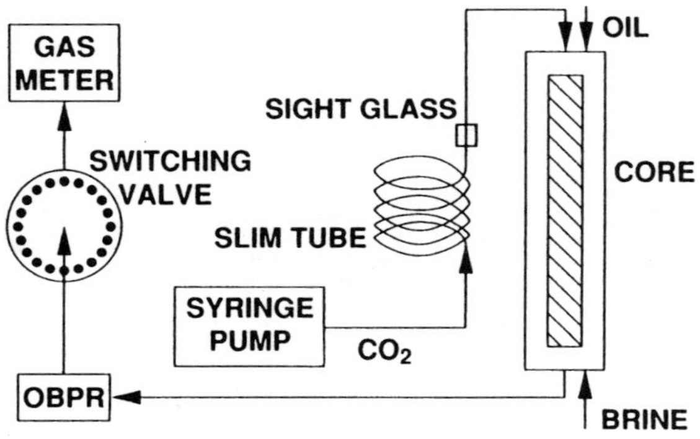

Fig. 1 – A schematic flow diagram for the near-miscible coreflood apparatus

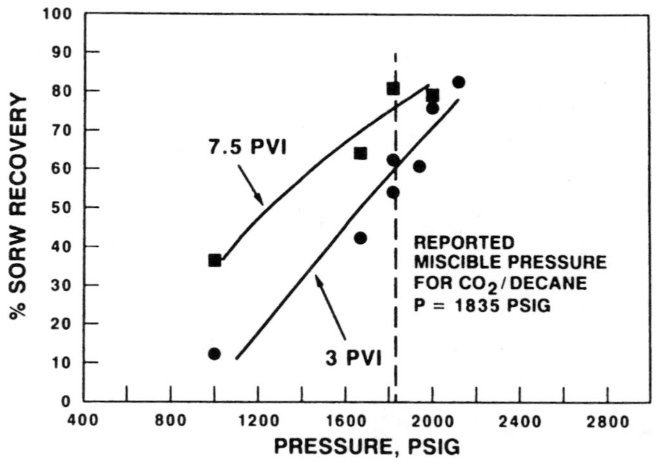

Fig. 2 – Effects of pressure, and PVI on tertiary flood oil recoveries in System 1: CO_2/Decane/Berea

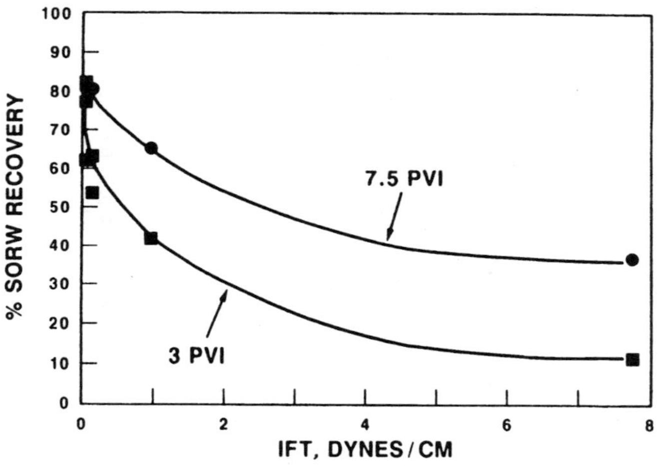

Fig. 3 – Effect of IFT, and PVI on tertiary flood oil recoveries in System 1: CO_2/Decane/Berea core at 160°F

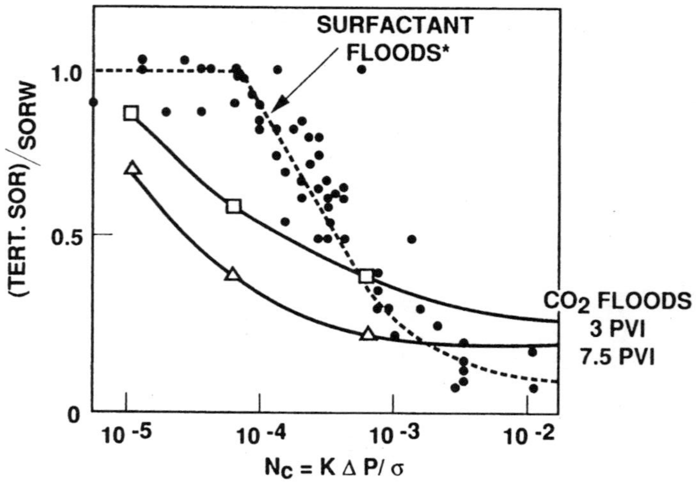

* SOLID SYMBOLS ARE SURFACTANT-FLOOD DATA FROM TABER, McMILLEN & FOSTER, GUPTA & TRUSHENSKI. (see ref. 8)

Fig. 4 – Capillary number correlation in Berea core: comparing CO_2 floods to other low IFT waterfloods

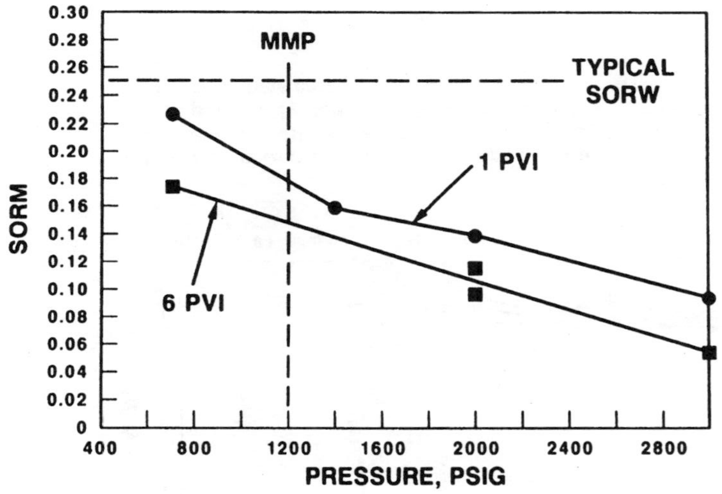

Fig. 5 – Effects of pressure, and PVI on tertiary flood residual oil saturations in System 2: CO_2/West Texas crude/Mixed-wet carbonate core at 100°F

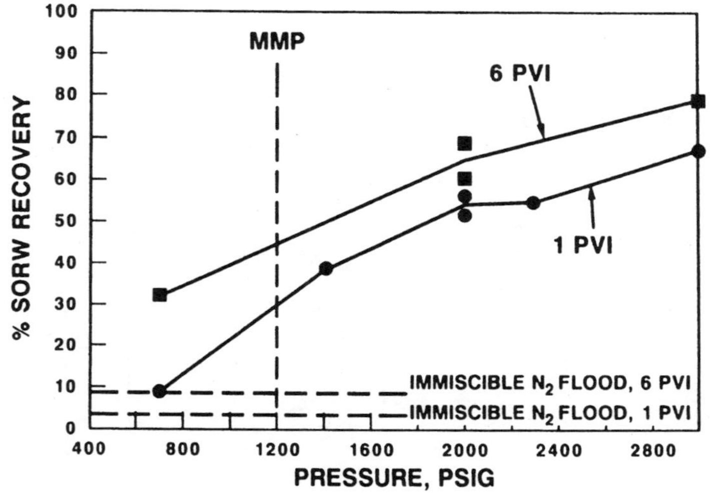

Fig. 6 – Effects of pressure, and PVI on tertiary oil recoveries in System 2: CO_2/West Texas crude/Mixed-wet carbonate core at 100°F

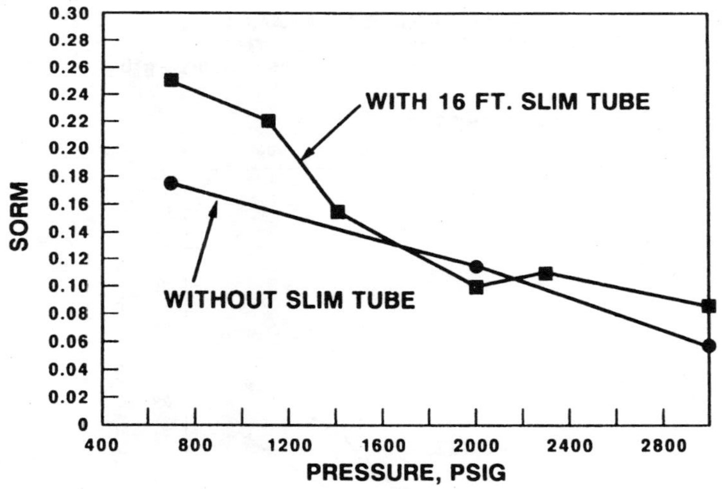

Fig. 7 – Effect of pre-equilibration using a slim-tube on tertiary flood residual oil saturation in System 2: CO_2/West Texas crude/Mixed-wet carbonate core at 100°F after 6 PVI

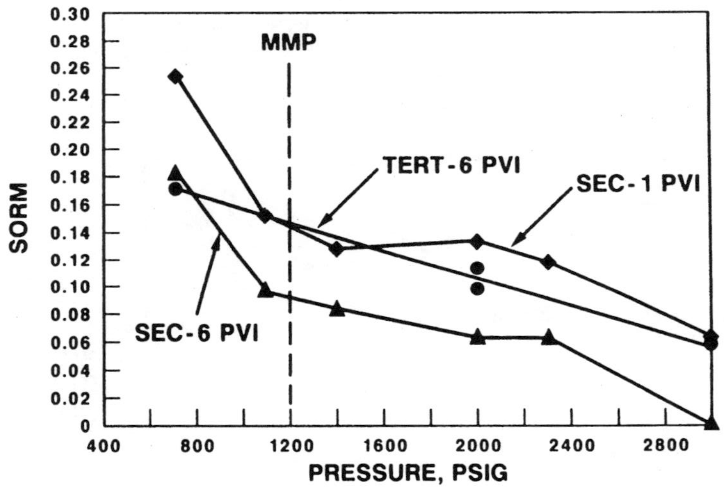

Fig. 8 – Effect of initial saturation on CO_2 flood residual oil saturation in System 2: CO_2/West Texas crude/Mixed-wet carbonate core at 100°F. This is a comparison of secondary and tertiary coreflood results.

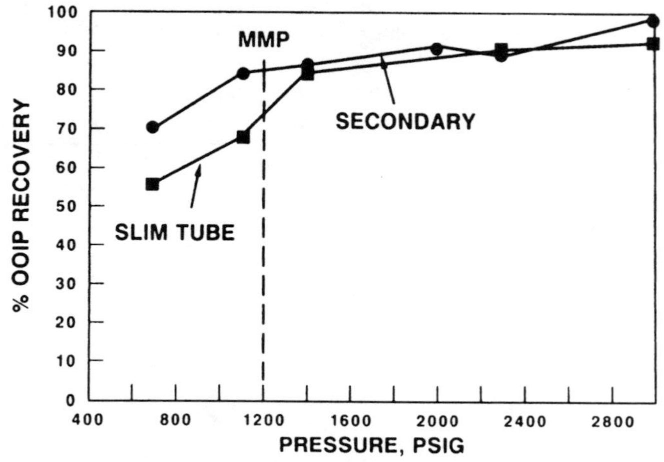

Fig. 9 – Comparison of pressure sensitivity on oil recovery for secondary CO_2 corefloods and slim-tube tests after 1.5 PVI in System 2: CO_2/West Texas crude

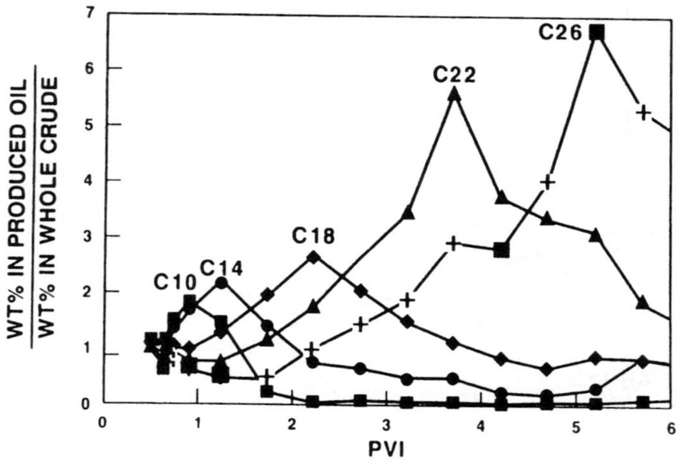

Fig. 10 – Normalized produced oil composition as a function of PVI for a tertiary CO_2 flood at 3000 psig in System 2: CO_2/West Texas crude/Mixed-wet carbonate core

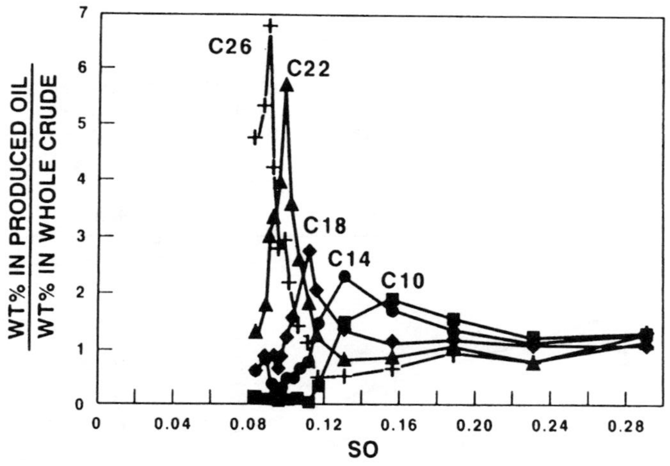

Fig. 11 – Normalized produced oil composition as a function of average oil saturation for a tertiary CO_2 flood at 3000 psig in System 2: CO_2/West Texas crude/Mixed-wet carbonate core

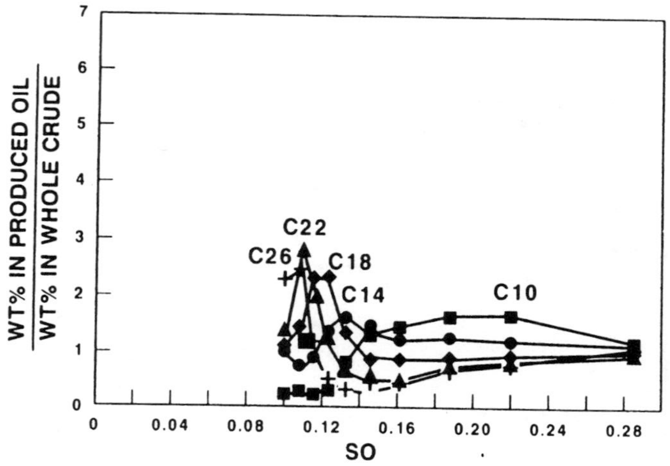

Fig. 12 – Normalized produced oil composition as a function of average oil saturation for a tertiary CO_2 flood at 2000 psig in System 2: CO_2/West Texas crude/Mixed-wet carbonate core

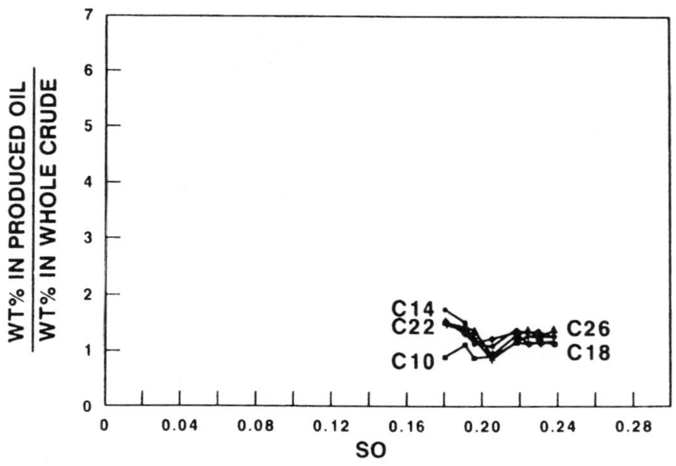

Fig. 13 – Normalized produced oil composition as a function of average oil saturation for a tertiary CO_2 flood at 700 psig in System 2: CO_2/West Texas crude/Mixed-wet carbonate core

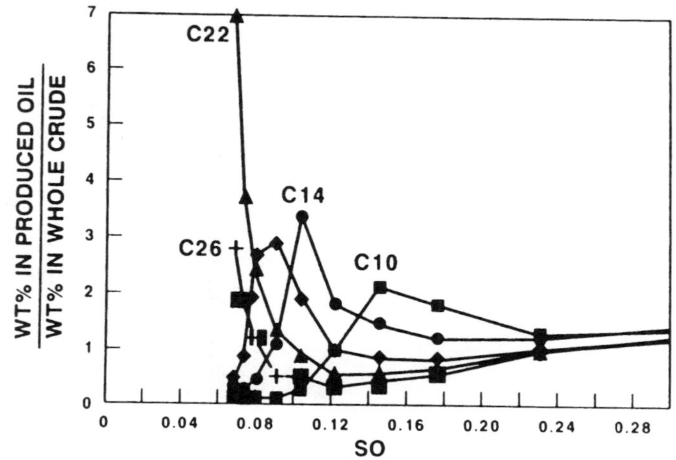

Fig. 14 – Normalized produced oil composition as a function of average oil saturation for a secondary CO_2 flood at 2000 psig in System 2: CO_2/West Texas crude/Mixed-wet carbonate core

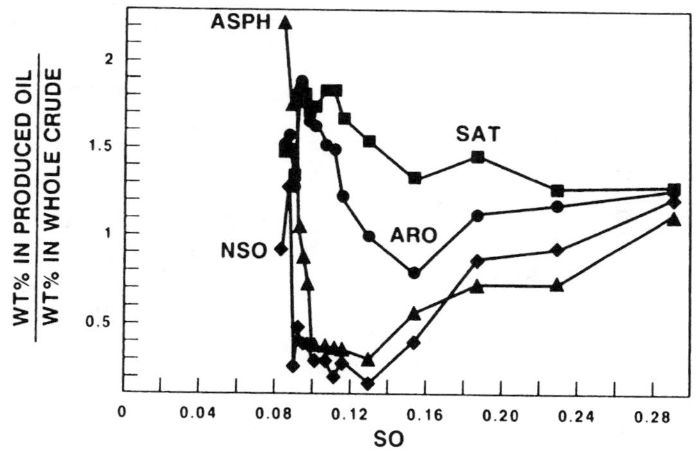

Fig. 15 – Normalized produced oil fractions by HPLC as a function of average oil saturation for a tertiary CO_2 flood at 3000 psig in System 2: CO_2/West Texas crude/Mixed-wet carbonate core

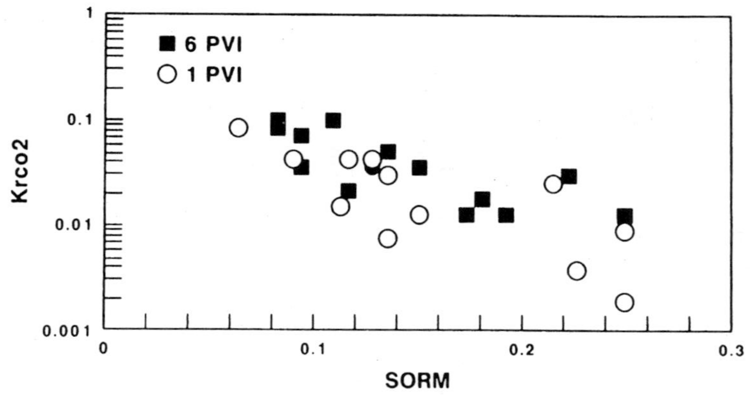

Fig. 16 – CO_2 flood end-point permeability as a function of residual oil saturation in a mixed-wet carbonate

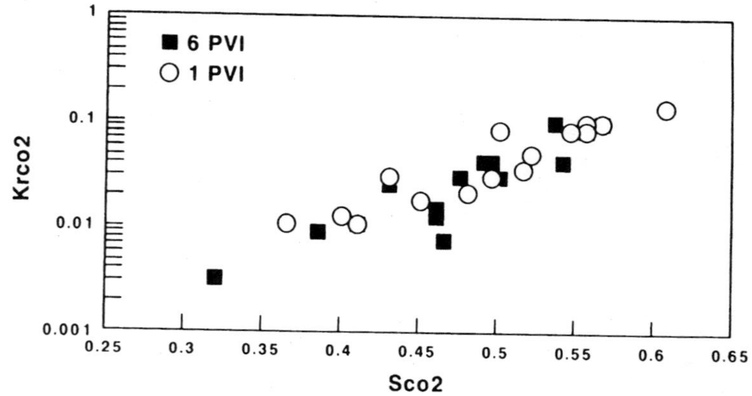

Fig. 17 – CO_2 flood end-point permeability as a function of CO_2 saturation in a mixed-wet carbonate

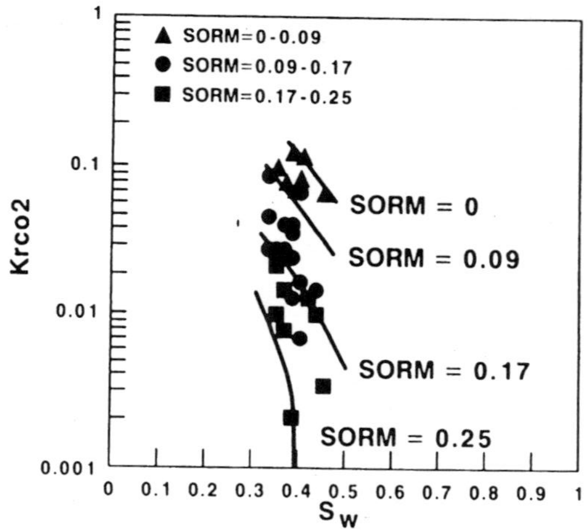

Fig. 18 – CO_2 flood end-point permeability as a function of water saturation and residual oil saturation in a mixed-wet carbonate

SPE

Society of Petroleum Engineers

SPE 22652

Mechanisms of Miscible Oil Recovery: Effects of Pore-Level Fluid Distribution

D. Stern, Exxon Production Research Co.
SPE Member

This paper was prepared for presentation at the 66th Annual Technical Conference and Exhibition of the Society of Petroleum Engineers held in Dallas, TX, October 6–9, 1991.

ABSTRACT

A series of tertiary multiple-contact miscible CO_2 corefloods have been carried out to study effects of flow rate, core length, oil viscosity, wettability, WAG ratio, and initial water saturation on displacement mechanisms. The goal of these experiments was to identify important recovery mechanisms that influence performance of reservoir-condition, multiple-contact miscible floods.

In these experiments, a bank of miscibly-displaced oil is produced first, followed by oil that is initially lighter than the original oil, and becomes increasingly heavy as the flood progresses. The implication is that some oil is initially bypassed and later recovered by extraction.

The amount of bypassing is not sensitive to flow rate or core length, but increases as the solvent/oil viscosity ratio decreases. Based on these results, it is concluded that pore-level effects such as capillarity-induced bypassing and dispersion are major bypassing mechanisms. Microvisualization experiments, in which fluids are hardened in-place, provide direct confirmation that pore-level bypassing takes place.

Viscous fingering (macroscopic bypassing) may also take place, but it is not thought to be the dominant bypassing mechanism under the conditions of this study.

Wettability also has a significant effect on coreflood performance, particularly at high WAG ratios. High WAG ratios result in less oil recovery by extraction. In water-wet rock, this is a large effect, with almost no extraction at high WAG ratios. In mixed-wet rock, however, the effect of WAG ratio is much smaller, and significant quantities of oil are recovered by extraction regardless of WAG ratio. This is likely the result of wall-coating oil films in the mixed-wet rock, which improve contact between oil and solvent.

References and illustrations at end of paper.

INTRODUCTION

Most field-scale miscible floods are designed using a numerical simulator, which requires data on relative permeability and final oil saturation. The simulator is used to predict performance of several process designs, and the injection scheme with the best economic potential is chosen. Because input data can have a large influence on predicted field performance, it is important to develop laboratory procedures that provide representative measurements of residual oil saturations and mobilities. These procedures provide the basis for predicting field performance from results of laboratory experiments. Our approach has been to interpret laboratory data in terms of pore-level process mechanisms. Mechanisms can then be incorporated into simulators that are used to estimate field performance. By focusing on process mechanisms under field conditions, we hope to ensure that our laboratory measurements are representative of swept regions in the field.

This paper presents results of a series of multiple-contact miscible displacements of oil by CO_2. This is a preliminary study to determine the important mechanisms that govern performance of reservoir-condition miscible displacements. A survey of the literature on this topic indicates that multiple-contact miscible displacements may be influenced by three bypassing mechanisms: dispersive and capillarity-induced pore-level bypassing and macroscopic viscous fingering.

To better understand the relative importance of these mechanisms under reservoir conditions, experiments were conducted to examine the effects of wettability, WAG ratio, initial water saturation, flow rate, core length and oil viscosity on mobility and oil recovery in laboratory CO_2 floods. By comparing the composition of the produced crude oil to that of the original crude, it is possible to infer the extent of bypassing, and relate that to recovery mechanisms. In addition, microvisualization experiments, which allow direct observation of pore-level fluid distributions, were conducted to better understand bypassing mechanisms.

The remaining portion of this paper is divided into four sections. The first section, entitled BACKGROUND, reviews some of the pertinent literature, leading to a more detailed discussion of important displacement mechanisms. The next section, EXPERIMENTS, describes experimental procedures used to obtain repeatable measurements of oil recovery and CO_2 mobility in laboratory cores. Experimental results are described and analyzed in the RESULTS and DISCUSSION sections.

BACKGROUND

In using laboratory data to predict field performance, several issues remain unsettled. For example, the effect of contact time on the development of miscibility has not been resolved. In a vaporizing gas drive, miscibility develops when light crude oil components mix with the solvent. As a result, contact time, which can influence the extent of mass transfer, could have a strong influence on flood performance. Although workers from Amoco[1-5] maintain that miscibility development can require ten feet, most laboratory measurements are carried out in cores that are 1-2 feet in length and are assumed representative of field behavior. In addition, laboratory slim-tube measurements give values of the miscible residual oil saturation (S_{orm}) close to zero, while field and coreflood measurements give values that are substantially higher.

Another question that often arises concerns predicting the mobility of oil and CO_2 in field-scale floods. Typically, the Todd-Longstaff mixing model[6] is used in simulators, but this has no physical basis, and limited experimental data are available to verify the formulation. The mixing model has been tested only against experimental data from single-phase first-contact miscible displacements. In contrast, tertiary CO_2 floods involve displacement of water and oil, and are multiple-contact miscible. Thus, the mixing model has not been completely verified against realistic multiple-contact-miscible experimental data.

The first step in addressing these issues is to define mechanisms that determine performance of multiple-contact miscible displacements; results of experiments aimed at this are presented here. This information can then be incorporated into simulators and used to better determine factors that influence field-scale performance.

Literature Review

Many experimental studies are documented in the literature concerning mechanisms of multiple-contact miscible displacements. A review of this work is presented to define in some detail important physical processes influencing miscible displacements.

Much of the experimental work has focused on the effects of process conditions such as flow rate and core length. Extensive data on contact-time effects were obtained by researchers at Amoco[1-5] in Berea cores ranging in length from 2 to 32 feet. In most experiments, the core did not contain any mobile water. Effects of core length are not well understood: Shelton and Schneider[1] observed only a small effect of core length in cores between 1 and 8 feet in length, while Yellig[2] and Negahban et al.[3] found that recovery is higher in longer cores. Negahban et al.[3] concluded that cores longer than 10 feet are required to develop miscibility. This is similar to conclusions reached earlier by Rathmell, et al.[7]

In secondary floods, Shelton and Schneider[1], and Tiffin and Kremesec[4] both observed increases in S_{orm} with flow rate, and suggested that this results from increased viscous fingering at higher flow rates, which leads to bypassing of oil. Watkins[8], at ARCO, reported the same trends in his experimental data and reached similar conclusions. Here viscous fingering refers to macroscopic bypassing caused by core-scale heterogeneity. Jones[9] on the other hand, found that, in tertiary floods, S_{orm} decreases with increasing flow rate, and suggested that this resulted from capillarity-induced bypassing. The capillary entry pressure is higher in small pores, so large pores are entered first, resulting in pore-level bypassing. In secondary floods, where water is not mobile, this effect is absent.

Gardner and Ypma[10] have also observed bypassing. They suggested that an interaction between flow patterns and phase behavior gives rise to the observed behavior. They hypothesized that macroscopic viscous fingers form as outlined above, bypassing some of the oil. Then, the lighter components in the bypassed oil are recovered by extraction, leaving behind a viscous residual that is immobile and largely insoluble in the flowing CO_2.

Orr and coworkers[11,12] suggested that pore-level bypassing could produce the same effect. Micromodel experiments[11] verify that effects analogous to those described by Gardner and Ypma[10] can occur on a pore level: the solvent develops a preferred path through high conductivity pores, and is unlikely to displace oil from other, lower conductivity pores. This effect leads to dispersion in first-contact miscible floods. Like Jones, they note that capillary pressure effects can also influence the extent of bypassing.

Important Bypassing Mechanisms

A clear understanding of the factors that influence dispersive and capillarity-induced bypassing, and viscous fingering provides a basis for interpreting experimental results. This section compares in detail these mechanisms, and briefly discusses experiments that can be used to discern their relative importance.

Dispersive Bypassing

Dispersive bypassing results from a distribution of pore sizes, and occurs in single-phase flow. The mixing zone between oil and solvent, once miscibility develops, will be subject to single-phase dispersive mixing; unfavorable viscosity ratios between solvent and oil will increase the extent of bypassing. Sahimi et al.[13], based on Monte Carlo simulations of tracer flow through regular networks of cylindrical pores, suggested that two factors influence the distribution of pathways and residence time seen by a fluid particle as it flows through the pore space. First, as a result of the distribution of pore radii, pathways of varying conductivity are present, giving rise to a distribution of path lengths and velocities. This mechanism is unchanged by flow rate. Second, as a result of the laminar velocity distribution in the pores and limited contact time at pore junctions, mixing is not complete at pore junctions. This results in a broader distribution of residence time, especially at high flow rates where contact time at pore junctions is short.

Bypassing from dispersion will increase as the contact time at pore junctions decreases. Thus, dispersive bypassing will depend on flow rate, but not on core length, since the contact time at pore junctions does not depend on core length.

Capillarity-Induced Bypassing

Capillarity-induced bypassing, described by Jones[9], and Campbell and Orr[11], is similar in nature to dispersive bypassing. It occurs in tertiary displacements because solvent must displace water to mobilize and recover oil. In water-wet rock, solvent displaces water from the largest pores first because their entry pressure is lower. The result is that during formation of the oil bank, the solvent develops preferred high conductivity paths, leading to pore-level bypassing. In mixed-wet rock, the solvent will first enter smallest oil-wet pores, and develop preferred pathways through them; however, wall-coating oil films present in mixed-wet rock may improve contact between oil and solvent. Thus, capillarity-induced bypassing may depend on the rock wettability, but is likely to occur in both mixed-wet and water-wet rock. This type of bypassing occurs when the oil is mobilized. Like dispersive bypassing, it will be evident at the back of the oil bank.

Capillarity-induced bypassing will be reduced as viscous forces in the solvent bank increase relative to capillary pressure forces: increases in flow rate or solvent viscosity are therefore expected to reduce bypassing by this mechanism. Note that this effect of flow rate is different from that expected for dispersive single-phase bypassing. Core length influences capillarity-induced bypassing to the extent that lower water saturations are usually seen in longer cores as a result of a reduction in the extent of capillary end effects. In the absence of significant capillary end effects, core length will not influence bypassing by this mechanism.

Viscous Fingering

Viscous fingering has been widely studied. (For recent discussion of this effect, see References 14 and 15.) It results from macroscopic-scale heterogeneities coupled with the mobility contrast between solvent and oil. As a result of core-scale heterogeneities, the solvent finds a preferred path through the core. Once this pathway develops, most solvent flows through it, and not much additional oil is recovered. Viscous fingering is mitigated by mixing between solvent and oil in the direction of flow, which reduces the mobility contrast at the displacement front, thereby eliminating some of the driving force for bypassing. In addition, mixing in the direction transverse to flow acts to smear out regions of high solvent saturation, reducing to some extent the amount of bypassing. The extent of viscous fingering will depend on, in addition to the viscosity ratio, the dimensionless parameters D_l/UL, D_t/D_l, and W/L, where D_t and D_l are the transverse and axial dispersion coefficients, U is the Darcy velocity, L is the core length, and W is the core width.

The result is that bypassing from viscous fingering increases as the contact time decreases because there is less time for mixing. Thus, bypassing from viscous fingering will increase as flow rate increases or core length decreases. The effects of core length and flow rate will not be equivalent, since the dispersion coefficients depend on flow rate, and the aspect ratio (L/W) changes with length but not flow rate. These trends are based on the results of simulations of single-phase first-contact miscible displacements. The same trends are expected in multiple-contact miscible displacements; however, without detailed simulations, the magnitude of these effects cannot be determined for tertiary multiple-contact miscible displacements. As a result, it will be difficult to determine from experimental data alone whether viscous fingering occurs.

Experimental observation of flow rate and core length effects, then, can give some indication of the relative importance of capillarity-induced, dispersive, and macroscopic viscous bypassing. If recovery increases with flow rate, it is likely that capillary pressure effects dominate. A decrease in recovery with flow rate, on the other hand, implies that either dispersive bypassing or viscous fingering dominate. If recovery is independent of core length, it is likely that either capillarity-induced or dispersive bypassing dominate; an increase in recovery with core length may indicate that viscous fingering has a significant effect.

The effects of oil viscosity may also give some indication of the relative importance of these three mechanisms. Viscous fingering and dispersive bypassing are expected to increase with oil viscosity, while capillarity-induced bypassing is a much weaker function of oil viscosity.

EXPERIMENTS

In this section, the experimental apparatus is described, rock and fluid properties are given, and experimental procedures are outlined.

Coreflood Experiments

<u>Equipment</u>

Floods were conducted in one-inch-square cores 12 or 40-inches long. The cores were coated with epoxy and cast in stainless-steel core holders using a lead-bismuth alloy. The core holders were equipped with internal taps, allowing measurement of pressure-drop as a function of position in the core. Positive displacement pumps were used to displace reservoir fluids from floating piston cells through the core. The pressure at the core outlet was controlled within ~ 1 psi with a CLAMAR oscillating back-pressure regulator. Produced fluids at atmospheric pressure were routed by means of a rotary 20-port switching valve (Valco) to test tubes where the liquids were collected. Gas production was monitored with a CLAMAR gasometer. An HP1000 computer was used to monitor pressure drop and gas production, and control the switching valve.

Analyses of the produced crude oil were carried out using a HP5880 gas chromatograph equipped with a flame ionization detector and a 25 m X 0.31 mm ID capillary column. The liquid phase consisted of 5% phenylmethylsilicone. Detailed procedures for the analysis are not presented; they are well known throughout the industry.

Materials

Experiments were conducted using Berea sandstone and San Andres Outcrop Carbonate. Berea sandstone is well characterized; a relatively homogeneous rock, it consists mainly of quartz (90%) and clays (4-6%) with minor quantities of calcite cement and iron-containing compounds. It was used as received, and assumed water-wet. Permeability in the sandstone ranged from 250 to 650 md.

Carbonate core material was obtained from a surface outcrop of the San Andres formation located in Last Chance Canyon in the western Guadalupe Mountains. It is a grainstone, comprised of pelloids of dolomite roughly 100 to 500 microns in diameter. Some calcite cement is present, but porosity is generally high, about 18-20%. SEM studies of pore casts prepared from outcrop samples indicate the pelloids have some limited microporosity, but most of the porosity is interparticle. The carbonate rock was rendered mixed-wet by aging in crude oil for one week at connate water saturation. Its permeability was between 130 and 230 md. Connate water saturations are listed in Table 3; they range from 0.25 to 0.4.

Oil used here was from the Means field in west Texas and from the Fullerton field; both oils were degassed and filtered prior to use. The degassed Means oil viscosity under flood conditions is 12 cp, and its minimum miscibility pressure is roughly 1200 psi. The degassed Fullerton crude viscosity under the conditions employed here is 4 cp. The minimum miscibility pressure for Fullerton crude is less than 1500 psi, which is the bubble point for the live crude. Slim tube tests have not been conducted with dead Fullerton oil.

GC analysis indicates both oils are mainly paraffinic in composition. Table 1 lists the density of each oil along with its composition determined from high performance liquid chromatography (HPLC). Shown is the fraction of each oil that falls into each of five categories: saturated; aromatic; nitrogen, sulfur, and oxygen-containing (NSO); asphaltenes; or unrecoverable by HPLC. The last three of these categories, when combined, are referred to as polar compounds.

Three main differences are noted between oils: Fullerton oil consists of lower molecular weight compounds, as evidenced by the chromatographs, and is somewhat more paraffinic, as indicated in Table 1. The two oils contain similar quantities of polar compounds, but the Fullerton oil contains fewer NSO's and asphaltenes.

Brine contained 3.5 wt% total dissolved solids, consisting of 0.8 wt% $CaCl_2$ with the remainder NaCl. It was prepared from deionized water, then filtered and degassed before use.

Procedures

Floods were carried out at 2000 psi and 100°F, corresponding to reservoir conditions at Means. Brine-filled cores were oil flooded to connate water saturation. With Berea cores, this consisted of injecting 10 pore volumes (PV) of oil at a frontal (intersticial) velocity of 100 ft/day. In the carbonate cores, oil was recirculated through the core for seven days with the pressure-drop set at 100 psi. These procedures gave connate water saturations in the range of 20-40%, and in the case of the carbonate rock, resulted in a mixed-wet core.

Waterfloods and WAG floods were carried out vertically with fluids injected at the top at a frontal velocity of 1 ft/day unless otherwise specified. Typically, 1-2 PV of brine were injected during the waterflood, and 3-4 PV CO_2 were injected during the miscible flood. At the end of each miscible flood, the core was flushed with 10 PV of trichloroethane to remove the miscible residual oil, dried with air, evacuated and resaturated with brine to give a brine-filled core. The effluent from the trichloroethane flush was retained, and the solvent was allowed to evaporate until no further change was seen. This gave an estimate of S_{orm} that could be compared to that derived from CO_2 flood production data.

Microvisualization

Microvisualization, using a technique developed by Yadov et al.[16], involved conducting the miscible flood using substitute fluids that were hardened in-place. Then, thin sections were prepared, and fluid distributions were observed. Significant effort has been devoted to showing that the hardening process does not influence the fluid configuration,[17,18] giving confidence that observed fluid distributions are representative of the styrene/water configuration before hardening. These experiments were carried out at atmospheric pressure using styrene as the oil in three-inch long Berea cores. Detailed procedures are given in Reference 16.

The miscible flood involved injecting styrene containing 0.5 wt% nile red dye to miscibly displace styrene containing 0.4 wt% diphenyl anthracene. This flood began at waterflood residual oil saturation.

Thin sections were observed using a Zeiss Photo Microscope III in the UV fluorescence mode. Using the filters described in Table 2, the oil appeared bright blue, solvent light green, and water dark brown or black. The rock appeared black. Water and rock could be distinguished using transmitted light.

RESULTS

In this section, results of 16 coreflood experiments and 1 microvisualization experiment are described. Results of corefloods are summarized in Table 3, which lists the conditions for each flood, the rock permeability and the final oil (S_{orm}) and water (S_w) saturations determined by material balance. In each experiment, we also determined fluid mobilities and conducted compositional analysis of the produced oil.

The remaining portion of the RESULTS section first discusses the repeatability of the experiments, then presents results of experiments in which the flow rate and core length were varied. These experiments were undertaken to better understand how measurements depend on contact time, with the aim of determining how these results will scale to the field. The next set of experiments examined the effects of oil viscosity, with the intent of providing insight into the bypassing mechanism. Finally, floods at varying wettability, WAG ratio, and initial water saturation were carried out to examine the importance of various displacement mechanisms under realistic CO_2 flood conditions. Also presented are the results of mobility measurements in those experiments. Finally, results of microvisualization experiments to verify the presence of bypassing are presented.

Repeatability of Experiments

Inspection of Table 3 indicates that, in carbonate cores, some variation in S_{orm} is seen from experiment to experiment, even in the same core. (Floods SAC-1, SAC-2, SAC-3, SAC-4, SAC-5 SAC-7, and SAC-8 were conducted in the same core.) Some variation is also seen in permeability, but this is not thought to significantly influence recovery or relative permeability.

The error in material balance, defined as the volume of oil initially present minus the volume of oil recovered in the CO_2 flood and the trichloroethane flush, averaged roughly ± 0.024 PV. This is similar to flood-to-flood variation in S_{orm}, seen in Table 3.

Effects of Core Length

For each set of experiments, oil recovery results, which are a gross measure of process performance are presented first, then compositional data are presented to help diagnose displacement mechanisms and clarify reasons for any variation in recovery.

Figure 1 shows the dependence of S_{orm} on core length in Berea cores. These are the results of floods B-1, B-4, and B-6. The values measured here were determined from the volume of oil produced during the CO_2 flood, and represent the oil remaining in the core after 3 pore volumes of CO_2 injection. For comparison, data from References 1 and 3 are also shown. Those floods were also conducted in Berea cores, but they started at connate water saturation. The values of S_{orm} from Reference 1 were determined after several pore volumes of CO_2 injection, while those from Reference 3 were determined after 1.2 PV. The floods reported in Reference 3 used west Texas oil with a viscosity of 3 cp. Reference 1 reports results of experiments using an oil with viscosity of 0.75 cp. Both sets of experiments were carried out at pressures well above the MMP.

Examination of Figure 1 shows that, for a given oil, flow rate, and initial water saturation, core length has only a small effect on S_{orm}, in cores ranging in length from 1 to 32 feet. This implies that miscibility develops over distances short compared to one foot. This contrasts with simulator predictions by Negahban et al.[3], which predicted much higher values of S_{orm} in cores shorter than 8 feet. Our conclusion is further supported by the results of experiments conducted by Shyeh-Yung[19], in which CO_2 was injected into a slim tube filled with oil and the slim tube effluent was injected into the core. In those experiments, S_{orm} was not strongly influenced by the presence of the slim tube at pressures above the minimum miscibility pressure (MMP). Thus, we believe that miscibility develops quickly over a distance that is smaller than the shortest core used (12 inches).

It is also evident from comparing results of experiment B-1 and B-4 that core orientation has little impact on oil recovery, indicating that gravity override is not influencing the results of horizontal experiments, and that the stabilizing effect of gravity is not significant in the vertical floods at 1 ft/day.

Compositional data provide more information about displacement mechanisms. This is illustrated in Figures 2 and 3, which depict the results of compositional analysis of the produced oil in experiments B-1 (12-inch core) and B-4 (40-inch core). The weight fraction of a given oil component, normalized to the weight fraction of that component in the original crude oil, is plotted against cumulative oil recovery. The plots in Figures 2 and 3 show the normalized weight fraction of normal alkanes containing 10, 15, 20, and 25 carbon atoms in the oil produced during CO_2 slug floods. We chose to plot concentrations of alkanes because they are the predominant chemical species present. Flood conditions are listed in Table 3.

The presence of bypassing can be inferred from these graphs. When oil is displaced directly, it will have the same composition as the original crude and the normalized weight fraction will be close to unity for all components. Deviation from unity indicates that some components are preferentially recovered.

Examination of Figure 2 indicates that until roughly 40% of the oil has been recovered, the normalized weight fraction is unity for all components, implying that this oil is recovered by direct miscible displacement. As more oil is recovered, the normalized weight fraction of C_{10} increases, and the normalized weight fraction of heavier components decreases. Eventually, the normalized weight fraction of C_{10} declines almost to zero, and the C_{15} concentration increases. As more oil is recovered, the C_{15} concentration also declines, and, after 65% of the oil has been recovered, increases are seen in concentrations of components containing 20 and 25 carbon atoms. Thus, as more oil is recovered, the produced oil is initially lighter than the original crude, and becomes increasingly heavy. This progressive increase in molecular weight of produced oil results from recovery of bypassed oil by extraction. The more volatile components are extracted first; as they are depleted, heavier components are recovered.

These results indicate that a bank of miscibly displaced oil is produced first, followed by oil recovered by extraction. The size of the miscible bank does not depend on core length; evidence of extraction is first seen at the same recovery level in both cores. However, the final recovery is slightly higher in the longer core. Based on this, it is concluded that the amount of bypassing is similar in these two floods, but slightly more oil is extracted from the longer core, as a result of the longer contact time. Based on the length effect observed here and the presence of a bank of miscibly displaced oil, the transition region between oil and solvent, where miscibility develops, is thought to be small compared to the core length.

Effect of Flow Rate

In Figure 4, the effect of flow rate on S_{orm} is illustrated for floods conducted in 1-foot-long, water-wet Berea cores with Means crude oil. These results were obtained in experiments B-1, B-2, and B-3. Our experiments show little dependence of S_{orm} on flow rate, with a shallow minimun at 1 ft/day. Also shown are results presented in Reference 8. Results from Reference 8 were obtained in secondary floods using a West Texas oil whose viscosity was 1.55 cp. Those experiments were conducted at pressures well above the MMP. It is seen that in the experiments presented in Reference 8, S_{orm} increases sharply at flow rates less than 1 ft/day and levels off at higher flow rates. Thus, the results presented in Reference 8 show a much stronger flow rate dependence between 0.3 and 1 ft/day than the results presented here.

The weak dependence of S_{orm} on flow rate in tertiary floods conducted here is thought to result from the combined effects of capillarity-induced bypassing, and dispersive or macroscopic viscous bypassing. Since capillarity-induced bypassing decreases with flow rate, its effect mitigates the effects of bypassing from viscous fingering and dispersion which tend to increase with flow rate.

The contrast between results presented here and those from Reference 8 likely results from differences in initial water saturation. Corefloods presented in Reference 8 started at connate water saturation with the result that capillarity-induced bypassing would not occur.

Figure 5 shows the compositional data from flood B-2 (0.3 ft/day). Comparing these results to those in Figure 2 (from experiment B-1 (1.0 ft/day)) illustrates the effects of flow rate. These results indicate that the onset of bypassing is evident slightly earlier at the lower flow rate consistent with the hypothesis that lower viscous forces result in more bypassing at the lower flow rate.

Figure 6 shows the effect of flow-rate on S_{orm} in the mixed-wet carbonate rock in experiments SAC-1, SAC-3, and SAC-4, conducted at 5, 1, and 0.3 ft/day. Here, as in Berea cores, little effect of flow rate is seen.

The composition of produced crude, shown in Figures 7 and 8 for experiments SAC-3 and SAC-4, is similar in both floods. In both cases, extraction is evident after roughly 40% of the oil is recovered, the C_{10} is almost depleted after 65% of the oil is recovered, and extraction of C_{20} and C_{25} is evident after roughly 65% of the oil is recovered. This indicates that the level of bypassing is only a weak function of flow rate, consistent with the hypothesis that opposing effects of dispersion and capillarity act to dampen changes in the amount of pore-level bypassing with flow rate.

Effects of Crude Oil Viscosity

To examine crude oil viscosity effects, we have compared results of experiments with Fullerton crude oil to those from experiments conducted with Means Crude. Since the pressure is substantially higher than the MMP for either crude oil, it is thought that any difference in flooding behavior will result from differences in viscosity, rather that the differences in oil composition.

It is found that, in floods where lower viscosity crude oil is used, less bypassing is seen. Figures 9 and 10 show the composition of produced oil from floods in Berea sandstone and San Andres carbonate when oil from the Fullerton field was used. Figure 9 shows the results of experiment B-7, while Figure 10 shows the results of experiment SAC-9. These are to be compared to Figures 2 and 7, which show compositional data for experiments B-1 and SAC-3, in which Means crude oil was used. In both cases, the onset of extraction occurred later when Fullerton crude was used than when Means crude was used. This is thought to result from the lower viscosity of the Fullerton crude (4 cp compared to 12 cp). Less contrast in mobility between the solvent in the preferred flow path and the bypassed oil reduces the extent of bypassing.

The viscosity contrast between solvent and oil is expected to significantly influence viscous fingering as well as dispersive bypassing, but will have a weaker effect on capillarity-induced bypassing. Thus, these results support the contention that, in addition to capillarity-induced bypassing, dispersive bypassing and/or viscous fingering also take place.

Results presented so far indicate that significant quantities of oil are bypassed in CO_2 miscible floods as a result of the viscosity contrast between CO_2 and oil coupled with heterogeneity and capillarity. Based on consideration of mechanisms of bypassing, we hypothesize that pore-level bypassing has a strong influence on the performance of miscible CO_2 floods. At this point, however, we cannot rule out the presence of larger-scale, more macroscopic bypassing.

Effects of WAG Ratio

In Tables 4 and 5, S_{orm} is listed as a function of WAG ratio for floods conducted in water-wet Berea and mixed-wet San Andres carbonate cores. These tables show results from experiments B-4 and B-5, and experiments SAC-3, SAC-5, SAC-6, SAC-7, and SAC-8. In water-wet rock, the well-known waterblocking effect[22-24] is seen: less oil is recovered at WAG ratios greater than zero. The presence of water reduces contact between oil and solvent. That is, some of the oil is trapped by water; the capillary pressure is not high enough for solvent to force water from those pores.

Analysis of the crude oil produced in these floods provides more information about the mechanisms of bypassing. In Figure 11, the composition of oil produced during the WAG=3 flood(B-5) in a Berea core is shown; this should be compared to the composition of the oil produced during the WAG=0 flood (B-4), shown in Figure 3. It is seen that at higher WAG ratios, extraction is largely eliminated, reducing the total oil recovery. The implication is that higher water saturations prevent mass transfer between bypassed oil and flowing solvent.

Examination of Table 5 indicates that in mixed-wet rock, the effect of WAG ratio is much smaller. The presence of wall-coating oil films as hypothesized by Salathiel[25] are thought to provide improved contact between oil and solvent, reducing the effect of water saturation. Figure 12 further illustrates the effects of wettability and WAG ratio; compositional data are shown for flood SAC-7, conducted in mixed-wet rock at

WAG=3. Comparison with Figure 7 indicates that extensive extraction is seen at WAG=3 and WAG=0.

A final result shown in Table 5 comes from a secondary WAG=3 flood (SAC-8). Oil recovery was similar to that seen in WAG=0 tertiary floods and higher than that seen in tertiary WAG=3 floods; this is thought to result from improved initial connectivity of the oil in these floods, and the absence of any capillary pressure effects. Compositional data, shown in Figure 14, can be compared to data from experiment SAC-7, replotted in terms of OOIP recovery in Figure 13. This indicates that more oil is recovered by direct displacement in the secondary flood, consistent with this hypothesis. Shyeh-Yung[19] has obtained similar results in near-miscible WAG=0 floods with Means crude.

In Figure 15, the relative permeabilities determined from the experiments shown in Table 5 are plotted as a function of CO_2 saturation along with data obtained by Shyeh-Yung[19] in near-miscible floods. We also show the oil relative permeabilities to oil determined at the end of the oil flood. Good agreement is seen between results obtained here and those obtained by Shyeh-Yung. CO_2 relative permeability is found to be a function only of the CO_2 saturation, consistent with a three-phase relative permeability formulation such as that proposed by Stone. This differs from the mixing model formulation typically used. The mixing model would estimate mobilities based on the total oleic phase saturation (i.e., the sum of CO_2 and oil saturations). In our floods, this gives mobilities 2-3 times larger than those seen experimentally. It is evident that the mixing model does not adequately account for the presence of S_{orm}, which reduces the permeability to CO_2.

Microvisualization

Presented here is the result of a single miscible microvisualization experiment in Berea sandstone, aimed at verifying the presence of pore-level bypassing. Other microvisualization experiments were stopped after the waterflood. These experiments verified that the crude-soaked carbonate was in fact mixed-wet, and that the Berea was water-wet.

As discussed in the EXPERIMENTAL section, miscible microvisualization experiments involve displacing styrene containing one dye with styrene containing a different dye. Thus, these are density- and viscosity-matched displacements. They will be influenced by dispersive and capillarity-induced bypassing. Displacements were conducted in Berea, at 1 ft/day. In thin sections subjected to UV light, oil fluoresces bright blue, rock grains are black, solvent is a very pale green, and water appears brown. Under transmitted light, water appears dark reddish-brown, and rock is a paler orange.

Figure 16 shows results of the microvisualization experiment described above. This photomicrograph shows a thin section taken behind the displacement front. Shown is an image obtained when the sample was illuminated with visible transmitted light, and subjected to UV light. Both transmitted light and UV fluorescence from the thin section are shown.

In this black and white photomicrograph, oil and solvent are not distinguishable, although the difference is easily seen in color photomicrographs. Shown along with the photomicrograph is a tracing of the photograph, where the oil, solvent, water, and rock have been assigned different tones.

The image in Figure 16 shows an oil blob and a solvent blob separated by a small-diameter pore containing water. This is an example of capillarity-induced bypassing. Other photographs, not shown here, give evidence that dispersive bypassing also occurs; both dyes are observed in different parts of a single oleic-phase blob.

DISCUSSION

Results of miscible floods presented here give clear evidence of the interaction between fluid flow and phase behavior suggested by Gardner and Ypma[10] and Orr and students[11,12]. We have determined the composition of the produced oil; the results of those experiments are consistent with the following scenario. A portion of the oil is recovered by direct displacement. The rest of the oil is bypassed, and a portion of the bypassed oil is recovered by extraction. The composition of the extracted oil changes with time; more volatile components are recovered first.

The amount of bypassing is not a strong function of core length or flow rate, but increases with the viscosity ratio between oil and solvent. The viscosity ratio dependence is indicative of a viscous instability which could result from either pore-level or macroscopic heterogeneities. The effects of velocity and core length are consistent with pore-level bypassing influenced by both dispersion and capillary pressure effects.

Other evidence of pore-level bypassing comes from microvisualization experiments, which show the presence of isolated blobs of oil behind the miscible solvent front. Thus, data presented here show indications of pore-level bypassing; it is not possible to determine from our data whether macroscopic bypassing (viscous fingering) also occurs.

We also find that high water saturations act to reduce the amount of extraction that occurs in both water-wet and mixed-wet rock, but the effect is much less pronounced in mixed-wet rock. This is attributed to the presence of wall-coating oil films in the mixed-wet rock, which act to improve contact between oil and solvent.

In hopes of clarifying the relative importance of viscous fingering, we have carried out some linear stability calculations. Using the procedure suggested by Christie and Bond[14] and Lee et al.[15], the transverse dimension of the smallest unstable finger has been calculated under conditions corresponding to our floods in mixed-wet carbonate cores. Smaller fingers than these will be smeared out by dispersion; larger fingers will grow over the course of the displacement. Thus, if the smallest stable finger is larger than the core dimension in the transverse direction (in this case, 1 inch), macroscopic bypassing is not likely to occur.

Results show that, at flow rates lower than 5 ft/day, the finger size is larger than the transverse dimension of the core, indicating that macroscopic bypassing is not likely at lower flow rates. Based on this analysis, it is concluded that although macroscopic bypassing may occur, it is likely that pore-scale bypassing is more important. Based on the observations and analysis discussed above, further experiments to determine the effects of viscous fingering may be warranted. An obvious experiment involves obtaining in situ images of saturation distributions, using either NMR or X-Ray CT. A second set of experiments would entail conducting floods in cores of varying radial dimension; this would give a rough indication of the length-scale of bypassing.

Another question that arises from the results presented here is whether extraction effects will be observed in the field. To examine this question, it may be appropriate to pursue two approaches. First, the composition of oil produced in field displacements could be monitored as a function of CO_2 throughput. A separate approach would involve developing a simulator that accounts for the effects shown here, and using that simulator to predict field-scale performance. Orr and Dai[12] have developed a simulator for this purpose; it may serve as a basis for further work.

CONCLUSIONS

1. High values of S_{orm} often seen in corefloods result from bypassing. Although light ends are subsequently recovered from bypassed oil, S_{orm} is still higher than that seen in first-contact-miscible floods because heavy ends are left behind.

2. Miscibility develops over relatively short distances, with the result that corefloods in 1 to 2 ft-long cores can be used to obtain representative measurements of S_{orm}.

3. In tertiary floods, bypassing is thought to result mainly from capillary pressure effects and dispersion (i.e., pore-level effects).

4. Macroscopic viscous fingering may also be present, but is thought to be a smaller effect under the conditions used here. Further experiments have been suggested to provide better insight into this effect.

5. Possibly as a result of the presence of thin wall-coating oil films, waterblocking is a small effect in mixed-wet rock. In water-wet rock, extraction of bypassed oil is eliminated at high WAG ratios, resulting in lower oil recovery.

ACKNOWLEDGEMENTS

The author thanks B. K. Boseck and J. M. Grobelny, who performed the corefloods, for their careful work. Thanks are also due to N. R. Daniel, who carried out the GC analysis, and D. I. Jezek, who prepared thin sections. J. P. Smith suggested some of the dyes used in the microvisualization experiments. Finally I thank J. J. Shyeh-Yung and T. R. Reppert for their helpful discussions.

REFERENCES

1. Shelton, J. L., and Schneider, F. N., "The Effects of Water Injection on Miscible Flooding Methods Using Hydrocarbons and CO_2", SPEJ (June 1975) 217-226.

2. Yellig, W. F., "Carbon Dioxide Displacement of a West Texas Reservoir Oil", SPEJ (Dec. 1982), 805-815.

3. Negahban, N., Shiralkar, G. S., and Gupta, S. P., "Simulation of the Effects of Mixing in Gasdrive Core Tests of Reservoir Fluids", SPERE (Aug. 1990), 402-408.

4. Tiffin, D. L., and Kremesec, V. J., "Mechanistic Study of Gravity-Assisted CO_2 Flooding", SPERE (May 1988), 524-533.

5. Kremesec, V. J., and Sebastian, H. M., "CO_2 Displacements of Reservoir Oils From Long Berea Cores: Laboratory and Simulation Results", SPERE (May 1988) 496-504.

6. Todd, M. R., and Longstaff, W. J., "The Development, Testing, and Application of a Numerical Simulator for Predicting Miscible Flood Performance", JPT (July 1972) 874-882.

7. Rathmell, J. J., Stalkup, F. I., and Hassinger, R. C., "A Laboratory Investigation of Miscible Displacement by Carbon Dioxide," Paper SPE 3483 presented at the 1971 Annual Fall Meeting, New Orleans, Oct. 3-6.

8. Watkins, R. W., "A Technique for the Laboratory Measurement of Carbon Dioxide Unit Displacement Efficiency in Reservoir Rock", Paper SPE 7474 presented at the 1978 SPE Annual Technical Conference and Exhibition, Houston, Oct. 1-4.

9. Jones, S. C., "Some Surprises in the Transport of Miscible Fluids in the Presence of a Second Immiscible Phase", SPEJ (Feb. 1985) 101-112.

10. Gardner, J. W., and Ypma, J. G. J., "An Investigation of Phase-Behavior/ Macroscopic Bypassing Interaction in CO_2 Flooding", SPEJ, (Oct. 1984) 508-520.

11. Campbell, B. T., and Orr, F. M., "Flow Visualization for CO_2/Crude Oil Displacements", SPEJ (Oct. 1985) 665-678.

12. Dai, K. K., and Orr, F. M., "Prediction of CO_2 Flood Performance: Interaction of Phase Behavior with Non-Uniform Flow," SPERE (Nov. 1987) 531-542.

13. Sahimi, M., Hughes, B. D., Scriven, L. E., and Davis, H. T., "Dispersion in Flow Through Porous Media - 1. One-Phase Flow," Chem. Eng. Sci. (1986) 41, 2103-2122.

14. Christie, M. A., and Bond, D. J., "Detailed Simulation of Unstable Processes in Miscible Flooding", SPERE, (Nov. 1987) 514-522.

15. Lee, S. T., Li, K. M. G., and Culham, W. E., "Stability Analysis of Miscible Displacement Processes", Paper SPE/DOE 12631 presented at the SPE/DOE Symposium on Enhanced Oil Recovery, Tulsa, Apr. 15-18.

16. Yadov, G. D., Dullien, F. A. L., Chatzis, I., and MacDonald, T. F., "Microscopic Distribution of Wetting and Non-wetting Phases in Sandstones During Immiscible Displacements", SPERE (May 1987) 137-147.

17. Sincock, K., J., and Black, C. J. J., "Validation of Water/Oil Displacement Scaling Criteria Using Microvisualization", Paper SPE 18284 presented at the 1988 SPE Annual Technical Conference and Exhibition, Houston, Oct. 2-5.

18. McKellar, M., and Wardlaw, N. C., "A Method of Viewing 'Water' and 'Oil' Distribution in Native-State and Restored-State Reservoir Core", AAPG Bull. (1988) 72, 765.

19. Shyeh-Yung, J. J., "Mechanisms of Miscible Oil Recovery: Effects of Pressure on Miscible and Near-Miscible Displacements of Oil by Carbon Dioxide," Paper SPE 22651, presented at the 1991 Annual Technical Conference and Exhibition, Dallas, Oct. 6-9.

20. Raimondi, P., and Torcaso, R. H., "Distribution of the Oil Phase Obtained on Imbibition of Water", SPEJ, (Mar. 1964) 49-55.

21. Tiffin, D. L., and Yellig, W. F., "Effects of Mobile Water on Multiple Contact Miscible Gas Displacements", SPEJ (June 1983) 447-455.

22. Spence, H. P., and Ostrander, J. F., "Comparison of WAG and Continuous Enriched-Gas Injection as Miscible Processes in Sadlerochit Core", Paper SPE 11962 presented at the 58th Annual SPE Conference and Exhibition, San Francisco, Oct. 1983.

23. Huang, E. T. S., and Holm, L. W., "Effect of WAG Injection and Rock Wettability on Oil Recovery During Carbon Dioxide Flooding", SPERE (Feb. 1988), 119-129.

24. Lin, F. C., and Huang, E. T. S., "The Effect of Rock Wettability on Waterblocking During Miscible Displacement", SPERE (May 1990) 205-212.

25. Salathiel, R. A., "Oil Recovery by Surface Film Drainage in Mixed-Wettability Rocks", JPT, (Oct. 1973) 1216.

SI METRIC CONVERSION FACTORS

inches x 0.0254 = m
feet x 0.3048 = m
PSI x 6.895 = kPa
md x 10^{-15} = m^2
ft/day x 3.528 x 10^{-6} = m/s

Table 1: Properties of Oils Used Here

Oil	Density, g/cm^3	Composition, wt% (by HPLC)				
		Saturates	Aromatics	NSO†	Asphaltenes	Unrecovered
Means	0.91	31	25	13	7	23
Fullerton	0.84	45	17	1	4	33

† Nitrogen/Sulfur/Oxygen-containing compounds.

Table 2: Optical Filters for UV Microscopy

Barrier Filter	Excitation Filter	Beam Splitter
> 365 nm	420 mn	> 420 nm

Table 3: Coreflood Results

Flood ID #	Absolute Permeability (md)	Core Length (inches)	Core Orientation#	Intersticial Velocity (ft/day)	WAG Ratio	S_{orw}	S_{cw}	S_{orm}	S_{wf}**
B-1	240	12	V	1.0	0.0	0.38	0.21	0.11	0.31
B-2	540	12	V	0.3	0.0	0.28	0.31	0.13	0.37
B-3	685	12	V	5.0	0.0	0.40	0.27	0.15	0.34
B-4	485	40	H	1.0	0.0	0.40	0.32	0.10	0.37
B-5	430	40	H	1.0	3.0	0.33	0.35	0.20	0.57
B-6	562	80	H	1.0	0.0	0.41	0.30	0.10	0.35
SAC-1	163	12	V	5.0	0.0	0.37	0.23	0.12	0.46
SAC-2	122	12	V	1.0	0.0	0.29	0.41	0.10	0.51
SAC-3	138	12	V	1.0	0.0	0.37	0.24	0.09	0.43
SAC-4	161	12	V	0.3	0.0	0.34	0.26	0.11	0.45
SAC-5	233	12	V	1.0	0.18	0.28	0.26	0.06	0.57
SAC-6	110	12	V	1.0	1.0	0.28	0.40	0.15	0.69
SAC-7	160	12	V	1.0	3.0	0.32	0.25	0.12	0.66
SAC+-8	234	12	V	1.0	3.0	--	0.36	0.08	0.61
B*-7	490	12	V	1.0	0.0	0.31	0.35	0.08	0.41
SAC*-9	162	12	V	1.0	0.0	0.30	0.33	0.07	0.43

+ Secondary Flood.
* Fullerton Crude; all other floods used Means crude.

#SAC - San Andres Carbonate
B - Berea Sandstone
H - Horizontal
V - Vertical

** Water Saturation at the End of CO_2 Flooding

Table 4: Oil Recovery in Water-Wet Berea Sandstone

1" x 1" x 40" core
1 ft/day intersticial velocity
100°F
2000 psi

Flood ID	WAG Ratio	S_{orm}	S_{wf}^{+}	S_{orw}	S_{cw}	Comparative Mobility++
B-4	0	0.10	0.37	0.40	0.320	30-50
B-5	3	0.18	0.57	0.33	0.350	0.35

\+ Water Saturation at the End of CO_2 Flooding

++ $\frac{\text{Mobility During } CO_2 \text{ Flood}}{\text{Water Mobility at } S_{orw}}$

Table 5: Oil Recovery in Mixed-Wet San Andres Carbonate

1"x1"x12" core
1 ft/day intersticial velocity
100°F
2000 psi

Flood ID	WAG Ratio	S_{orm}	S_{wf}^{**}	S_{orw}	S_{cw}	Comparative Mobility++
SAC-3	0+	0.09	0.43	0.37	0.24	4.0
SAC-5	0.18+	0.06	0.57	0.38	0.26	1.0
SAC-6	1.0+	0.15	0.69	0.28	0.40	0.7
SAC-7	3.0+	0.12	0.66	0.32	0.25	0.4
SAC-8	3.0*	0.08	0.61	--	0.36	--

\+ Tertiary Floods

* Secondary Floods

** Water Saturation at the End of CO_2 Flooding

++ $\frac{\text{Mobility During } CO_2 \text{ Flood}}{\text{Water Mobility at Sorw}}$

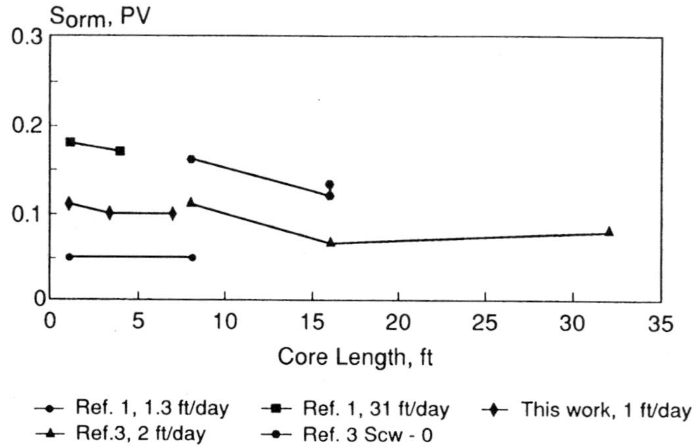

Figure 1. Dependence of S_{orm} on Core Length, in Experiment B-1, 4, 6; compared to that seen in Reference 8.

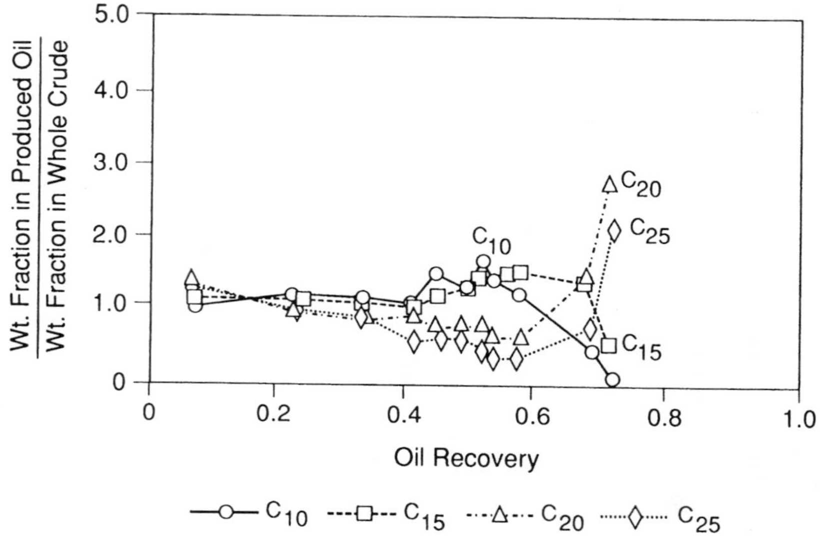

Figure 2. Composition of Produced Oil in Experiment B-1 (1 ft Berea core, 1 ft/day).

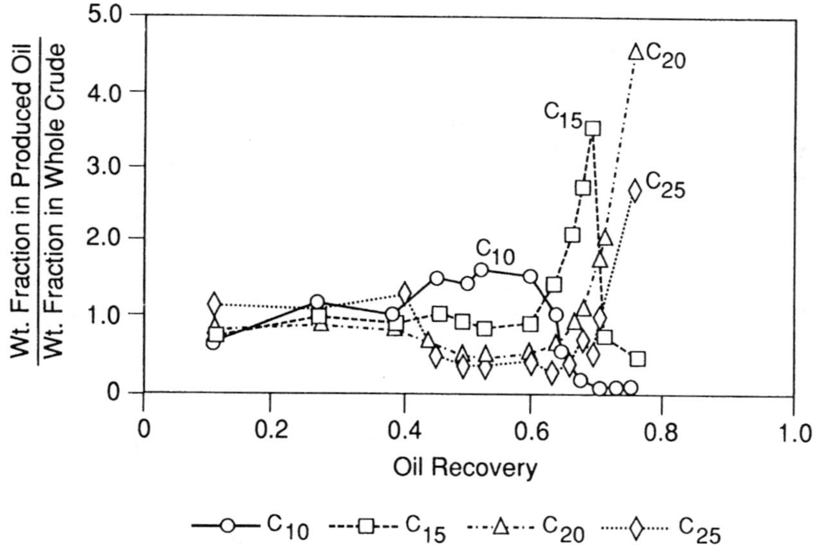

Figure 3. Composition of Produced Oil in Experiment B-4 (40-inch Berea core, 1 ft/day).

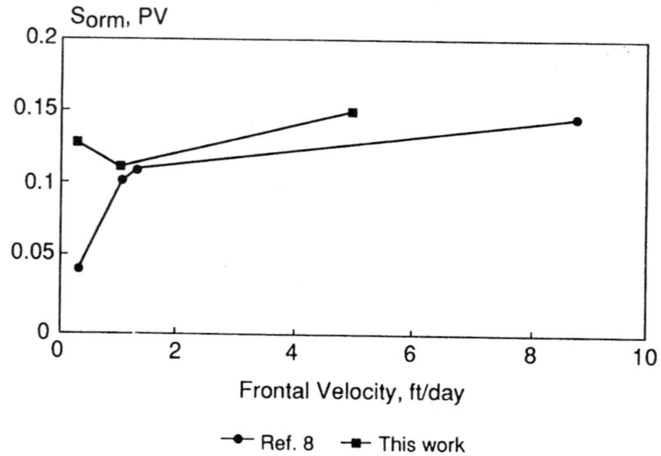

Figure 4. Effect of Flow Rate on S_{orm} in Experiments B-1, 2, and 3; compared to that seen in Reference 8.

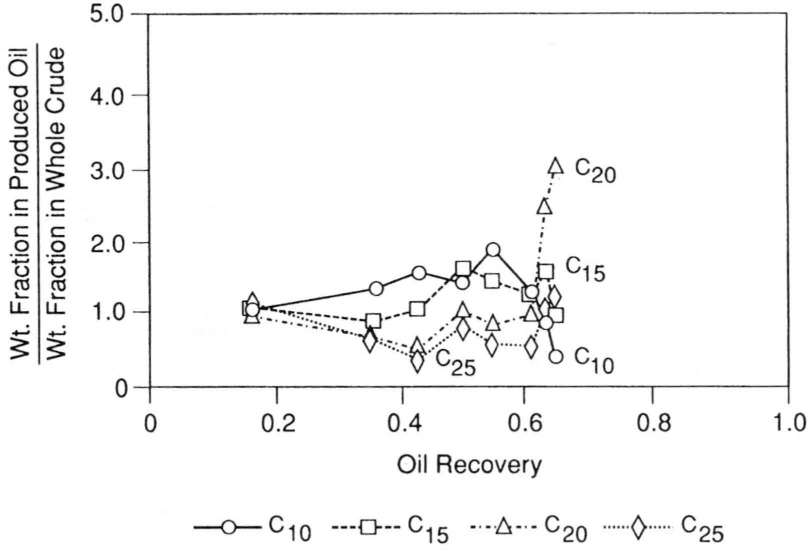

Figure 5. Composition of Produced Oil in Experiment B-2 (1 ft Berea core, 0.3 ft/day).

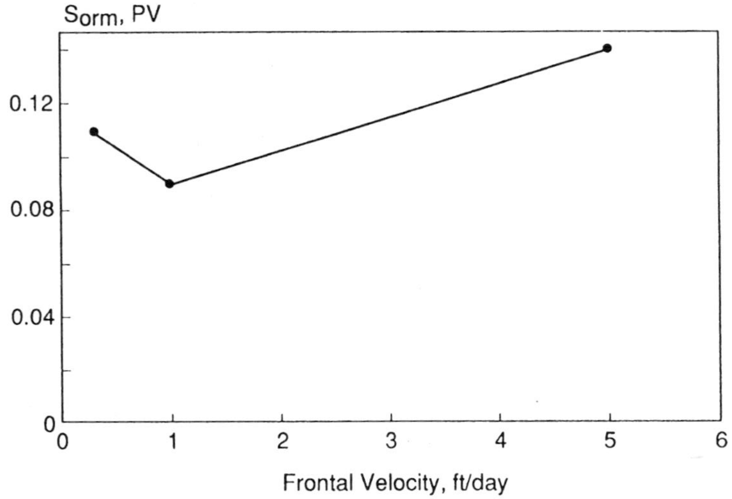

Figure 6. Effect of Flow Rate on S_{orm} in Experiments SAC-1, 3, 4.

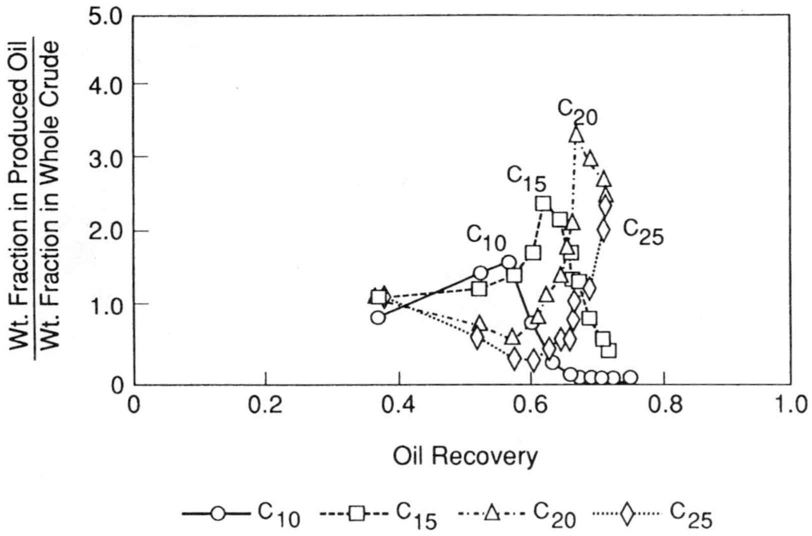

Figure 7. Composition of Produced Oil in Experiment SAC-3 (San Andres core, 1 ft/day).

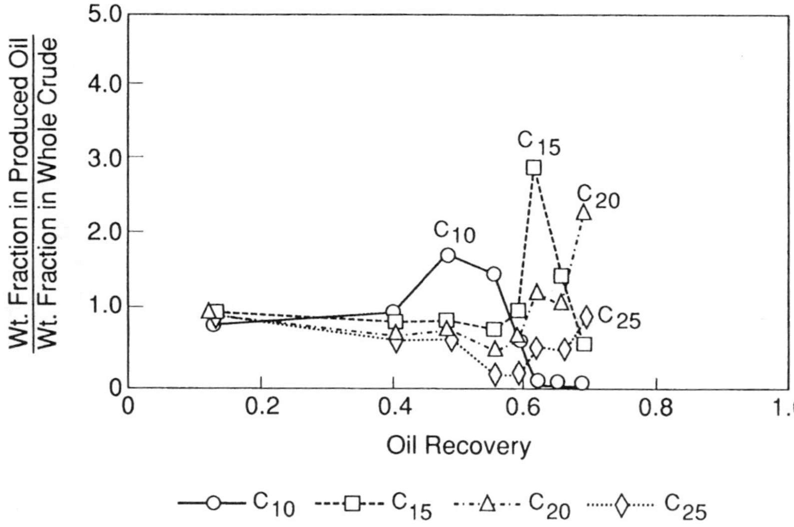

Figure 8. Composition of Produced Oil in Experiment SAC-4 (San Andres core, 0.3 ft/day).

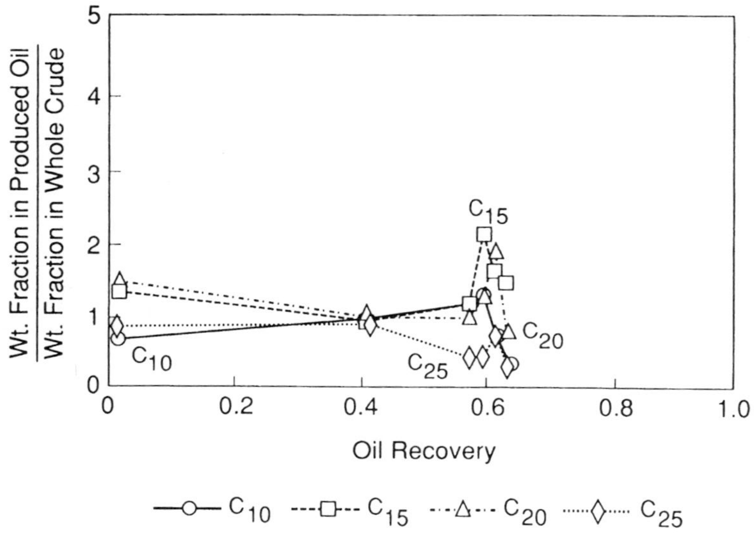

Figure 9. Composition of Produced Oil in Experiment B-7 (Berea core, Fullerton crude).

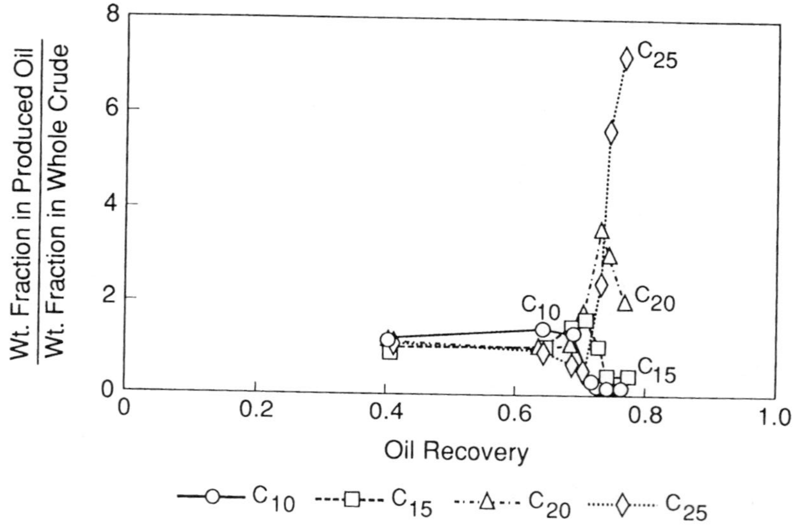

Figure 10. Composition of Produced Oil in Experiment SAC-9 (San Andres core, Fullerton crude).

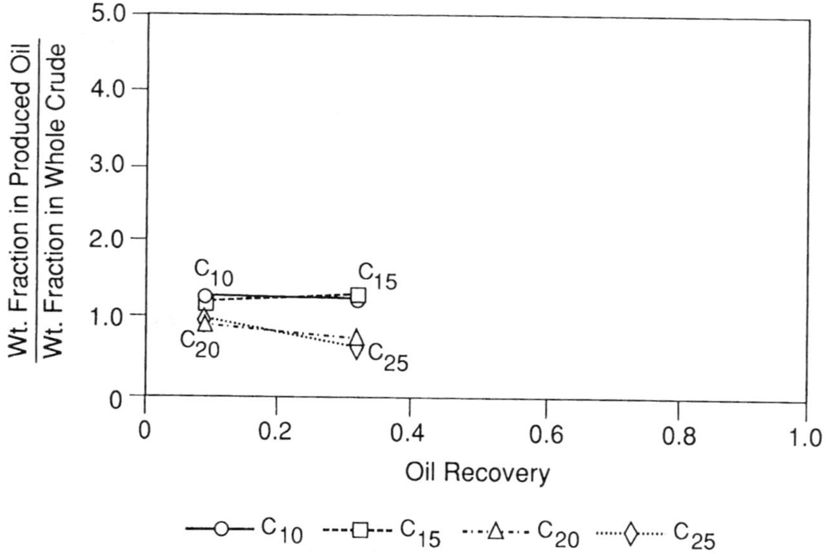

Figure 11. Composition of Produced Oil in Experiment B-5 (40-inch Berea core, WAG = 3).

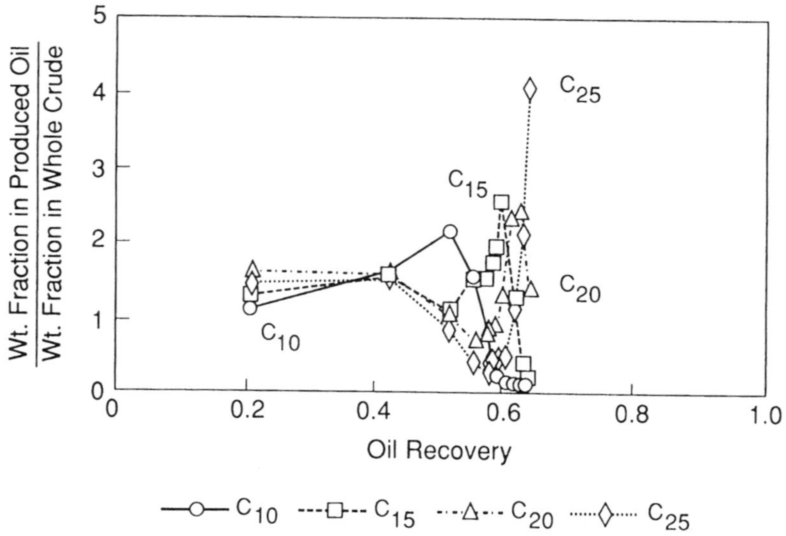

Figure 12. Composition of Produced Oil in Experiment SAC-7 (San Andres core, WAG = 3).

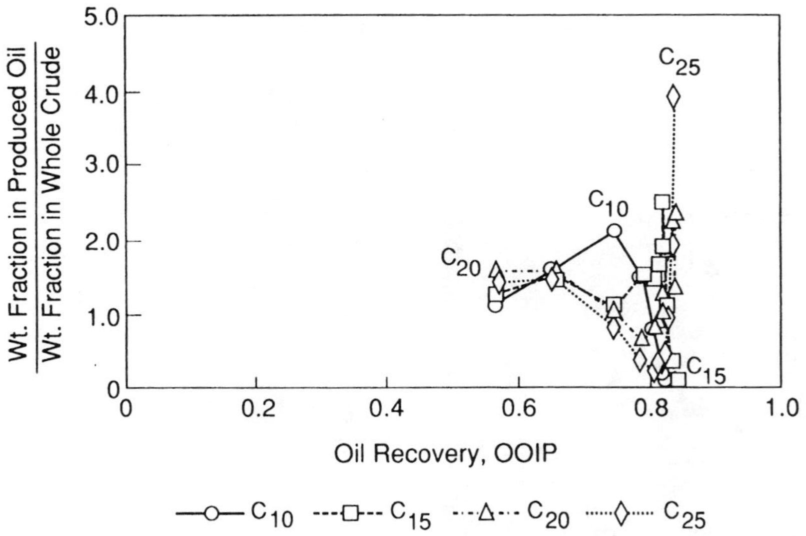

Figure 13. Composition of Produced Oil in Experiment SAC-7, with Recovery in % OOIP.

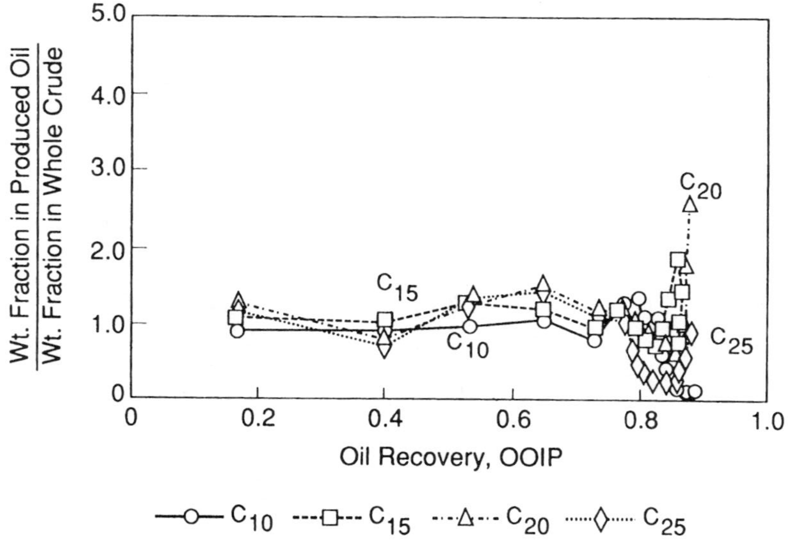

Figure 14. Composition of Produced Oil in Experiment SAC-8, with Recovery in % OOIP (San Andres core, secondary flood, WAG = 3).

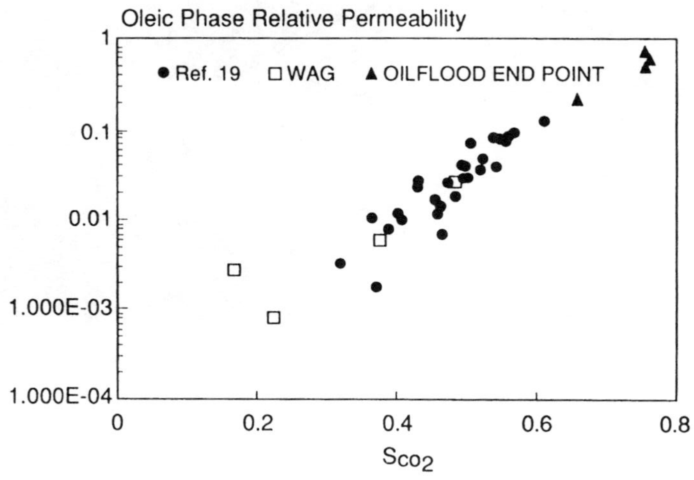

Figure 15. Relative Permeabilities Determined in Experiments SAC-3, 5, 6, 7; Compared to Those Reported in Reference 19.

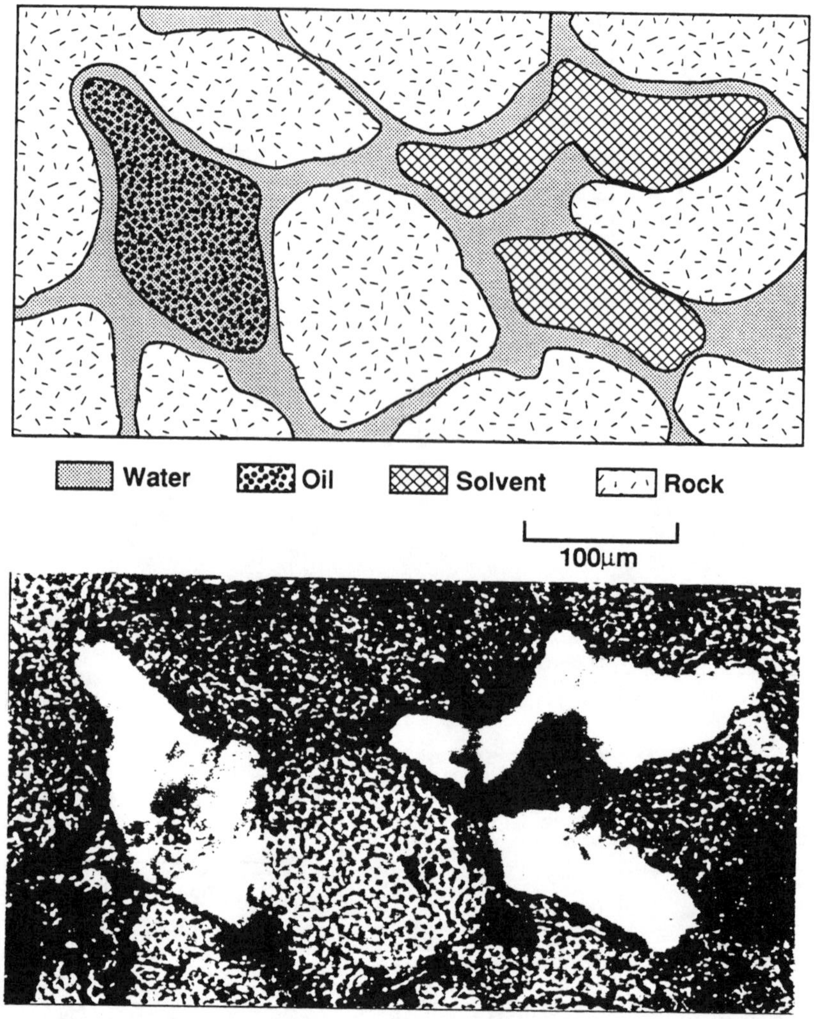

Figure 16. Image Obtained in a Miscible Microvisualization Experiment.

Annotated Bibliography

Field Case Histories—Performance

Flanders, W.A. and DePauw, R.M.: "Update Case History: Performance of Two Freds Tertiary CO_2 Project," Paper SPE 26614 presented at the 1994 SPE Permian Basin Oil and Gas Recovery Conference, Midland, Texas, 16–18 March.

This paper is an update of the performance of the Two Freds tertiary project in the Delaware sandstone. The project was developed in two stages: the east side, begun in 1974 and now essentially complete, and the west side expansion, started in the early 1980's. This field had high initial water saturations: 46.2% in the east side and 36.6% in the west side. The waterflood did not respond adequately in the east side, and this side of the field was essentially at its waterflood economic limit at the start of CO_2 injection. The west side had a moderate waterflood response and was still under waterflood at the start of CO_2 injection. CO_2 injection into the east side was approximately 57% of the mobile pore volume (PV). This was followed by injection of exhaust gas, by exhaust-gas water-alternating-gas (WAG) injection, and finally by water injection only. In the west side, a slug of CO_2 was followed by recycle-gas WAG injection. At the time the paper was written, a slug size of approximately 44% of mobile PV had been injected. Incremental tertiary recovery from the east side was 15.8% of the original oil in place (OOIP)—18.6% of the oil in place (OIP) at the start of injection and 107% of the primary plus secondary recovery. Net utilization of CO_2 was 8 Mcf CO_2/bbl incremental oil. Oil recovery in the west-side flood was approximately 10% of OIP at the start of injection, and net CO_2 utilization was 11.2 Mcf/bbl. The paper summarizes procedures used to monitor field performance and to estimate slug size and production lost to overinjection. The authors concluded that controlling injection on a pattern basis was critical for optimizing project economics and that exhaust-gas WAG had no detrimental effect on ultimate recovery or oil-production rate.

Harpole, K.J., and Hallenbeck, L.D.: "East Vacuum Grayburg San Andres Unit CO_2 Flood Ten-Year Performance Review: Evolution of a Reservoir Management Strategy and Results of WAG Optimization," paper SPE 36710 presented at the 1996 SPE Annual Technical Conference and Exhibition, Denver, Colorado, 6–9 October.

After 10 years of CO_2 miscible injection, the flood is projected to recover approximately 10% of the OOIP from the process and approximately 1.5% from infill drilling associated with the CO_2 flood. CO_2 injection began in 1985 in an area of approximately 5,000 acres containing an estimated 260 million STB. Initial development consisted of 80-acre inverted nine spots. Original design called for injection of approximately a 30% HCPV slug at a fixed 2:1 WAG. Large WAG cycles of approximately 2% HCPV per CO_2 half–cycle were implemented. As the project progressed, several problems were identified and addressed

1. Reservoir pressure in part of the Upper San Andres declined to less than miscibility pressure because initial injection policy did not allow high-capacity injectors to make up injection for less effective injectors and because of the high ratio of producers to injectors. Corrective action required increasing the total gas injection rate and drilling additional injectors to the Upper San Andres.

2. Poor sweep efficiency and CO_2 utilization resulted from some injectors and producers aligned on a northeast-southwest permeability trend, from predominant injection into the Lower San Andres, and from high-permeability thief zones. To improve sweepout, a program of well conversions, pattern realignment, and infill drilling was initiated. In addition, selective interval fracture stimulations were performed where a large contrast existed between Upper and Lower San Andres permeabilities.

3. The large CO_2 WAG cycles caused large swings in gas production. CO_2 half-cycles were reduced, and WAG ratios were tailored for each pattern on the basis of HCPV, performance, and pattern maturity.

Masoner, L.O. and Wackowski, R.K.: "Rangely Weber Sand Unit CO_2 Project Update," *SPERE* (August 1995) 203; *Trans.*, AIME, **299.**

The Rangely Weber Sand Unit is the largest field in the Rocky Mountains in daily and cumulative production. The field was discovered in 1933 and unitized in 1957. Peripheral water injection was started in 1958, fieldwide pattern waterflooding in 1969, and miscible CO_2 injection in 1986. Approximately 2 years before CO_2 injection was started, an effort was made to increase reservoir pressure to more than the minimum miscibility pressure (MMP) of 2,900 psi. Early response to CO_2 injection was characterized by significant incremental-oil production and early gas breakthrough. Gas-production volumes were greater than expected, necessitating construction of additional CO_2-compression capacity. The development plan calls for total CO_2 injection of 40% HCPV. At the time the paper was written, incremental recovery was 2.3% OOIP in the processed area, and a 6.8% OOIP cumulative tertiary recovery was expected. Although operational changes have been made, the project is considered economic. The original design called for a 1 : 1 WAG ratio, but WAG tapering to higher ratios was necessary because of the large gas production. WAG tapering was applied on the basis of incremental gas/oil ratio (GOR)—total-pattern gas production to total-pattern incremental oil. In a trial with 10 injection patterns, encouraging results were obtained for controlling gas production by decreasing WAG cycle size from 1.5 to 0.25% HCPV, and more than 90% of the WAG injectors were put on reduced half-cycles. Although WAG tapering and reduced half-cycle lengths have been effective in controlling gas production, vertical and areal sweep efficiencies remain the greatest challenges for the CO_2 project.

Stiles, L.H. and Magruder, J.B.: "Reservoir Management in the Means San Andres Unit," *JPT* (April 1992) 469.

This paper describes the evolution of reservoir management from primary recovery through waterflooding and implementation of a CO_2 flood. At the start of the CO_2 flood, existing wells included original 40- and 20-acre infill wells (waterflooding). Infill drilling to 10-acre spacing was an integral part of the CO_2 project so that patterns were 40-acre inverted nine spots with WAG injectors and center producers as the new wells. The 6-cp oil viscosity, 2,000-psi MMP, low formation-parting pressure, potential low injectivity, and potential CO_2 override were thought to be different from other large San Andres CO_2 projects. Project implementation began in 1983 and included 172 patterns covering 8,500 acres, or more than 80% of the OOIP. Operating objectives were (1) to complete producers and injectors in all floodable pay, (2) to maintain reservoir pressure above the MMP, (3) to maximize injection pressure below the fracture pressure, (4) to pump off producers, and (5) to maintain balanced injection/withdrawal by pattern. To help achieve these objectives, areal-flood balancing, vertical conformance, production, injection, and CO_2-produced/-injected ratio were monitored. While the infill drilling and pattern modification resulted in an initial production-rate increase in the project area from approximately 7,500 to 14,000 BOPD, CO_2 injection resulted in a sustained rate of 12,000 BOPD. Exact tertiary response is difficult to quantify because of simultaneous infill drilling and pattern modification. Of the original estimated reserves of 38 million STB, 16.6 million STB was associated with CO_2 injection and 21.4 million were associated with infill drilling and pattern modification. After 0.2 HCPV of CO_2 injection, CO_2 retention is 72%.

Field Case Histories—Operations

Fullbright, G.D. *et al.*: "Evolution of Conformance Improvement Efforts in a Major CO_2 WAG Injection

Project," paper SPE 35361 presented at the 1996 SPE/DOE Improved Oil Recovery Symposium, Tulsa, Oklahoma, 21–24 April.

The Rangely Weber Sand Unit is a giant oil field in northwestern Colorado. Field development has evolved through many stages, including infill drilling, waterflood expansions, and CO_2 injection. Conformance improvement has become a major strategy as CO_2 injection matures and production and oil cuts decline. A multidisciplinary team was developed to focus on the conformance issue. As with any effort to change the focus of an operation, conformance improvement requires a detailed study of scope of potential benefits, which then must be communicated to the decision makers. This paper summarizes the Rangely Conformance Improvement Team's effort, including (1) organizational approach to the effort, (2) strategic alignment of conformance improvement with operating-team priorities, (3) assessment of the state of conformance and the tools developed for problem identification and opportunity ranking, (4) conformance-improvement selection, (5) results to date, and (6) future conformance-improvement efforts and challenges.

Onan, D.D.: "Effects of Supercritical Carbon Dioxide on Well Cements," paper SPE 12593 presented at the 1984 SPE Permian Basin Oil and Gas Recovery Conference, Midland, Texas, 8–9 March.

Widespread implementation of large-scale miscible CO_2 injection projects has raised concerns regarding the performance and/or durability of hydrated cement located across producing and injection intervals in CO_2-affected wells. Carbonation is known to affect the microstructure of cement, altering its porosity and compressive strength. CO_2 reactivity is characterized by chemical composition and such variables as partial pressure, temperature, and relative humidity. However, the effects of these variables on Portland cement still are not clearly understood, giving rise to contradictory opinions. This led to the need for and development of a laboratory program to examine the effects of supercritical CO_2 on preset cement and the influence of carbonation on the early stages of the cement-hydration process. Results show that, after prolonged exposure to CO_2 at supercritical conditions, the hydration products formed in hydration of common Portland cement undergo decomposition into calcium carbonate and a siliceous residue. Cement samples exposed to lower extremes (temperature and pressure) of a supercritical CO_2 environment exhibited greater reactivity under dynamic conditions compared with those tested under static conditions. Increasing CO_2 pressure increased the degree of reaction, regardless of carbonation conditions used.

Schneider, B.D., Hogan, G.P. II, and Holt, W.N. Jr.: "Pulsed-Neutron Decay-Spectrum Data and Multi-Inflatable Packer Plugdown Assemblies Improve Oil Production Rates," *SPERE* (August 1997) 169.

A gravity-stable, vertical CO_2 miscible injection project was implemented in the 824-ft-thick Wellman Unit. The reservoir is completed in the Wolfcamp formation, which was deposited in a reef environment. CO_2 injection began in July 1983. CO_2 is injected into the crest of the reservoir to displace the oil vertically downward. The producing wells constantly are plugged down out of the "gassed-out" interval into the oil bank. The paper describes the use of a through-tubing logging tool to determine CO_2/oil/water contacts and the use of multiple-inflatable-packer/sliding-sleeve plugdown assemblies to achieve zonal isolation between gassed-out and producing intervals. Previous cement squeezes were unsuccessful because of poor detection of the gas/oil interface. In addition, isolating the target zone had previously been difficult because of poor wellbore integrity.

Analysis, Prediction, and Optimization

Brinkman, F. P. *et al.*: "Use of Full-Field Simulation To Design a CO_2 Miscible Flood," *SPEREE* (June 1999) 230.

This paper illustrates a method for using full-field reservoir simulation to evaluate CO_2 flooding in the carbonate reservoir of the Sharon Ridge Canyon Unit in west Texas. The method combines three different full-field 3D models, a very-finely-gridded geological model, a coarsely gridded waterflood model, and a fine-grid CO_2 flood model. In addition, four finely-gridded-type-pattern models representing one-quarter of an inverted nine-spot pattern are used to help assess gridding sensitivity for the CO_2 flood simulations. The geologic model contains more than 4 million cells that were 200 ft on a side and 1 ft thick. The hydrocarbon column averaged 115 ft and ranged up to 450 ft. The model gave an improved estimate of OOIP as well as an improved picture of stratigraphic heterogeneity. Properties of the geologic model were scaled up to a coarser $40 \times 79 \times 19$ model, which was used for primary depletion and waterflood history matching. After many gridblocks were made inactive around the edges of the field, the coarse-grid model had approximately 20,000 active cells. Areal gridblocks ranged in size from 5 to 10 acres with at least one gridblock between wells. To obtain a satisfactory history match, the geologic model permeabilities had to be increased by a factor of 5 to 10. The model highlighted areas of the field with high oil saturations as priority targets for CO_2 flooding. Saturation and pressure distributions from the waterflood model were used to initialize subsequent more-detailed CO_2-flood simulations. The CO_2-flood simulator had four components: oil, water, solution gas, and CO_2. Parameters entered into the simulator included MMP as a function of solution-gas content in the total gas and minimum oil saturation to CO_2 flooding. Four type models representing one-quarter of an inverted five spot were used to assess sensitivity to areal gridding as well as other important design parameters. On the basis of these studies, a 76x10x15 full-field model with approximately 63,000 active gridblocks was used for the final CO_2-flood evaluations. A 15-year prediction with the model took approximately 6 calendar days on the workstation. The model predicted that a 70% HCPV slug gave incremental recovery of approximately 13% OOIP, gross CO_2 utilization of 13 Mcf/STB, and net utilization of 4 Mcf/STB.

Chopra, A.K., Stein, M.H., and Dismuke, C.T. "Prediction of Performance of Miscible-Gas Pilots," *JPT* (December 1990) 1564.

A model was developed to predict field performance of miscible-gas-injection projects. A critical component of the model is inclusion of a solvent relative permeability, different from the oil relative permeability, in the presence of a final oil saturation. The model of CO_2 displacement was included within a reservoir simulator, operating in modified black-oil mode, and used to match a miscible-gas-injection pilot. A satisfactory match of both production- and injection-well performance (reduced injectivity in particular) was obtained. Pilot reservoir description, obtained from a match of primary and waterflood performance, was not altered to achieve the CO_2-flood match. Interested readers can follow up on the pilot description by referring to a later paper by one of the authors (Stein) that is included as one of the reprints in this volume.

Flanders, W.A., McGinnis, R.A., and Shatto, A.G.: "CO_2 EOR Economics for Small-to-Medium-Size Fields," paper SPE 26391 presented at the 1993 SPE Annual Technical Conference and Exhibition, Houston, 3–6 October.

CO_2 projects undertaken in large fields have benefited from economy of scale and substantial remaining life. Smaller, more mature fields may not have these advantages. However, because CO_2 source fields and pipeline distribution systems are now in place, smaller fields do not have to bear these costs and may be candidates for CO_2 EOR. This paper gives information for evaluating the economic viability of small-to-medium-size fields. It first presents production histories and reviews the similarities and differences of six typical projects: Seminole field and Wasson Denver Unit in the San Andres formation, North Cross and Cordona Lake in the Devonian formation, and Hansford Marmaton and Two Freds East in sandstones. Next, it reviews ranges of startup costs (e.g., conversion of wells, well workovers, installation of CO_2-injection equipment and lines, and modifications to production facilities and flowlines) and operating costs (e.g., well operation and gas and CO_2 recycling). Finally, it develops a base case that uses actual production data from a small CO_2 project as well as equipment and operating costs in the range of ongoing projects. Sensitivity to ranges of these factors is discussed. The paper concludes that economic viability of CO_2 EOR is very field specific. Although cost parameters become dated very quickly, the paper presents an enlightening discussion of factors that go into evaluating the economics of CO_2 projects in small fields where large, expensive simulation studies may not be warranted.

Masoner, L.O., Abidi, H.R., and Hild, G.P.: "Diagnosing CO_2 Flood Performance Using Actual Performance Data," paper SPE 35363 presented at the 1996 SPE/DOE Improved Oil Recovery Symposium, Tulsa, Oklahoma, 21–24 April.

The Rangely Weber Sand Unit in western Colorado is one of the largest CO_2 floods. Conducting day-to-day management of the more than 260-pattern flood with large-scale simulation is impractical. This paper discusses ways to analyze pattern data and performance measures with off-the-shelf software tools. Pattern performance is built by allocating a percent of the total production from each impacted offset producer and 100% of the injection with the objective of analyzing each pattern's financial and technical performance. Three methods of allocating production are described and evaluated: areal, streamtube model, and engineering judgment. Engineering judgment was thought to be the most accurate. The authors distinguish between "economic" incremental oil (the difference between CO_2-flood and waterflood recovery estimated by extrapolating waterflood performance based on waterflood-processing rate) and "technical" incremental oil (the difference from extrapolating waterflood performance based on WAG-processing rate). They conclude that the value of economic incremental oil is useful in post-mortem evaluation while value of technical incremental oil is necessary for estimating current benefits and making future decisions. The paper also examines allocated-pattern performance measures, such as injected-fluid volume, injection/production ratio, processing-rate index, cash flow per day, operating expense per barrel of oil, CO_2 retention and utilization, natural gas liquid recovery, GOR, WAG ratio, and half-cycle slug size. These measures yield insights into such things as allocation-factor quality, total and incremental cash flow and expense, optimal WAG ratios and cycle slug sizes, contacted PV, pattern sweep efficiency, commitment to conformance-improvement activities, and cessation CO_2 injection.

Phase Behavior

Fong, W.S. *et al.*: "Phase-Behavior Modeling Techniques for Low-Temperature CO_2 Applied to McElroy and North Ward Estes Projects," paper SPE 24184 presented at the 1992 SPE/DOE Enhanced Oil Recovery Symposium, Tulsa, Oklahoma, 22–24 April.

An approximate tuning procedure has been developed that uses a two-phase equation of state (EOS) to characterize the displacement behavior of CO_2 in the three-phase domain. The technique results in good prediction of laboratory slim-tube displacement tests and the MMP for both pure and impure CO_2 streams. Laboratory corefloods and field-scale displacements can also be modeled at a practical level of accuracy. With this pseudo-two-phase representation, complex three-phase flash calculations and the need for four-phase relative permeabilities are avoided. The approach was used to forecast the North Ward Estes CO_2 project, and reasonable agreement with observed field performance was obtained. An immiscible CO_2 huff 'n' puff test in the McElroy field was examined as well.

Hagedorn, K.D. and Orr, F.M. Jr.: "Component Partitioning in CO_2/Crude-Oil Systems: Effects of Oil Composition on CO_2 Displacement Performance," *SPE Advanced Technology Series* (April 1994) 177.

This paper addresses the importance of molecular structure on the development of CO_2 miscibility and presents dissociation, *K*, values determined from gas chromatography/mass spectrometry (GC/MS) for more than 250 components in CO_2/crude-oil mixtures. The data show that both molecular size and structure affect the way a molecule partitions into a CO_2-rich phase. Multiring aromatic compounds partition much less favorably than other components with similar size or GC elution times. GC/MS analysis of two crude oils with a similar carbon-number distribution but different slim-tube displacement behavior showed that the oil with a significant quantity of multiring aromatics produced much lower slim-tube recovery at a given pressure and had a correspondingly higher MMP. EOS compositional simulations indicate that, when oil contains significant quantities of multiring aromatic compounds, the properties of the multiring aromatics must be included as a separate component in an oil characterization to reproduce the behavior observed in single-contact partitioning experiments and slim-tube displacements.

Turek, E.A., Metcalfe, R.S., and Fishback, R.E.: "Phase Behavior of Several CO_2/West Texas Reservoir-Oil Systems," *SPERE* (May 1988) 505.

Measured phase equilibrium and volumetric properties are reported for several west Texas oil/CO_2 systems. Reservoir oils studied were from the San Andres, Grayburg, and Devonian Chert formations. Both static (single-contact) and multiple-contact measurements were conducted in a visual cell. Static data cover a wide range of CO_2 compositions and provide a general understanding of CO_2/reservoir-oil phase behavior. Multiple-contact measurements where the CO_2-rich phase is repeatedly put in contact with reservoir oil simulate phase behavior occurring at a flood front; multiple-contact measurements where the oil-rich phase is repeatedly put in contact with pure CO_2 simulate phase behavior exhibited by residual oil near a CO_2-injection well. The multiple-contact data cover narrow compositional paths encountered in displacement processes and serve as a basis for EOS evaluation. Phase-behavior trends common to all systems are discussed.

Wang, Y. and Orr, F.M. Jr.: "Calculation of Minimum Miscibility Pressure," paper SPE 39683 presented at the 1998 SPE/DOE Improved Oil Recovery Symposium, Tulsa, Oklahoma, 19–22 April.

This paper discusses a way to calculate accurately the MMP for multicomponent systems with an arbitrary number of components. The procedure uses 1D, dispersion-free theory (method of characteristics) for multicomponent mixtures. An accurate EOS is required. The paper describes the numerical approach for making the MMP estimates and shows excellent comparisons between calculation results and literature-reported MMP measurements.

Injectivity, Displacement Mechanisms, and Relative Permeability

Dria, D.E., Pope, G.A., and Sepehrnoori, K.: "Three-Phase Gas/Oil/Brine Relative Permeabilities Measured Under CO_2 Flooding Conditions," *SPERE* (May 1993) 143.

This paper reports the results of experimental measurements of three-phase CO_2/oil/brine relative permeabilities in water-wet Baker dolomite, a sedimentary grainstone dolomite with intergranular porosity, cut from an outcrop. The three-phase relative permeability measurements were made under steady-state injection of brine, decane, and CO_2 after the core was driven to residual water saturation and after a subsequent steady-state saturation was achieved from brine/oil injection. Measurements were made at 71°C and average core pressures of approximately 9.65 MPa to simulate phase behavior at reservoir conditions. The typical description of three-phase gas/oil/brine flow is that one liquid phase strongly wets the rock, the gas phase is totally nonwetting, and the second liquid phase is of intermediate wettability. This assumes that the relative permeabilities of the wetting and nonwetting phases depend only on their saturations and that the relative permeability of the intermediate wettability phase depends on two saturations. However, the results show that the relative permeability of each phase in this study depends only on the saturation of that phase and also that the relative permeabilities of the CO_2- and oil-rich phases are nearly identical. The results further show that significant differences exist between the three-phase gas/oil/brine relative permeabilities when the gas is CO_2 and when the gas is N_2. The authors speculate that this difference may be because the oil and gas phases are much more alike in the CO_2/oil system than in the N_2/oil systems.

Pande, K.K. and Orr, F.M. Jr.: "Interaction of Phase Behavior, Reservoir Heterogeneity, and Crossflow in CO_2 Floods," paper SPE 19668 presented at the 1989 SPE Annual Technical Conference and Exhibition, San Antonio, Texas, 8–11 October.

Much of the theoretical basis for understanding CO_2 floods is restricted to 1D flow in homogeneous media. This paper seeks to extend this understanding to heterogeneous media, although the form of the heterogeneity (two uniform layers in perfect communication with each other) is ideal. The results are based on an extension of the method of characteristics to media in vertical (quasistatic) equilibrium. Results indicate that displacements that would be multicontact miscible (MCM) in 1D media with essentially no two-phase flow would exhibit some two-phase flow in heterogeneous media because viscous crossflow causes the composition path to mix deeper into the multiphase region. The increased mixing caused by crossflow leaves a larger oil saturation to the CO_2 displacement than would be the case for MCM displacement in one dimension.

Patel, P.D., Christman, P.G., and Gardner, J.W.: "Investigation of Unexpectedly Low Field-Observed Fluid Mobilities During Some CO_2 Tertiary Floods," *SPERE* (November 1987) 507.

Phase behavior, inorganic precipitation, and wettability are investigated as possible reasons for the unexpectedly low field-observed mobilities during some CO_2 floods (in particular, the Denver Unit, Wasson field CO_2 pilot). The observed mobility was not a near-wellbore effect and probably played a major role in reservoir sweep; the low effective CO_2 relative permeability offset the detrimentally low CO_2 viscosity. Experimental and simulation studies, supplemented by literature data, led the authors to conclude that rock wettability could not be the root cause of these low fluid mobilities. Phase-behavior effects, while they may play a role, are not necessary to explain the injectivity behavior, and inorganic precipitates probably have little effect under the conditions investigated. Interested readers can investigate the subject of injectivity further by referring to a later paper by Christman (and coauthor Gorell) that appears in this volume.

Prieditis, J. and Brugman, R.J.: "Effects of Recent Relative Permeability Data on CO_2-Flood Modeling," paper SPE 26650 presented at the 1993 SPE Annual Technical Conference and Exhibition, Houston, 3–6 October.

This paper shows that modeling and prediction of tertiary CO_2-flood performance can be improved if the reported results of laboratory CO_2-displacement tests at representative reservoir conditions are used. The calculations are made by modifying the relative permeability treated in a mixing-parameter-type reservoir simulator (the authors acknowledge that some differences may exist in the way relative permeabilities influence results in compositional simulators that are often used). Specifically, the study investigates the influence of reduced CO_2 relative permeability, residual oil saturation to a miscible flood, residual CO_2 saturation, and water and oil relative permeability hysteresis. Four methods to represent CO_2 relative permeability were investigated: (1) same as the oil relative permeability; (2) saturation-weighted average of two-phase oil relative permeability in the presence of water, k_{row}, and two-phase gas relative permeability in the presence of oil, k_{rg}; (3) simple average of k_{row} and k_{rg}; and (4) equal to either k_{row} or k_{rg}, depending on which is the larger saturation. The calculations were made for a one-quarter five-spot pattern representative of west Texas floods. The mixing parameter was held constant at a value of two-thirds, and WAG was modeled by simultaneous injection. The authors concluded that the best predictions were made with the saturation-weighted method and if all mechanisms were included. When this is done, predicted oil recoveries were very close to the original mixing-parameter-model predictions; however, differences in predicted gas production and injectivity were significant. The presence of a large residual CO_2 saturation had a significant effect on predicted CO_2-flood performance during the final chase waterflood.

Wegener, D.C. and Harpole, K.J.: "The Determination of Relative Permeability and Trapped-Gas Saturation for Predictions of WAG Performance in the South Cowden CO_2 Flood," paper SPE 35429 presented at the 1996 SPE/DOE Improved Oil Recovery Symposium, Tulsa, Oklahoma, 21–24 April.

CO_2 relative permeability, trapped-gas saturation, and hysteresis effects are key parameters in determining injectivity and displacement efficiency in a miscible CO_2 WAG injection project. This paper is an experimental/numerical study to determine these parameters for the South Cowden field flood. A unique feature of this work is that the cores were screened with magnetic-resonance imaging before the experimental work to eliminate cores with gross heterogeneity. Trapped-gas saturation in the main South Cowden facies was found to be 20 to 25%. In addition, measured CO_2 relative permeabilities were significantly lower than oil relative permeabilities at comparable water saturations. The impact of values and an observed reduction in water relative permeability after CO_2 flooding was determined with numerical simulation.

SI Metric Conversion Factors

acre × 4.046 856	E − 01	= ha
bbl × 1.589 873	E − 01	= m^3
ft × 3.048*	E − 01	= m
ft^3 × 2.831 685	E − 02	= m^3
°F (°F − 32)/1.8		= °C
psi × 6.894 757	E + 00	= kPa

*Conversion factor is exact.

Bibliography

Papers appearing in this Reprint volume are indicated with an **[R]** after the entry, and papers included in the Annotated Bibliography are indicated with an **[A]** after the entry.

Alderman *et al.*: "Field Facilities for CO_2 Tertiary Floods: A Conceptual Design," paper SPE 11234 presented at the 1982 SPE Annual Technical Conference and Exhibition, New Orleans, 26–29 September.

Attanucci, V. *et al.*: "WAG Process Optimization in the Rangely CO_2 Flood," paper SPE 26622 presented at the 1993 SPE Annual Technical Conference and Exhibition, Houston, 3–6 October.

Bangia, V.K., Yau, F.F., and Hendricks, G.R.: "Reservoir Performance of a Gravity-Stable Vertical CO_2 Miscible Flood: Wolfcamp Reef Reservoir, Wellman Unit," *SPERE* (November 1993) 261; *Trans.*, AIME, **295.**

Beliveau, D., Payne, D.A., and Mundry, M.: "Waterflood and CO_2 Flood of the Fractured Midale Field," *JPT* (September 1993) 881.

Bellavance, J.F.R.: "Dollarhide Devonian CO_2 Flood: Project Performance Review 10 Years Later," paper SPE 35190 presented at the 1996 SPE Permian Basin Oil and Gas Recovery Conference, Midland, Texas, 27–29 March. **[R]**

Borling, D.C.: "Injection Conformance Control Case Histories Using Gels at the Wertz Field CO_2 Tertiary Flood in Wyoming," paper SPE 27825 presented at the 1994 SPE/DOE Improved Oil Recovery Symposium, Tulsa, Oklahoma, 17–20 April.

Bowser, P.D. *et al.*: "Innovative Techniques for Converting Old Waterflood Injectors to State-of-the-Art CO_2 Injectors," paper SPE 18976 presented at the 1989 SPE Rocky Mountain Region/Low Permeability Reservoirs Symposium, Denver, Colorado, 6–8 March.

Brinkman, F.P. *et al.*: "Use of Full-Field Simulation To Design a Miscible CO_2 Flood," *SPEREE* (June 1999) 230. **[A]**

Brokmeyer, R.J., Borling, D.C., and Pierson, W.T.: "Lost Soldier Tensleep CO_2 Tertiary Project, Performance Case History; Bairoil, Wyoming," paper SPE 35191 presented at the 1996 SPE Permian Basin Oil and Gas Recovery Conference, Midland, Texas, 27–29 March. **[R]**

Chang, Y-B. *et al.*: "CO_2 Flow Patterns Under Multiphase Flow: Heterogeneous Field-Scale Conditions," *SPERE* (August 1994) 208. **[R]**

Chang, Y-B., Coats, B.K., and Nolen, J.S.: "A Compositional Model for CO_2 Floods Including CO_2 Solubility in Water," paper SPE 35164 presented at the 1996 SPE Permian Basin Oil and Gas Recovery Conference at Midland, Texas, 27–29 March.

Chase, C.A. Jr. and Todd, M.R.: "Numerical Simulation of CO_2 Performance," *SPEJ* (December 1984) 597.

Chopra, A.K., Stein, M.H., and Dismuke, C.T.: "Prediction of Performance of Miscible-Gas Pilots," *JPT* (December 1990) 1564. **[A]**

Chou, S.I. *et al.*: "CO_2 Foam Field Trial at North Ward-Estes," paper SPE 24643 presented at the 1992 SPE Annual Technical Conference and Exhibition, Washington, DC, 4–7 October.

Christiansen, R.L.: "Gasflooding Experiments for the East Side of the Yates Field Unit," *SPERE* (February 1990) 14.

Christie, M.A., Muggeridge, A.H., and Barley, J.J.: "3D Simulation of Viscous Fingering and WAG Schemes," *SPERE* (February 1993) 19; *Trans.*, AIME, **295.**

Christman, P.G. and Gorell, S.B.: "Comparison of Laboratory and Field Observed CO_2 Tertiary Injectivity," *JPT* (February 1990) 226; *Trans.*, AIME, **289. [R]**

Collier, T.S.: "State-of-the-Art Installation of CO_2 Injection Equipment: A Case Study," paper SPE 17294 presented at the 1988 SPE Permian Basin Oil and Gas Recovery Conference, Midland, Texas, 10–11 March.

Creek, J.L. and Sheffield, J.M.: "Phase Behavior, Fluid Properties, and Displacement Characteristics of Permian Basin Reservoir Fluid/CO_2 Systems," *SPERE* (February 1993) 34.

Dai, K.K. and Orr, F.M. Jr.: "Prediction of CO_2 Flood Performance: Interaction of Phase Behavior With Microscopic Pore Structure Heterogeneity," *SPERE* (November 1987) 531; *Trans.*, AIME, **283.**

Dria, D.E., Pope, G.A., and Sepehrnoori, K.: "Three-Phase Gas/Oil/Brine Relative Permeabilites Measured Under CO_2 Flooding Conditions," *SPERE* (May 1993) 143; *Trans.*, AIME, **295. [A]**

El-Saleh, M.: "Analogy Procedure for the Evaluation of CO_2 Flooding Potential for Reservoirs in the Permian and Delaware Basins," paper SPE 35391 presented at the 1996 SPE/DOE Improved Oil Recovery Symposium, Tulsa, Oklahoma, 21–24 April.

Flanders, W.A. and DePauw, R.M.: "Update Case History: Performance of the Two Freds Tertiary CO_2 Project," paper SPE 26614 presented at the 1994 SPE Permian Basin Oil and Gas Recovery Conference, Midland, Texas, 16–18 March. **[A]**

Flanders, W.A., Stanberry, W.A., and Martinez, M.: "CO_2 Injection Increases Hansford Marmaton Production," *JPT* (January 1990) 68.

Flanders, W.A., McGinnis, R.A., and Shatto, A.G.: "CO_2 EOR Economics for Small-to-Medium-Size Fields", paper SPE 26391 presented at the 1993 SPE Annual Technical Conference and Exhibition, Houston, 3–6 October. **[A]**

Fleming, E.A., Brown, L.M., and Cook, R.L.: "Overview of Production Engineering Aspects of Operating the Denver Unit CO_2 Flood," paper SPE 24157 presented at the 1992 SPE/DOE Enhanced Oil Recovery Symposium, Tulsa, Oklahoma, 22–24 April. **[R]**

Folger, L.K. and Guillot, S.N.: "A Case Study of the Development of the Sundown Slaughter Unit CO_2 Flood, Hockley County, Texas," paper SPE 35189 presented at the 1996 SPE Permian Basin Oil and Gas Recovery Conference, Midland, Texas, 27–29 March.

Fong, W.S. *et al.*: "Phase Behavior Modeling Techniques for Low-Temperature CO_2 Applied to McElroy and North Ward Estes Projects," paper SPE 24184, presented at the 1992 SPE/DOE Enhanced Oil Recovery Symposium, Tulsa, Oklahoma, 22–24 April. **[A]**

Fox, C.E., Tugwell, J.O., and Morell, J.I.: "The Denver Unit CO_2 Flood Transforms Former Waterflood Injectors Into Oil Producers," paper SPE 27674 presented at the 1994 SPE Permian Basin Oil and Gas Recovery Conference, Midland, Texas, 16–18 March.

Fullbright, G.D. *et al.*: "Evolution of Conformance Improvement Efforts in a Major CO_2 WAG Injection Project," paper SPE 35361 presented at the 1996 SPE/DOE Improved Oil Recovery Symposium, Tulsa, Oklahoma, 21–24 April. **[A]**

Gardner, J.W., Orr, F.M. Jr., and Patel, P.D.: "The Effect of Phase Behavior on CO_2-Flood Displacement Efficiency," *JPT* (November 1981) 2067.

Hadlow, R.E.: "Update of Industry Experience With CO_2 Injection," paper SPE 24928 presented at the 1992 SPE Annual Technical Conference and Exhibition, Washington, DC, 4–7 October. **[R]**

Hagedorn, K.D. and Orr, F.M. Jr.: "Component Partitioning in CO_2/Crude Oil Systems: Effects of Oil Composition on CO_2 Displacement Performance," *SPE Advanced Technology Series* (April 1994) 177. **[A]**

Harpole, K.J. and Hallenbeck, L.D.: "East Vacuum Grayburg San Andres Unit CO_2 Flood Ten-Year Performance Review: Evolution of a Reservoir Management Strategy and Results of WAG Optimization," paper SPE 36710 presented at the 1996 SPE Annual Technical Conference and Exhibition, Denver, Colorado, 6–9 October. **[A]**

Hawkins, J.T. *et al.*: "SACROC Unit CO_2 Flood: Multidisciplinary Team Improves Reservoir Management and Decreases Operating Costs," *SPERE* (August 1996) 141.

Hazlett, R.D., Furr, M.J., and Navarro, R: "A Correlation for Miscible Flood Displacement Efficiency in the San Andres With NMR Relaxation," *SPE Advanced Technology Series* (April 1994) 185.

Healy, R.N., Holstein, E.D., and Batycky, J.P.: "Status of Miscible Flooding Technology," *Proc.*, 14th World Pet. Cong., Stavanger (1994) **2.**

Hervey, J.R. and Iakovakis, A.C.: "Performance Review of a Miscible CO_2 Tertiary Project: Rangely Weber Sand Unit, Colorado," *SPERE* (May 1991) 163.

Hill, W.J. *et al.*: "CO_2 Operating Plan, South Welch Unit, Dawson County, Texas," paper SPE 27676 presented at the 1994 SPE Permian Basin Oil and Gas Recovery Conference, Midland, Texas, 16–18 March.

Hsu, C-F., Morell, J.I., and Falls, A.H.: "Field-Scale CO_2-Flood Simulations and Their Impact on the Performance of the Wasson Denver Unit," *SPERE* (February 1997) 4. **[R]**

Hsu, C.-F. *et al.*: "Design and Implementation of a Grass-Roots CO_2 Project for the Bennet Ranch Unit," paper SPE 35188 presented at the 1996 SPE Permian Basin Oil and Gas Recovery Conference, Midland, Texas, 27–29 March.

Huang, E.T.S.: "The Effect of Oil Composition and Asphaltene Content on CO_2 Displacement," paper SPE 24131 presented at the 1992 SPE/DOE Symposium on Enhanced Oil Recovery, Tulsa, Oklahoma, 22–24 April.

Johns, R.T. and Orr, F.M. Jr.: "Miscible Gas Displacement of Multicomponent Oils," *SPE Journal* (March 1996) 39.

Johnston, J.R.: "Weeks Island Gravity-Stable CO_2 Pilot," paper SPE 17351 presented at the 1988 SPE/DOE Enhanced Oil Recovery Symposium, Tulsa, Oklahoma, 17–20 April.

Jonas, T.M., Chou, S.I., and Vasicek, S.L.: "Evaluation of a CO_2 Foam Field Trial: Rangely Weber Sand Unit," paper SPE 20468 presented at the 1990 SPE Annual Technical Conference and Exhibition, New Orleans, 23–26 September.

Khan, S.A., Pope, G.A., and Sepehrnoori, K.: "An Expert System for Miscible Gas Flooding," *SPECA* (February 1993) 11.

Khan, S.A., Pope, G.A., and Sepehrnoori, K.: "Fluid Characterization of Three-Phase CO_2/Oil Mixtures," paper SPE 24130 presented at the 1992 SPE/DOE Enhanced Oil Recovery Symposium, Tulsa, Oklahoma, 22–24 April.

Kleinstelber, S.W.: "The Wertz Tensleep CO_2 Flood: Design and Initial Performance," *JPT* (May 1990) 630; *Trans.*, AIME, **289.**

Kremesec, V.J. Jr. and Sebastian, H.M.: "CO_2 Displacements of Reservoir Oils From Long Berea Cores: Laboratory and Simulation Results," *SPERE* (May 1988) 496; *Trans.*, AIME, **285.**

Langston, M.V., Hoadley, S.F., and Young, D.N.: "Definitive CO_2 Flooding Response in the SACROC Unit," paper SPE 17321 presented at the 1988 SPE/DOE Enhanced Oil Recovery Symposium, Tulsa, Oklahoma, 17–20 April. **[R]**

Lee, K.H. and El-Saleh, M.M.: "A Full-Field Numerical Modeling Study for the Ford Geraldine Unit CO_2 Flood," paper SPE 20227 presented at the 1990 SPE/DOE Enhanced Oil Recovery Symposium, Tulsa, Oklahoma, 22–25 April.

Lin, M.T. *et al.*: "Simulation of Carbon Dioxide Flooding Using Horizontal Wells," paper SPE 24929 presented at the 1992 SPE Annual Technical Conference and Exhibition, Washington, DC, 4–7 October.

Linn, L.R.: "CO_2 Injection and Production Field Facilities Design Evaluation and Considerations," paper SPE 16830 presented at the 1987 SPE Annual Technical Conference and Exhibition, Dallas, 27–30 September. **[R]**

Magruder, J.B., Stiles, L.H., and Yelverton, T.D.: "Review of the Means San Andres CO_2 Tertiary Project," *JPT* (May 1990) 638; *Trans.*, AIME, **289.**

Masoner, L.O.: "Modeling CO_2 Recovery Using Solubility-Adjusted Saturations and a Two-Flow-Regime Geological Model," paper SPE 35453 presented at the 1996 SPE/DOE Improved Oil Recovery Symposium, Tulsa, Oklahoma, 21–24 April. **[A]**

Masoner, L.O. and Wackowski, R.K.: "Rangely Weber Sand Unit CO_2 Project Update," *SPERE* (August 1995) 203; *Trans.*, AIME, **299. [A]**

Masoner, L.O., Abidi, H.R., and Hild, G.P.: "Diagnosing CO_2 Flood Performance Using Actual Performance Data," paper SPE 35363 presented at the 1996 SPE/DOE Improved Oil Recovery Symposium, Tulsa, Oklahoma, 21–24 April.

Merchant, D.H., and Thakur, S.C.: "Reservoir Management in Tertiary CO_2 Floods," paper SPE 26624 presented at the 1993 SPE Annual Technical Conference and Exhibition, Houston, 3–6 October.

Merritt, M.B. and Groce, J.F.: "A Case History of the Hanford San Andres Miscible CO_2 Project," *JPT* (August 1992) 924.

Mizenko, G.J.: "North Cross (Devonian) Unit CO_2 Flood: Status Report," paper SPE 24210 presented at the 1992 SPE/DOEEnhanced Oil Recovery Symposium, Tulsa, Oklahoma, 22–24 April. **[R]**

Moffitt, P.D. and Zornes, D.R.: "Postmortem Analysis: Lick Creek Meakin Sand Unit Immiscible CO_2 Waterflood Project," paper SPE 24933 presented at the 1992 Annual Technical Conference and Exhibition, Washington, DC, 4–7 October. **[R]**

Müller, T. and Lake, L.W.: "Theoretical Study of Water-Blocking in Miscible Flooding," *SPERE* (November 1991) 445.

Negahban, S. and Kremesec, V.J. Jr.: "Development and Validation of Equation-of-State Fluid Descriptions for CO_2/Reservoir-Oil Systems," *SPERE* (August 1992) 363. **[R]**

Negahban, S., Shiralkar, G.S., and Gupta, S.P.: "Simulation of the Effects of Mixing in Gasdrive Core Tests of Reservoir Fluids," *SPERE* (August 1990) 402.

Nghiem, L.X. and Li, Y.K.: "Effect of Phase Behavior on CO_2 Displacement Efficiency at Low Temperatures: Model Studies With an Equation of State," *SPERE* (July 1986) 414.

Onan, D.D.: "Effects of Supercritical Carbon Dioxide on Well Cements," paper SPE 12593 presented at the 1984 SPE Permian Basin Oil and Gas Recovery Conference, Midland, Texas, 8–9 March. **[A]**

Pande, K.K. and Orr, F.M. Jr.: "Interaction of Phase Behavior, Reservoir Heterogeneity, and Crossflow in CO_2 Floods," paper SPE 19668 presented at the 1989 SPE Annual Technical Conference and Exhibition, San Antonio, Texas, 8–11 October. **[A]**

Pariani, G.J. *et al.*: "An Approach To Optimize Economics in a West Texas CO_2 Flood," *JPT* (September 1992) 984; *Trans.*, AIME, **293. [R]**

Patel, P.D., Christman, P.G., and Gardner, J.W.: "An Investigation of Unexpectedly Low Field-Observed Fluid Mobilities During Some CO_2 Tertiary Floods," *SPERE* (November 1987) 507. **[A]**

Perry, G.E.: "Weeks Island 'S' Sand Reservoir B Gravity-Stable Miscible CO_2 Displacement, Iberia Parish, Louisiana," paper SPE 10695 presented at the 1982 SPE/DOE Enhanced Oil Recovery Symposium, Tulsa, Oklahoma, 4–7 April.

Pittaway, K.R. and Rosato, R. J.: "The Ford Geraldine Unit CO_2 Flood: Update 1990," *SPERE* (November 1991) 410.

Power, M.T., Leicht, M.A., and Barnett, K.L.: "Converting Wells in a Mature West Texas Field for CO_2 Injection," *SPEDE* (December 1990) 306; *Trans.*, AIME, **289.**

Prieditis, J. and Brugman, R.J.: "Effects of Recent Relative Permeability Data on CO_2 Flood Modeling," paper SPE 26650 presented at the 1993 SPE Annual Technical Conference and Exhibition, Houston, 3–6 October. **[A]**

Prieditis, J., Wolle, C.R., and Notz, P.K.: "A Laboratory and Field Injectivity Study: CO_2 WAG in the San Andres Formation of West Texas," paper SPE 22653 presented at the 1991 SPE Annual Technical Conference and Exhibition, Dallas, 6–9 October. **[R]**

Ring, J.N. and Smith, D.J.: "An Overview of the North Ward Estes CO_2 Flood," paper SPE 30729 presented at the 1995 SPE Annual Technical Conference and Exhibition, Dallas, 22–25 October. **[R]**

Roper, M.K. Jr., Pope, G.A., and Sepehrnoori, K.: "Analysis of Tertiary Injectivity of Carbon Dioxide," paper SPE 23974 presented at the 1992 SPE/DOE Enhanced Oil Recovery Symposium, Tulsa, Oklahoma, 22–24 April.

Roper, M.K. *et al.*: "Interpretation of a CO_2 WAG Injectivity Test in the San Andres Formation Using a Compositional Simulator," paper SPE 24163 presented at the 1992 SPE/DOE Enhanced Oil Recovery Symposium, Tulsa, Oklahoma, 22–24 April.

Schneider, B.D., Hogan, G.P. II, and Holt, W.N. Jr.: "Pulsed-Neutron Decay-Spectrum Data and Multi-Inflatable-Packer Plugdown Assemblies Improve Oil Production Rates" *SPERE* (August 1997) 169. **[A]**

Schwanitz, B.J.: "A Combination of Logging Measurements for Quantitative Monitoring of CO_2 Floods," paper SPE 17276 presented at the 1988 SPE Permian Basin Oil and Gas Recovery Conference, Midland, Texas, 10–11 March.

Sebastian, H.M., Wenger, R.S., and Renner, T.A.: "Correlation of Minimum Miscibility Pressure for Impure CO_2 Streams," *JPT* (November 1985) 2076.

Sharma A.K. and Clements, L.E.: "From Simulator to Field Management: Optimum WAG Application in a West Texas CO_2 Flood: A Case History," paper SPE 36711 presented at the 1996 SPE Annual Technical Conference and Exhibition, Denver, Colorado, 6–9 October.

Shaw, E.C.: "A Simple Technique To Forecast CO_2 Flood Performance," paper SPE 23975 presented at the 1992 SPE Permian Basin Oil and Gas Recovery Conference, Midland, Texas, 18–20 March.

Shyeh-Yung, J-G.J.: "Mechanisms of Miscible Oil Recovery: Effects of Pressure on Miscible and Near-Miscible Displacements of Oil by Carbon Dioxide," paper SPE 22651 presented at the 1991 SPE Annual Technical Conference and Exhibition, Dallas, 6–9 October. **[R]**

Sides, W.S. III: "Injection Safety Valve Solutions for CO_2 WAG Cycle Wells," paper SPE 24000 presented at the 1992 SPE Permian Basin Oil and Gas Recovery Conference, Midland, Texas, 18–20 March.

Stadler, M.P., Stessman, A.N., and Orr, F.M. Jr.: "Prediction of CO_2/Crude-Oil Phase Behavior Using Supercritical Fluid Chromatography," paper

SPE 24111 presented at the 1992 SPE/DOE Enhanced Oil Recovery Symposium, Tulsa, Oklahoma, 22–24 April.

Stalkup, F.I. and Crane, S.D.: " Reservoir Description Detail Required To Predict Solvent and Water Saturation at an Observation Well," *SPERE* (February 1994) 35; *Trans.*, AIME, **297.**

Stein, M.H. *et al.*: "Slaughter Estate Unit CO_2 Flood: Comparison Between Pilot and Field-Scale Performance," *JPT* (September 1992) 1026. **[R]**

Stephenson, D.J., Graham, A.G., and Luhning, R.W.: "Mobility Control Experience in the Joffre Viking Miscible CO_2 Flood, " *SPERE* (August 1993) 183; *Trans.*, AIME, **295. [R]**

Stern, D.: "Mechanisms of Miscible Oil Recovery: Effects of Pore-Level Fluid Distribution," paper SPE 22652 presented at the 1991 SPE Annual Technical Conference and Exhibition, Dallas, 6–9 October. **[R]**

Stiles, L.H. and Magruder, J.B.: "Reservoir Management in the Means San Andres Unit," *JPT* (April 1992) 469. **[A]**

Stone, P.C., Steinberg, B.G., and Goodson, J.E.: "Completion Design for Water and CO_2 Floods," *SPEPE* (November 1989) 365; *Trans.*, AIME, **287.**

Surguchev, L.M., Korbøl, R., and Krakstad, O.S.: "Optimum Water-Alternate-Gas Injection Schemes for Stratified Reservoirs," paper SPE 24646 presented at the 1992 SPE Annual Technical Conference and Exhibition, Washington, DC, 4–7 October.

Tanner, C.S. *et al.*: "Production Performance of the Wasson Denver Unit CO_2 Flood," paper SPE 24156 presented at the 1992 SPE/DOE Enhanced Oil Recovery Symposium, Tulsa, Oklahoma, 22–24 April. **[R]**

Tiffin, D.L. and Kremesec, V.J.: "Mechanistic Study of Gravity-Assisted CO_2 Flooding," *SPERE* (May 1988) 524.

Turek, E.A., Metcalfe, R.S., and Fishback, R.E.: "Phase Behavior of Several CO_2/West Texas Reservoir-Oil Systems," *SPERE* (May 1988) 505. **[A]**

Wackowski, R.K. and Masoner, L.O.: "Rangely Weber Sand Unit CO_2 Project Update: Operating History," *SPERE* (August 1995) 203; *Trans.*, AIME, **299.**

Waggoner, J.R., Castillo, J.L., and Lake, L.W.: "Simulation of EOR Processes in Stochastically Generated Permeable Media," *SPEFE* (June 1992) 173; *Trans.*, AIME, **293.**

Walsh, M.P. and Lake, L.W.: "Applying Fractional Flow Theory to Solvent Flooding and Chase Fluids," *J. Pet. Sci. & Eng.* (July 1989) 281.

Wang Y. and Orr, F.M. Jr.: "Calculation of Minimum Miscibility Pressure," paper SPE 39683 presented at the 1998 SPE/DOE Improved Oil Recovery Symposium, Tulsa, Oklahoma, 19–22 April. **[A]**

Wegener, D.C., and Harpole, K.J.: "Determination of Relative Permeability and Trapped-Gas Saturation for Predictions of WAG Performance in the South Cowden CO_2 Flood," paper SPE 35429 presented at the 1996 SPE/DOE Improved Oil Recovery Symposium, Tulsa, Oklahoma, 21–24 April. **[A]**

White, J.T. and Benoit, R.L.: "A CO_2 Measurement and Control System," paper SPE 14288 pesented at the 1985 SPE Annual Technical Conference and Exhibition, Las Vegas, Nevada, 22–25 September.

Xue, G. *et al.*: "A Comparative Technical and Economic Analysis of Waterflood Infill Drilling and CO_2 Flood in West Texas Carbonate Reservoirs," paper SPE 27642 presented at the 1994 SPE Permian Basin Oil and Gas Recovery Conference, Midland, Texas, 16–18 March.